Thermal Energy

This book presents the essentials of thermal energy storage techniques along with recent innovations and covers in-depth knowledge of thermal energy applications. Different aspects of thermal energy storage systems are covered, ranging from fundamentals to case studies. Major topics covered include application of thermal energy in water heating, solar cooking and solar pond, thermal energy storage materials for indoor comfort in buildings, thermal management of battery, hydrogen production, reducing carbon footprints, and so forth.

Key features:

- Presents current research and technological updates along with applications and market scenarios in thermal energy storage, thermal management, and applications of thermal energy
- Explores sensible, latent, and thermochemical energy storage aspects
- Emphasizes the need and adequate utilization of abundant heat energy for clean energy perspectives
- Reviews use of thermal energy in hydrogen production, the oil and gas sector, along with market analysis
- Includes pertinent case studies

This book is aimed at researchers and graduate students in energy and mechanical engineering, energy storage, and renewables.

Thermal Energy
Applications, Innovations, and Future Directions

Edited by Amritanshu Shukla,
Atul Sharma, and Karunesh Kant

CRC Press
Taylor & Francis Group
Boca Raton New York London

CRC Press is an imprint of the
Taylor & Francis Group, an **informa** business

First edition published 2024
by CRC Press
2385 NW Executive Center Drive, Suite 320, Boca Raton FL 33431

and by CRC Press
4 Park Square, Milton Park, Abingdon, Oxon, OX14 4RN

CRC Press is an imprint of Taylor & Francis Group, LLC

ISBN: 978-1-032-38542-6 (hbk)
ISBN: 978-1-032-38544-0 (pbk)
ISBN: 978-1-003-34555-8 (ebk)

DOI: 10.1201/9781003345558

Typeset in Times
by Apex CoVantage, LLC

Contents

PART I   *Thermal Energy: Storage*

PART II   Thermal Energy: Advancements in Applications

Navendu Misra, Saurabh Pandey, Amritanshu Shukla,
and Atul Sharma

Saurabh Pandey, Abhishek Anand, Amritanshu Shukla,
and Atul Sharma

PART III   Thermal Energy: Global Scenario and Future Directions

R. Kumar, Anil Kumar, and Atul Sharma

Shubham Raina, Krishma Kumari, Deepak Pathania, and Richa Kothari

Chapter 13 Solid Particle Thermal Energy Storage for High-Temperature
CSP Applications 262

Karunesh Kant and R. Pitchumani

Chapter 14 Thermal Management of Batteries by Thermal Energy
Storage Materials 275

Fathimathul Faseena A M and A. Sreekumar

Preface

Fast global economic development has led to rapidly increasing energy demand. Today in terms of gigajoules of energy consumption per person per year, the US consumes about 300, Japan about 170, the EU about 150, China close to 100, India 20, Nigeria 5, and Ethiopia 2. To grow from Nigeria to China is a 20-fold increase just on per capita terms. This shows the pressing need to develop and efficiently utilize clean and abundantly available energy resources worldwide. However, conventional fossil fuel energy sources are limited, and their increased usage adversely affects ecology due to the emission of harmful gasses, which are supposed to be responsible for climate change and environmental pollution. Further, conventional energy sources have several disadvantages, such as their unbalanced geographical distribution in different countries, which increases investment costs in transport and security for countries with low resources, and their direct impact on the environment, signified by global warming and its impacts on the ecosystem. The need for clean, cheap, and sustainable energy sources becomes a priority to overcome the challenges mentioned earlier. Using renewable energy sources (solar, wind, hydro, etc.) decreases the demand for traditional energy resources and decentralizes the energy supply. Solar energy is abundantly available but intermittent in nature. Due to intermittency in the energy supply, energy storage devices are required to reduce the gap between energy demand and supply. The energy storage systems store excess energy to be used later in deficiency periods. Thermal energy storage through the sun is one of its best options. Thermal energy enhances the performance of solar thermal systems by absorbing or releasing heat energy. It can be utilized for several applications, such as concentrating solar power (CSP) plants, thermal management of photovoltaics, buildings, electronics cooling, etc.

In this connection, the present book offers up-to-date coverage of the fundamentals as well as recent advancements in thermal energy storage techniques, materials, characterization, and their technological applications. It also covers recent innovations and in-depth knowledge of thermal energy applications. Different aspects of thermal energy storage systems are covered, from fundamentals to case studies. The proposed book intends to provide a comprehensive perspective to the researchers, scientists, engineers, technologists, students, policy-makers, etc., who wish to learn more about thermal energy storage systems and applications. We hope readers find this book interesting and that it succeeds in serving as a text and reference material for researchers in academia as well as in industry, enabling them to get an overall research update on latent heat-based thermal energy storage technology.

We would like to convey our appreciation to all the contributors, academic and research collaborators, who have always been a source of inspiration to work within this field of study. We would also like to thank all team members at CRC Press for their kind support and great effort in bringing the book to fruition. Last but not least, the editors are sincerely thankful to their family members, whose unending support and affection have always been a driving force in making any professional assignment a success!

Amritanshu Shukla, Atul Sharma, and Karunesh Kant

About the Editors

Amritanshu Shukla, Ph.D., is currently working as a Professor in Physics at Lucknow University, Lucknow, India. He has also served as Head of the Division of Basic Sciences and Humanities and as Chief Vigilance Officer of the Rajiv Gandhi Institute of Petroleum Technology, Jais, Amethi, Uttar Pradesh, India. His research interests include theoretical physics, nuclear physics, and physics of renewable energy resources, including thermal energy storage materials. He has published more than 100 research papers in various international journals and international and national conference proceedings. He has also written five international books, including *Energy Security & Sustainability: Present and Future* (CRC Press/Taylor & Francis), *Sustainability through Energy-Efficient Buildings* (CRC Press/Taylor & Francis), and *Low Carbon Energy Supply – Trends, Technology, Management* (Springer Nature). He has delivered invited talks at various national and international institutes and is involved with several national as well as international projects and active research collaborations in India and abroad (such as with the University of Lund, Sweden; Kunshan University, Taiwan; Université Clermont Auvergne, France; Virginia Tech University, USA) on the topics of his research interests. Formerly he has worked at some international research institutes, namely the Institute of Physics Bhubaneswar (Department of Atomic Energy, Government of India); Physical Research Laboratory Ahmedabad (Department of Space, Government of India); University of Rome/Gran Sasso National Laboratory, Italy; and University of North Carolina Chapel Hill, USA. He was awarded the prestigious young scientist award in 2004 for attending a meeting with Nobel Laureates held at Lindau, Germany.

Atul Sharma is currently a Professor in Physics at the Rajiv Gandhi Institute of Petroleum Technology (set up through an Act of Parliament by the Ministry of Petroleum and Natural Gas as an Institute of National Importance on the lines of Indian Institutes of Technology). He has worked as a Scientific Officer at Devi Ahilya University, India; as a Research Assistant at the Korea Institute of Energy Research, Daejon, South Korea; and as a Visiting Professor at the Department of Mechanical Engineering, Kun Shan University, Tainan, Taiwan, Republic of China. Dr. Sharma completed his MPhil and Ph.D. from the School of Energy and Environmental Studies, Devi Ahilya University, Indore, Madhya Pradesh, India. Dr. Sharma has published several edited books from various well-known international publishers and research papers in various international journals and conferences. He completed several projects as principal investigator with funding from agencies, including the Department of Science and Technology, New Delhi, and the Council of Science and Technology, Uttar Pradesh, India. He is working on the development and applications of phase change materials, green buildings, solar water heating systems, solar air heating systems, and solar drying systems. Further, he served as a reviewer for many national and international journals, project reports, and book chapters.

Karunesh Kant is currently working as a Research Associate at the Department of Mechanical Engineering, Virginia Tech, Blacksburg, VA, USA. He received his bachelor's and master's in mechanical engineering and Ph.D. in energy, specializing in thermal energy storage. He was awarded the prestigious Bhaskara Advanced Solar Energy Fellowship by Indo-US Science and Technology Forum and the Department of Science and Technology, Government of India to undertake a part of his doctoral thesis work at the Advanced Materials and Technologies Laboratory, Virginia Tech. After his doctorate, he joined the Energy Technology Group, Department of Mechanical Engineering, Eindhoven University of Technology, Netherlands, as a postdoctoral fellow in September 2018, where he worked on thermochemical energy storage material for thermal energy storage. He received a postdoctoral fellowship at University Clermont Auvergne (UCA), Clermont, France, in October 2020 to work on the mathematical modeling of nano-enhanced phase change materials for thermal energy storage. At the UCA, he also worked on multi-3D particle tracking velocimetry for indoor airflow study in large enclosures. After completing his fellowship at UCA, he joined the Advanced Materials and Technologies Laboratory, Department of Mechanical Engineering, Virginia Tech, USA as a research associate in September 2021. He is currently working on multiple projects funded by the US Department of Energy. He is listed in the top 2% of global researchers based on the data given by Elsevier and Stanford University, USA. He is an Early Career Editorial Board member of Elsevier's *e-Prime* journal. He served as guest editor of *Energies* journal, published by MDPI, and *Solar Energy* journal, published by Frontiers in Energy Research. He is a reviewer of the various international journals published by Elsevier, Wiley, ACS, Taylor & Francis, and MDPI. He has published 35 peer-reviewed journal articles, 13 book chapters in internationally reputed books, and eight conference papers and posters. His publications got more than 2,000 citations and 20 of them h indexed.

Contributors

Fathimathul Faseena A M
Solar Thermal Energy Laboratory,
  Department of Green Energy
  Technology
Pondicherry University
Puducherry, India

Abhishek Anand
Institute of Power Engineering
Universiti Tenaga Nasional
Kajang, Selangor, Malaysia

Mohamad Aramesh
Mechanical and Automotive Engineering,
  School of Engineering
RMIT University
Bundoora, Victoria, Australia

Rabeb Ayed
The Research and Technology
  Centre of Energy
Hamem Lif, Tunisia

Sara Baddadi
The Research and Technology
  Centre of Energy
Hamem Lif, Tunisia

Salwa Bouadila
The Research and Technology
  Centre of Energy
Hamem Lif, Tunisia

Frank Bruno
Future Industries Institute
University of South Australia
Mawson Lakes, South Australia, Australia

Ross Flewell-Smith
Future Industries Institute
University of South Australia
Mawson Lakes, South Australia, Australia

Alireza Gorjian
Electrical Engineering Department,
  Faculty of Engineering
Bu-Ali Sina University
Hamedan, Iran

Shiva Gorjian
Biosystems Engineering
  Department, Faculty of
  Agriculture
Renewable Energy Department, Faculty
  of Interdisciplinary Science and
  Technology
Tarbiat Modares University
Tehran, Iran

Pragya Gupta
Advanced Materials Research
  Laboratory, Department of Polymer
  and Process Engineering
Indian Institute of Technology Roorkee,
  Saharanpur Campus
Saharanpur, India

Rhys Jacob
Future Industries Institute
University of South Australia
Mawson Lakes, South Australia,
  Australia
Forschungszentrum Jülich GmbH
Institute of Energy and Climate
  Research, Structure and Function of
  Materials (IEK-2)
Jülich, Germany

Karunesh Kant
Advanced Materials and
  Technologies Laboratory,
  Department of Mechanical
  Engineering
Virginia Tech
Blacksburg, VA

Richa Kothari
Department of Environmental
    Sciences
Central University of Jammu,
    Rahya-Suchani (Bagla) Samba,
    Jammu
Jammu and Kashmir, India

Anil Kumar
Department of Mechanical Engineering
    and Centre for Energy and
    Environment
Delhi Technological University
Delhi, India

R. Kumar
Department of Mechanical Engineering
Delhi Technological University
Delhi, India

Krishma Kumari
Department of Environmental
    Sciences
Central University of Jammu,
    Rahya-Suchani (Bagla) Samba,
    Jammu
Jammu and Kashmir, India

Pradip K. Maji
Advanced Materials Research
    Laboratory, Department of Polymer
    and Process Engineering
Indian Institute of Technology Roorkee,
    Saharanpur Campus
Saharanpur, India

Muhamad Mansor
Institute of Power Engineering
Universiti Tenaga Nasional
Kajang, Selangor, Malaysia

Navendu Misra
Non-Conventional Energy Laboratory
Rajiv Gandhi Institute of Petroleum
    Technology
Jais, Amethi, India

Hamed Mokhtarzadeh
Biosystems Engineering Department,
    Faculty of Agriculture
Tarbiat Modares University
Tehran, Iran

Abhijith M T
Department of Green Energy
    Technology, Madanjeet School of
    Green Energy Technologies
Pondicherry University
Puducherry, India

Saurabh Pandey
Non-Conventional Energy Laboratory
Rajiv Gandhi Institute of Petroleum
    Technology
Jais, Amethi, India

Deepak Pathania
Department of Environmental
    Sciences
Central University of Jammu,
    Rahya-Suchani (Bagla) Samba,
    Jammu
Jammu and Kashmir, India

R. Pitchumani
Advanced Materials and Technologies
    Laboratory, Department of
    Mechanical Engineering
Virginia Tech
Blacksburg, VA

B. K. Purohit
Department of Chemical Engineering
B. V. Raju Institute of Technology
Narsapur, Medak, Telangana, India

Shubham Raina
Department of Environmental
    Sciences
Central University of Jammu,
    Rahya-Suchani (Bagla) Samba,
    Jammu
Jammu and Kashmir, India

Bahman Shabani
Mechanical and Automotive
    Engineering, School of Engineering
RMIT University
Bundoora, Victoria, Australia

Atul Sharma
Non-Conventional Energy Laboratory
Rajiv Gandhi Institute of Petroleum
    Technology
Jais, Amethi, India

Shane Sheoran
Future Industries Institute
University of South Australia
Mawson Lakes, South Australia,
    Australia

Amritanshu Shukla
Department of Physics
Lucknow University
Lucknow, India
Rajiv Gandhi Institute of Petroleum
    Technology
Jais, Amethi, India

Kamran Soleimani
Department of Biosystems Engineering,
    Faculty of Agriculture
Ferdowsi University of Mashhad
Mashhad, Iran

A. Sreekumar
Department of Green Energy
    Technology, Madanjeet School of
    Green Energy Technologies
Pondicherry University
Puducherry, India

Yaghuob Molaie
Biosystems Engineering Department,
    Faculty of Agriculture
Tarbiat Modares University
Tehran, Iran

.

Acknowledgments

Before starting this project in June 2022, we were assured that in order to develop this type of book, at least one year would be needed. We considered bringing an edited book to share knowledge, development, and scientific advancement in thermal energy storage (applications, innovations, and future directions) with a large group of interested readers. Therefore, we thought that we could connect with well-known researchers worldwide and request them to contribute their recent leading-edge research experience in the form of book chapters for this edited volume. As in the past, we had already worked on such projects and had an excellent experience with them. It felt so simple at the beginning; however, we were wrong, and we faced many difficulties in completing this project in the last stage. After completing this project, we now understand the issue better, as the authors who contributed to this book are from the scientific and academic community with prior commitments. However, we received great encouragement and enthusiasm from our authors and reviewers. We are honored to have all the contributors who have been very supportive, dedicated, and responsive throughout our interactions. All of us are highly thankful to all our passionate authors and reviewers!

We would like to express our sincere gratitude to the many persons who saw us through this book: to all those who provided support, talked things over, read, wrote, offered comments, and allowed us to quote their remarks. This book would never have taken off without the generous support of the authors of the book chapters. It would never have been completed without the cooperation and administrative and editorial assistance of Dr. Gagandeep Singh and Ms. Aditi Mittal of CRC Press. Our heartfelt thanks to all these dedicated and cooperative individuals!

This book would not have been possible without the support and encouragement of our families. Words cannot express our gratitude for their kind support and encouragement. There is no doubt that we are very excited, enthusiastic, and gratified about the outcome. We also feel sorry about unwittingly neglecting them on many occasions, especially during weekends. We dedicate this book to our families for all their patience and moral support.

Amritanshu Shukla, Atul Sharma, and Karunesh Kant

Part I

Thermal Energy

Storage

1 Introduction to Thermal Energy and Its Storage Materials

Pragya Gupta, Pradip K. Maji, and Atul Sharma

1.1 INTRODUCTION

Thermal energy storage (TES) is indeed demanding for its efficient utilization in various economic aspects. It acts as a bridge between energy demand and supply (Garg, Mullick and Bhargava, 2012). The storage of thermal energy is the capability of materials to stock thermal energy, which will be utilized in future for different applications such as maintaining the thermal comfort of occupants for heating or cooling the building, industrial processes such as metallurgical transformations, and solar power generation. In the current scenario, depletion of natural resources and greenhouse gaseous emissions are major concerns all over the globe. Thermal energy storage enhances the efficiency and sustainability of the energy transmission network and contributes significantly to overall energy conservation (Fernandes *et al.*, 2012). Thermal energy storage could be accomplished by heating, cooling, melting, freezing, or evaporation of the material to accumulate or release the heat (Pielichowska and Pielichowski, 2014). Thermal energy storage is categorized as chemical, latent, and sensible heat storage systems, which depend upon the characteristics or processing of material.

1.2 SENSIBLE HEAT STORAGE MATERIAL

The temperature varies with respect to the energy stored in the sensible heat storage system. The sensible storage material must have a high specific heat capacity to absorb a high amount of energy, which is unaffected during thermal cycling for long-term stability and compatibility with the storage container (Demirbas, 2006). The specific heat, temperature difference, and sensible heat storage material mass influences sensible heat as per eq. 1:

$$Q = \int_{T_i}^{T_f} mC_p dT \qquad (1)$$

Where Q denotes the amount of sensible heat (J), M signifies the mass of sensible heat stored material (kg), C_p indicates the specific heat of sensible heat stored material (J/kg.K), and T_i and T_f specify initial and final temperatures of sensible heat stored material (K).

DOI: 10.1201/9781003345558-2

The heat can be absorbed or released by conduction, convection, and radiation during the process (Pielichowska and Pielichowski, 2014). Liquid and solid constituents can be appropriate for sensible heat storage. The material should be non-flammable, non-corrosive, and have no chemical decomposition during heat absorbance and release phenomenon.

1.2.1 LIQUID THERMAL STORAGE MATERIAL

A liquid storage medium has the advantage of being easily circulated, which enables it to transport heat as per the requirement. An active system is one where the storage material can be easily circulated. The buoyancy force contributes to the formation of a promising temperature gradient through the storage material due to the density difference carried out by heating liquid. A few liquids can store sensible heat effectively.

1.2.1.1 Water

Water is a cheap, readily available thermal storage media available around the world. It can be used in a wide temperature range. The high specific heat of water is beneficial to utilize in a low-temperature application. However, it requires an insulated and high-pressure vessel due to the high vapor pressure of water at a high temperature range. For instance, the sensible heat energy of water is 250 kJ/kg at 60°C.

For thermal comfort, energy flow, and basic household applications, hot water is generally employed in the radiators. Hot water is mainly used in radiators for thermal comfort, energy transportation, and essential household applications. For instance, a solar energy system could employ water as a means of energy storage and transportation. As a result, water may be currently the most popular storage option for solar-based space heating and hot water. Due to buoyancy forces, temperature drift occurred at the topmost and lowermost end of the tank, so thermal stratification or a thermal gradient can be established in a solar-powered thermal storage system. The removal of mixing during storage allows for stratification, which has two benefits: (1) if the energy is provided towards the output temperature as collected instead of a reduced storage temperature, it can be used more efficiently, and (2) if the temperature of the intake fluid to the receiver is reduced, then the amount of energy stored may increase.

Steel, aluminum, reinforced concrete, fiber glass, and other materials are used to construct water storage tanks. Polyurethane, glass wool, or mineral wool are used to insulate the tanks. The tanks utilized range in size from several hundred liters to several thousand cubic meters. The application of subsurface natural aquifers for large-scale storage applications has also been explored by researchers. Basically, the aquifers exist as geological growths that contain groundwater and can be used to store heat for a long time. The sand or gravel that has been saturated with water serves as the storage medium in aquifers. Aquifer sizes typically range from tens of thousands to millions of cubic meters; therefore, there is sufficient storage space available. The periodic storing that is thermal energy between hot and cold weather is a good application for the particular category. Aquifer stowing is not promising for small loads, such as particular homes, due to its bulky nature. Several experimental and theoretical research projects on aquifer thermal energy storage have been performed due to its extensive acceptance as a beneficial technology. Still, additional comprehensive computer simulation research is necessary to identify the ideal structural alignment due to the abundance of various methods as well as bulky characteristics. Aquifer storage is attractive now because of its cheap cost, high input or output rate, and enormous capability.

1.2.1.2 Storage in Salty Water

Large amounts of solar energy can be captured and stored inexpensively and easily using solar ponds as low-temperature (50–95°C) thermal energy reservoirs. It reflects a high prospect for electric power production, industrial heat supply, and applications including space heating and cooling. Four fundamental characteristics can be used to categorize solar ponds, such as convecting/non-convecting, partitioned/non-partitioned, gelled/ungelled, and isolated separator and collector/in-pond collector. Yet, the salt gradient solar pond without convection is popular in the current state of research. In the case of a solar pond, salt-containing water or sea water is used to generate a density gradient due to the concentration difference with depth from the surface. The salts that are regularly utilized in the pond are sodium chloride and magnesium chloride. A dark or black bottom on a salt gradient pond allows sun radiation to be absorbed, increasing the water's temperature up to 95°C as a result. Without disturbing the upper layers, it is simple to extract the thermal energy trapped in the pond's lower levels.

1.2.1.3 Mineral Oil

Concentrated solar power plants transmit thermal energy by using mineral oil as a fluid. It allocates the heat to the boiler taken from the receiver, and steam is produced to power the turbine. A similar method will be able to collect heat for use in the night in highly insulated storage tanks. Heat transmission fluid also assists as energy storage in a direct system. It reduces the costs by diminishing the requirement for heat exchangers. The liquid form of mineral oil can be exploited at high temperatures up to 400°C and reduced vapor pressure in comparison to water. Mineral oil cannot freeze in pipes in the night, unlike molten salts, necessitating the application of an antifreeze system. However, molten salt amalgamates by means of low melting points was found as a mineral oil substitute because molten salts are less expensive than mineral oil. In recent development, concentrated solar power has been employed as an indirect system, in which molten salt combinations serve as sensible heat materials and mineral oil serves as a heat transfer fluid.

1.2.1.4 Molten Salts

These types of salts are now common thermal energy storage material. They are very economic and have a high energy storage density due to their higher density in comparison to other liquid thermal storage material. Molten salts have a lower vapor pressure than water and can be employed in liquid form up to 400°C. It permits the system to work at high temperature, enhancing the Rankine cycle performance. Molten salts have a required low melting point at room temperature to facilitate their liquid form throughout the process and require a minimum antifreeze agent at night due to the unavailability of solar energy. The cons are that pure molten salts typically reflect melting temperatures above 200°C (Alva *et al.*, 2017). In the present era, it is common to utilize salt composites to lower the melting point to below 100°C while maintaining high temperature more than 500°C. It will work as a heat transfer fluid, while an antifreeze system is safer to work with any freezing risk, or else, the heat transmission fluid will be made of mineral oil. The fact that molten salts are extremely corrosive and oxidizing agents make it difficult to encompass a higher temperature are one of their downsides. Additionally, they have a low thermal conductivity and a 6% volume change throughout the melting process.

1.2.1.5 Liquid Metals and Alloys

Pristine metal and its alloy presented a melting point over 300°C, although having low melting points, but utilized as sensible thermal storage materials. It also reflects higher working temperatures and thermal conductivity in addition with minimum vapor pressure. However, it has disadvantages such as expensive cost. Additionally, it could need an environment free from oxygen and oxides to diminish corrosion.

1.2.2 Solid Thermal Storage Material

This type of thermal storage material is inexpensive and widely accessible. It does not have a vapor pressure problem. Therefore, there is no requirement for pressure vessels and no leakage issues because it works at ambient pressure. The solid materials are only capable of passive heat storage due to their difficulty in being circulated, hence they require a heat transfer fluid such as air, water, or mineral oil to discharge or collect heat as per application by means of loosely bounded solid material. The direct contact between air and solid material increases the efficiency for heat transfer at the time of charging and discharging processes. The major drawback of sensible solid material is that the heat transfer fluid cannot maintain temperature over a long period because the temperature of the solid storage material reduces throughout its period of discharging. There are few solid materials that can be used in low temperature applicability, such as space heating as well as industrial waste heat recovery.

1.2.2.1 Rocks

Rocks can be crushed in a range of 4–5 cm and utilized in a packed bed after being heated with air for thermal transmitting. During the charging, heated air passes through spaces of the packed rocks as well as heats the rock, and at the time of discharging process, cooled air passes through the spaces of the packed rocks and absorbs the heat from the rock. The efficiency of thermal transmission between air and rocks can be increased by availability of contact surface area. The low thermal conductivity and a small interfacial surface are desirable to minimize thermal float during storage. Rocks have the benefits of being non-toxic, non-flammable, extremely affordable, and widely accessible in nature. Some disadvantages of rocks include the requirement for a high rate of air flow and pressure difference (Hänchen, Brückner and Steinfeld, 2011).

1.2.2.2 Concrete

A solid material that has the capability to store heat such as concrete is simple to work with. The higher mechanical strength of concrete does not demand a container to hold it. One method is to design a heat exchanger utilizing to concrete with heat transfer fluid to run pipes carrying heat transmitted with a concrete block. After several sequences of thermal expansion in addition to contraction, fracture formation is one of the major problems that arises at high temperatures.

1.2.2.3 Sand

Easily accessible small particles such as silica sand can be employed as a solid sensible heat storage material. The silica sand ranges between 0.2–0.5 mm and has been used as a solid packed bed system while air has been used as a heat transfer fluid. Small grain particles will have a higher packing density. Basalt gravel, for example, has a size of about 0.4 mm. For the solar collection, the sand particles flow through gravitational force from an inlet to capture solar heat energy. The hot sand can be placed in an insulated storage tank after being heated by concentrated sun rays by means of fall in the tower. When the temperature of the gravel has reached around 700–1000°C, it can generate vapor to power Rankine cycle power plants.

1.2.2.4 Bricks

To decrease the rate of electricity used for space heating, bricks in building walls contain heat. The bricks can be heated with less expensive power overnight during off-peak hours, and they are able to accumulate thermal energy. The stored energy in bricks can be released either naturally through convection or radiation or otherwise artificially by forced convection powered by a device in the daytime (Farid *et al.*, 2004). During the high demand of electricity, thermal energy keeps a building comfortable without using extra consumption to minimize the peak load.

1.3 LATENT HEAT STORAGE MATERIALS

Latent heat storage material absorbs or releases heat during heating or cooling due to the latent heat of fusion. During the absorbing or releasing process, the phase of material changes with a minute variation in the temperature, which is termed as phase change material. Thermal energy or heat belongs to phase change material stated as $Q = m \cdot L$, where m signifies mass of latent heat of material and L denotes latent heat fusion of material. Frequently, the solid-liquid phase change technique is employed throughout the process. Phase change from liquid to gas acquires the largest heat in the latent form, although the storage is complicated and unrealistic because the significant volume of storage materials changes by evaporation (Cárdenas and León, 2013). When solid-solid materials have been exploited, there is no requirement for encapsulation because of the absence of fluid, which entirely eliminates the possibility of spillage. The materials, having high thermal conductivity and high latent heat, are desirable for heat storage application (Farid *et al.*, 2004). The material should melt uniformly with negligible subcooling, with a melting point close to a thermal energy storage system with a prerequisite operational temperature range, and is chemically inert, inexpensive, non-toxic, and non-corrosive. In contrast to sensible heat, heat transfer fluid temperature remains consistent over time because the storage material temperature is constant through the discharging process. Temperature alteration among thermal storage as well as release is smaller for materials that store latent heat (Farid *et al.*, 2004). In comparison with the specific heat, the latent heat of fusion should be high. As an illustration, the "latent heat of fusion" of sodium nitrate, which takes a specific heat of 1.1 kJ/kg. K., is approximately 172 kJ/kg. The latent heat storage materials benefit from having a high energy storage density as a result of this significant differential. The phenomenon decreases the volume of the heat storage container, which decreases the surface area in its exterior walls and diminishes heat loss.

1.3.1 ORGANIC

Several domestic and industrial applications, including space heating in buildings, electronic devices, refrigeration and air conditioning, solar air/water heating, textiles, automobiles, food, and the space industry, have successfully tested and used organic materials and their eutectic blends (Xu, Li and Chan, 2015). Consistent melting without phase separation is possible for organic material; however, the organic material exhibits low heat conductivity from 0.1 to 0.35 W/m K, which is why a higher surface area is enhanced at the rate of heat transfer. Furthermore, organic phase change materials have use under investigation for high-temperature applications due to their low melting point.

1.3.1.1 Paraffin

Straight n-alkanes chains (CH_3-(CH_2)-CH_3) are arranged to form paraffin waxes (Sharma *et al.*, 2015). Due to the high cost of pure paraffin waxes, only technical-grade paraffin has been utilized as a phase change material. Technical-grade paraffin waxes are inexpensive, have a limited melting point range between −10°C to 67°C, and have enough density for thermal storage (200 kJ/kg or 150 MJ/m^3) (Farid *et al.*, 2004; Pielichowska and Pielichowski, 2014). The paraffin wax experiences infinitesimal sub-cooling, no phase segregation, and is chemically inert. The waxes are non-toxic, odorless, durable, affordable, accessible, environmentally harmless, and non-corrosive. Their applications are constrained due to their low heat conductivity (0.2 W/m K) (Pielichowska and Pielichowski, 2014). The distillation of crude oil produces commercial-grade paraffin. The majority of paraffin-based phase change materials are blends with saturated hydrocarbon with varying carbon. The hydrocarbon chain length increases with the paraffin melting and fusion temperatures. The properties of material will be tailored via combining physically different paraffins. Technical paraffin along with pure paraffin have high thermal cycle ability even after 1000–2000 cycles. Paraffins are attuned with metal containers since it does not encourage decomposing. The chemical structure of paraffin wax, however, is analogous to long chain polymer, which is why it interacts with some polymeric containers (Pielichowska and Pielichowski, 2014). Nowadays, researchers are using prepared composite with paraffin to enhance its internal properties. The micro phase change material was prepared by using paraffin-based core material, while methanol-improved melamine formaldehyde- and graphene-based shell material, which is schematically presented by Figure 1.1. The authors reported that the incorporation of graphene enhances the thermal response sensitivity of synthesized material (Su *et al.*, 2017).

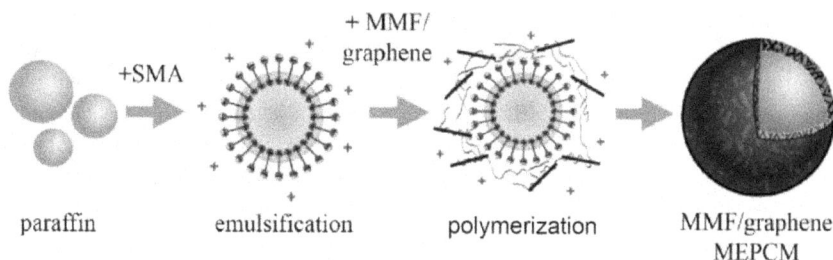

+ MMF/
graphene

+SMA

paraffin   emulsification   polymerization   MMF/graphene
                                                                  MEPCM

FIGURE 1.1 Schematic presentation of graphene-improved melamine formaldehyde-based, paraffin-encapsulated latent heat storage material.

Through interfacial polycondensation and surface self-assembly, microparticles with a crystallized titania shell and a specialized paraffin (i.e., n-eicosane) core coupled using graphene nanosheets have been developed. In a typical preparation method, formamide and n-eicosane were stirred. Formamide was employed to disperse graphene nanosheets. Pluronic 104, a non-ionic surfactant, and a suspension containing graphene were then poured into the flask. Subsequently, formamide containing water was then introduced dropwise into the emulsion. A greyish dispersion was obtained by continually stirring the reactant mixture. To develop a crystalline titania shell, sodium fluoride was successively poured into the flask with stirring. In the end, n-eicosane-titania-graphene microspheres are received by the filtration process. The microspheres are then repeatedly treated with deionized water, ethanol, as well as diethyl ether, and then desiccated in a vacuum before being used again. The prepared composite depicted an enthalpy of phase change as 160 J/g and an increment in the thermal conductivity (Liu, Wang and Wu, 2017). For the application of photothermal transformation, regenerated chitin, polyurea, and graphene oxide-abased shell and paraffin core have been prepared by Maithya *et al.* (2021).

1.3.1.2 Fatty Acids

The fatty acids consist of optimum thermal and physical characteristics of thermal storage material at a lower temperature with a common structure of $(CH_3 (CH_2)_{2n}$-COOH. Some common fatty acids include capric, oleic, lauric, d-lattic, myristic, palmitic, stearic acids, etc. In comparison to paraffin, fatty acids have higher melting and boiling temperatures and negligible volume changes during phase transitions. They also exhibit repeatable melting and freezing behavior. However, they are costlier than commercial paraffin, corrosive, and have a stingy odor. In general, the carbon of fatty acids is directly proportional to melting and freezing temperatures, melting heat, and degree of crystallization (Pielichowska and Pielichowski, 2014). Fatty acids such as lauric, palmitic, and stearic acids have been utilized as a phase change material for solar stills (Kateshia and Lakhera, 2022).

1.3.1.3 Ester

Ester can be produced using acid by substituting an alkyl (-O) group for a hydroxyl (-OH) group. A limited range of temperatures is required for the solid-liquid transition of esterified fatty acid that have been depicted in a specified temperature range. The eutectic mixture of fatty acid has formed without subcooling. The phase transition temperature for eutectic mixtures of esters are nearby ambient temperature with high transition enthalpy (Farid *et al.*, 2004). Esters are classified into two types depending on where they come from. The esterification of bio-based acids and alcohols falls into one group, while another group includes animal fat. Almost all esters have melting points that remain within the range of 18–30°C and have a high latent heat. The flame-retardant and thermal conductivity of esters can be used directly for thermal energy storage. Through life cycle analysis, Fabiani *et al.* (2020) evaluated expired palm oil from the food business and indicated that it exhibited excellent thermal stability, less risk to human health, as well as minimal leak issues. Because of this, it can be assumed in terms of a potential resource for future thermal energy storage. A granular panel composed of natural clay and cellulose fiber has been combined with non-cocoa vegetable fat retrieved from a confectionery business by Boussaba *et al.* (2021). The blended vegetable fat reflected excellent thermal and chemical stability and more than 2,000 thermal cycles. The addition of vegetable fat somewhat diminished the supporting matrix mechanical properties. In order to develop a composite based on fatty acid esters, Sari et al. (Sarı and Karaipekli, 2012) combined erythritol tetrapalmitate, erythritol tetrastearic acid, diatomite, and expanded perlite. Gypsum with bio-originated phase change material was found to have better thermal insulation features in comparison with pristine gypsum (Serrano *et al.*, 2015). Esters have a great deal of potential for creating thermal energy storage, according to the studies mentioned here. There is a huge demand for fatty acid esters because of widespread viable use for the polymer, powder, paint, and smart clothing industries.

1.3.1.4 Alcohols

Polyalcohols, polyols, also referred to as sugar alcohols, are regarded as intermediate temperature (90–200°C) phase change material. Until now, researchers have not explored alcoholic phase change material due to its drawbacks of supercooling. The latent heat of fusion for alcohols such as xylitol, erythritol, and mannitol is around 300 kJ/kg, which performs considerably better as compared with materials of the same family. Other alcoholic materials in the family comprise stearyl alcohol, inositol, dulcitol, xylitol, cetyl alcohol, mannitol, erythritol, and lauryl alcohol, etc. The chemical instability of inositol, D-mannitol, is due to its reactivity with oxygen, and the low cycle stability of galactitol makes it demotivated for its commercial use. Some candidates, though, have shown a lot of promise. For instance, erythritol can store more than 300 J/g of latent heat and has a very high latent heat storage capacity.

1.3.1.5 Glycols

The dimethyl ether with a hydroxyl on the end is composed of polyethylene glycol. The chemical structure is $HO-CH_2-(CH_2-O-CH_2-)_n-CH_2-OH$. Polyethylene glycol is also termed as polyoxyethylene and polyethylene oxide. It exists soluble in organic compounds as well as water. It exists in different grades as per molecular weight. Polyethylene glycols are affordable, non-flammable, non-toxic, thermally and chemically stable, non-corrosive, and their melting temperature and latent heat increases with molecular weight (Sarier and Onder, 2012). The major drawbacks are poor thermal conductivity similar to other organic compounds.

1.3.2 Inorganic

The inorganic material reflected high thermal conductivity and stability after a long-term thermal cycle. The inorganic material contains high volumetric thermal density but is prone to phase separation and sub-cooling (Sharma *et al.*, 2015). It also dissolves inconsistently and is highly corrosive.

1.3.2.1 Salt Hydrates

The working temperature for salt hydrates is in the range of 30–50°C. The salts are inorganic and have the general formula AB_nH_2O. The salt becomes dehydrated as it transitions phases. However, a significant figure of hydrate that contains high latent heat and an appropriate melting temperature and the ability to produce phase change melts inconsistently. The crystalline salt that is created during the dehydration process is not dissolved by the water that is released during the phase transition. Phase separation and sedimentation take place in containers as a result of the salt's and water's different densities. In terms of real-world applications, this is a severe technological issue. The addition of gelling or thickening agents is one technique used to stop the heavier phase from segregating and sedimenting (Pielichowska and Pielichowski, 2014). Due to the thickening agent's ability to raise the viscosity of hydrate and aid in holding molecules together, the addition of polymer or cellulose as a gel material to salt prevents sedimentation. The considerable supercooling of salt hydrates is another drawback. This is because of their low ability to form nuclei, thus nucleating chemicals are added to the mixture to solve the issue. Borax, charcoal, and other nucleating chemicals, among others, aid in overcoming sub-cooling but also slow down heat transfer by diminishing thermal conductivity. Some popular salt hydrates, such as $MgCl_2 \cdot 6H_2O$, $NH_4Al(SO_4)_2 \cdot 12H_2O$, $Mg(NO_3)_2 \cdot 6H_2O$, $Ba(OH)_2 \cdot 8H_2O$, CH_3COONa, $Na_2S_2O_3 \cdot 5H_2O$, $Na_2HPO_4 \cdot 12H_2O$, $Zn(NO_3)_2 \cdot 6H_2O$, $CaBr_2 \cdot 6H_2O$, $Na_2CO_3 \cdot 10H_2O$, $Na_2SO_4 \cdot 10H_2O$, $LiNO_3 \cdot 3H_2O$, $CaCl_2 \cdot 6H_2O$, $KF \cdot 4H_2O$, $LiClO_3 \cdot 3H_2O$, are reported by researchers (Dixit *et al.*, 2022).

1.3.2.2 Salts

In the case of molten salt, salts with somewhat lower melting temperatures are used as sensible liquid storage media, but for applications involving high temperatures greater than 100°C, they can also serve as latent heat storage materials. Salts have an extensive series of melting temperatures and are classified into numerous subgroups, including nitrates, carbonates, hydroxides, and chlorides, among others.

1.3.2.3 Metals and Metal Alloys

Higher conductivity, exceptional stability, dependability, and recurrence are characteristics of metals and their alloys (Pielichowska and Pielichowski, 2014). They have the highest heat of phase transition with respect to volume otherwise mass. They can store a huge amount of energy as a result. During phase shift, their volume changes very little and their vapor pressure is minuscule. Metals can, however, compete with salts when volume is a top priority.

1.3.2.4 Eutectic

A mixture of two or more constituents, such as organic/organic, organic/inorganic, and inorganic/inorganic, is called a eutectic. Each of them undergoes a consistent phase shift, meaning that it has a distinct melting and freezing temperature and, during crystallization, combines to generate a variety of component crystals. There is typically no probability of component separation in eutectics because they typically melt and freeze simultaneously.

1.4 COMPOSITE PHASE CHANGE MATERIALS

The thermal conductivity of phase change material has been enhanced by researchers to solve the issue of poor thermal conductivity as a result of dissolving high thermally conductive material into the phase change material, such as metals, graphite, or carbon. The term "composite phase change material" refers to such a composition. Phase transition materials that are both organic and inorganic can be used to produce composites. The phase change material ought to be compatible with the additional thermal conductive materials. For instance, the additional conductive material needs to be corrosion resistant if the phase change material corrodes in the presence of a few inorganic salts.

1.4.1 GRAPHITE COMPOSITES

Graphite is also utilized in various ways. When Pincemin *et al.* (2008) investigated the effectiveness of phase change material composites made of eutectic $NaNO_3$/ KNO_3 with thermally conductive graphite dispersed throughout the molten salt, they discovered a 14-fold increase in thermal conductivity, leading to an effective conductivity of 9 $Wm^{-1}K^{-1}$. The effectiveness of phase change material composites made of expanded graphite and n-docosane paraffin was investigated by Sari and Karaipekli (2007). Thermal conductivities of the composite phase change material showed that the paraffin (0.22 W/m K) thermal conductivity increased by 81.2%, 136.3%, 209.1%, and 272.7%, respectively, with mass fractions of 2%, 4%, 7%, and 10% expanded graphite.

1.4.2 GRAPHITE FLAKES OR NATURAL GRAPHITE

Graphite reflected a strong thermal conductivity and a well-aligned crystal structure. It has carbon sheets that are piled on top of one another, combined by weak van der Waals forces instead of the strong covalent bonds that hold carbon atoms together. As a result, graphite flakes are tightly packed and have a high density. Graphite is also inexpensive to produce and requires very little processing.

1.4.3 EXPANDED NATURAL GRAPHITE

The chemical treatment and thermal exfoliation processes have been used for the synthesis of this material created from natural graphite. When reacting with different acids, weak bonds are broken, and reactant molecules have amalgamated among graphite stacks. The reactant molecules force graphite assembly to enlarge to preserve the sheet. A compound has been intercalated. The graphite intercalated compound is heated to cause the graphite layers to rapidly expand after being washed in water, dried, and then intercalated. The heat treatment is carried out at temperatures above 500°C in an air atmosphere. During the creation of the intercalated compound, the water is integrated between the layers of carbon. The expansion is caused by the water's almost immediate evaporation during the heat treatment procedure. The graphene layers are dispersed throughout each natural graphite particle, giving rise to a high level of porosity. A solidified graphite matrix with high porosity and thermal conductivity can be created by arranging these worms into a bed and then compressing it.

1.4.4 EXPANDED GRAPHITE POWDER

Expanded graphite can be grounded into a fine powder. Compared to natural graphite, it is less compact and has a susceptibility to form networks even at low concentration. Therefore, as compared with natural graphite, less expanded graphite powder is needed to achieve a specified level of heat conductivity. However, it costs more than natural graphite to produce due to the associated treatments required.

1.4.5 NANO COMPOSITES

Different nanostructures have been incorporated into phase change material as additives for thermal improvement. Different types of nanoparticles were investigated by Khodadadi, Fan, and Babaei (2013) as potential materials for thermal conductivity boosters. The advantage of nanoparticles is their incredibly small size. That is why they act as a fluid and prevent clogging while flowing through pipes. In general, it has been discovered that carbon-based nanostructures produce superior thermal improvements over metallic and metal oxide materials.

A critical assessment of experimental research with respect to the bulk features of phase change material on the nanoparticles distribution was conducted by Kibria *et al.* (2015). Sarier *et al.* (2011) have developed intercalated composite with n-hexadecane as a phase change material and in the base matrix of Na-montmorillonite. The intercalation between both components have been done surfactant, as shown in Figure 1.2. The prepared composite depicted high heat storage as well as release capacity.

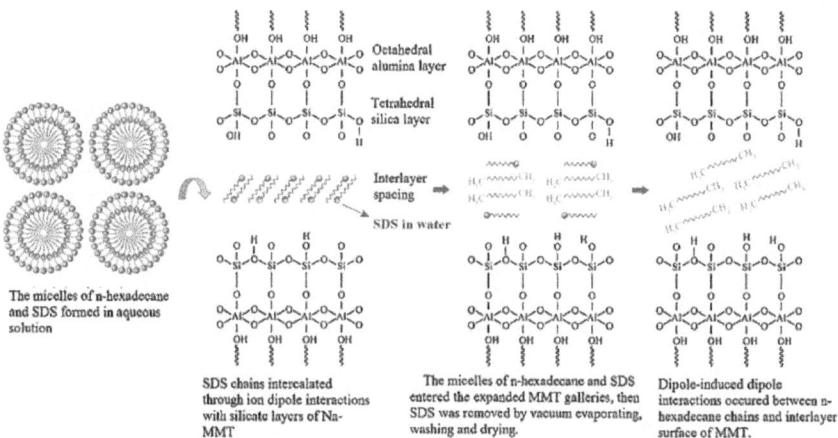

FIGURE 1.2 Intercalated structure of n-hexadecane and Na-montmorlionite (Sarier et al., 2011).

1.4.6 Form-Stable Composite

A composite can be considered as form stable when it can retain an identical structure in a solid phase while its temperature exceeds more than the melting point of the material. Besides encapsulation to assist for accelerated heat transfer, phase change material can be molded into the desired shapes in a form-stable composite with a greater surface area to volume ratio. Here, increasing thermal conductivity is not a top priority; the major goal is to enhance the phase change material contact surface area with the heat transfer fluid. To prevent leakage of melted paraffin, Sari (2004) had investigated a form-stable composite of paraffin wax and polyethylene composites, in which the paraffin assists as a thermal storing media, but polyethylene acts as a supportive matrix. It was possible to achieve a 24% increase in thermal conductivity.

1.5 MICROENCAPSULATED PHASE CHANGE MATERIALS

Microcapsules are defined as constituent with sizes between 10 and 1000 μm that have a core substance encased in a coating or shell. Commercial uses include textiles, adhesives, cosmetics, pharmaceuticals, and architecture, which all frequently use microencapsulation (Sharma et al., 2015). The requisite shape, diameter, and high mechanical properties of shell, permeability, and thermal strength are the basic properties of microencapsulated phase change materials. The microcapsules containing pouches, tubes, spheres, panels, and other containers serve as heat exchangers directly. The shell protects the liquid phase change materials from the environment and prevents decomposition. It expands the area of contact for heat transmission. The shell should have enough mechanical strength to stabilize capsules as well. The microcapsules come in a variety of sizes and forms. The method of microencapsulation is rather expensive. The best materials for microcapsule shells include urea-formaldehyde resin, melamine-formaldehyde resin, and polyurethanes.

1.6 COACERVATION

Electrostatic interaction due to contrast-charged particles or ions causes coacervation when two or more contrast particles are dispersed in a media of phase change material. Spherical droplets are made up of the concentrated colloidal particles, which are kept together by electrostatic forces. For instance, a combination of paraffin, gelatin, and gum arabic can be added to synthesized micron-sized particles. During microencapsulation, the phase change material based core is paraffin, and the shell is made of gelatin and gum arabic.

1.6.1 Suspension Polymerization

A method based on suspension polymerization can produce microcapsules having polymer shell and a phase change material based core. Using mechanical agitation, such as vigorous stirring, a monomer or combination of monomers and initiator is distributed in an aqueous media of phase change material in this method. Heat can be used to break down the initiator into free radicals, and the free radicals then cause the monomerization of other molecules to begin. Chemical factors such as viscosity, density, and interfacial tension, equipment aspects (i.e., porosity, pore and pipe diameter), and operating conditions (i.e., flow and shear rate and pulsation) affect the droplet size of microcapsules.

1.6.2 EMULSION POLYMERIZATION

In the method, emulsion is prepared by polymerization of monomer with surfactants in liquid medium. The aqueous solution disperses the phase change material drops that work by means of an oil phase. Surfactants with hydrophilic and hydrophobic sides arrange themselves at the interface of liquid droplets of phase change material and aqueous solution monomers. Both the hydrophilic and hydrophobic components make contact with the liquid droplet of phase change material. As a result, the initiator causes the monomer molecules to begin polymerizing around the sphere-shaped droplet. n-Eicosane as a core and phenol-formaldehyde-based shell have synthesized by oil-in-water emulsion (Liu *et al.*, 2022). The microencapsulation process has been done by in-situ polymerization process.

1.6.3 POLYCONDENSATION

A phase change material incorporated microcapsule can be prepared via regulating the phase change liquid temperature and by using a polymer as an outer layer. The monomer molecules combine, with the loss of smaller molecules as byproducts such as water or methanol in the polycondensation process. Microencapsulation can be accomplished by polycondensing a shell material with core as the phase change material. Rapid polymerization process and negligible absorbency of the products are benefits of interfacial polycondensation (Liang *et al.*, 2009). The explanation for interfacial polycondensation is as follows. The droplets of the core material have been formed. The exterior layer of droplets generate the microsphere shell, and the reactive monomers begin to form oligomers. When the previously prepared oligomers encompass and become hydrophobic at the droplet interface, a thin layered film synthesizes over the droplet. The outer layer turns out to be a shell as a result of polycondensation, which also causes a miniature shell to form throughout droplets. The monomer for the preparation of a shell can be polyurea, polyurethane, polyester, polyamide, and amine resin, etc. By means of interfacial polymerization using ethylenediamine and toluene-2, 4-diisocyanate, for example, monomer in an emulsion system, phase change material such as butyl stearate can be microencapsulated, as reported by Liang *et al.* (2009). Toluene-2,4-di-isocyanate and ethylenediamine were procured for synthesis of polyurea microcapsules and butyl stearate serving as the main component. The synthesized material reflected the properties, such as 29°C of phase change temperature and latent heat of 80 J/g. Zhang and Wang had prepared microencapsulated n-octadecane with polyurea as an outer layer for thermal regulation application (Zhang and Wang, 2009). The compound has been prepared by polymerization by using oil-based monomer (i.e., toluene-2, 4-diisocyanate) and water-based amines (i.e., ethylene diamine, diethylene triamine), as presented by Figure 1.3. The microcapsule reflected an optimized outcome at 70/30 ratio of core

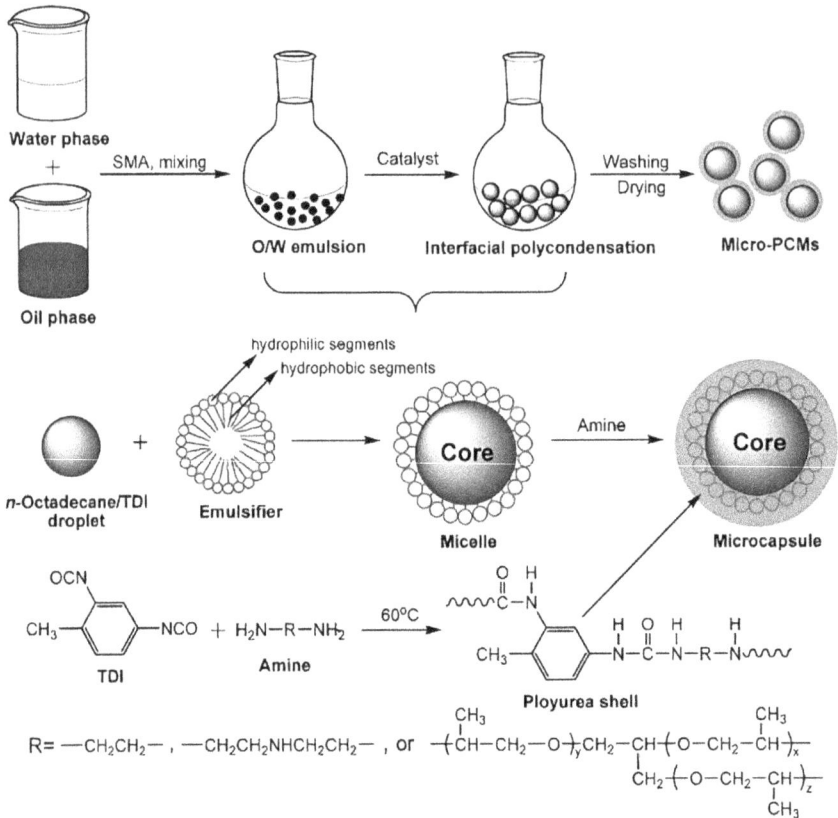

FIGURE 1.3 Graphic representation of microencapsulated n-octadecane with polyurea by the polycondensation process (Zhang and Wang, 2009).

and shell material Zhu *et al.* (2019) reported nanoencapsulated material by taking *n*-octadecane as a core material and silica-graphene as a shell material. The shell material was synthesized by hydrolysis and the polycondensation process. In this process, the polymerization of tetraethoxysilane and methyltrimethoxy with the self-assembly of graphene take place, and the nanoencapsulated n-octadecene-silica-graphene composite shell have been synthesized. The method is displayed in Figure 1.4. Nano phase change material has been formed in the oil-in-water miniemulsion in which cethyltriammonium bromide acts as a surfactant to stabilize it. As a result, the outer surface of the generated miniemulsion droplets becomes positively charged. With the help of sodium dodecyl sulphate, graphene is consistently distributed in water and has an opposite charges surface. Electrostatic contact causes graphene to self-assemble on the interface of the miniemulsion droplets as soon as the graphene dispersed is dropped into it. The interfacial hydrolysis and polycondensation are catalyzed to form silica networks after the addition of aqueous ammonia. The silica-graphene composite shell thereby prepared the emulsion droplets. The authors claimed that thermal conductivity was enhanced 132.9% due to the addition of graphene.

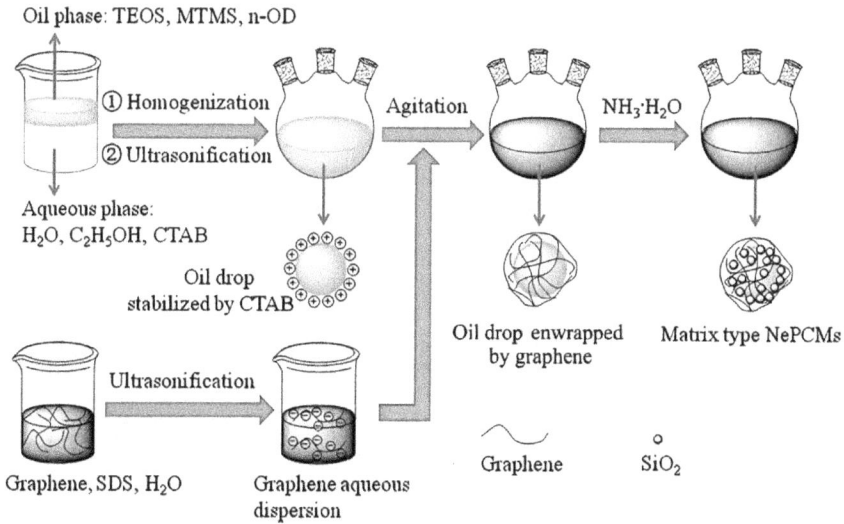

FIGURE 1.4 Schematic illustration of nanoencapsulated n-octadecane with silica-graphene by polycondensation process (Zhu *et al.*, 2019).

1.6.4 POLY-ADDITION

An oil-based monomer and an aqueous-based monomer are the basis for the technique for preparing phase change material filled microcapsules. In order to initiate the reaction by using monomers, it may be necessary to disperse an oil-based solution into an additional aqueous-based solution. Quickly, a polyurea membrane develops on the surface of the phase change material distributed oil phase droplets. The biodegradable and solvent-free phase change material were described by Fu *et al.* (2017). The synthesized substances are characterized for accelerated thermal cycling,

1.7 THERMOCHEMICAL HEAT STORAGE MATERIALS

The microcapsules incorporated with phase change material have been prepared by interfacial poly-addition reaction between oil-phase and aqueous-phase monomers. In order to prepare phase change material, a monomer of oil-phase solution may be dispersed into a monomer of aqueous-phase solution and an initiator be added to begin the poly-addition process between the two monomers. At the outer layer of the phase change material distributed oil droplet, a membrane forms immediately. In the present time, the sensible and latent heat–based inventions are launched in the market, however, researchers are improving and transforming the thermochemical heat storage technology to increase its efficiency. The thermochemical heat storage materials contain higher energy density than latent and sensible storage materials. While thermochemical storage materials are worked at ambient temperature, the process temperature is required for latent and sensible storage temperature. Therefore, the minimum amount of insulation and heat loss is needed for thermochemical storage.

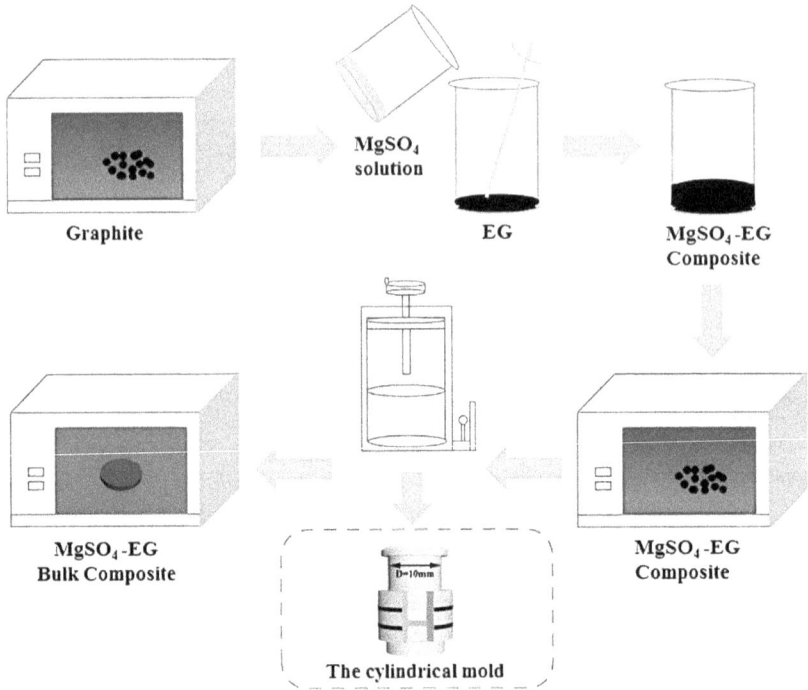

FIGURE 1.5 Preparation method of bulk composite (Miao *et al.*, 2021).

Thermochemical processes have negligible heat loss, hence they are hypothetically inexhaustible.

While latent heat and sensible heat storage materials have simple technologies, other ones have a complex technical methodology. The choice of process and analysis for several parameters, such as reversibility, rate, and operating parameters (i.e., pressure, temperature, and kinetics), are essentials for development of a thermochemical storage system (Pardo *et al.*, 2014).

The following criteria must be encountered by an appropriate thermochemical storage material: (1) The endothermic reaction needed to store heat must take place below 1273 K. It is recommended that the exothermic reaction needed to recover heat take place above 773 K. (2) The capacity for storing the heat can be maximized via high reaction enthalpy and low molar volume. (3) The chemical byproducts of the reactions are manageable. The environment and chemical molecules should not interact. (4) Magnesium sulfate and hydroxyapatite-based thermochemical heat storage materials have been synthesized by the impregnation process. (5) The composite material reflected stability over both hydration and dehydration cycles (Nguyen *et al.*, 2022). (6) However, the use of expanded graphite enhances thermal and mass transfer characteristics of magnesium sulfate (Figure 1.5). (7) The polymethylhydroxysiloxane and magnesium sulfate-based composite foam have been synthesized by Calabrese *et al.* (2019). Foam based on composite reflected mechanical stability by increasing magnesium sulfate.

1.8 CONCLUSION

The broad applications of energy storage materials are crucial for developing clean and renewable energy for society. While they hold significant promise in terms of improving thermal properties, relevant theories and research have also generated various promising findings. The preparation technologies are advancing recently.

REFERENCES

Alva, G. *et al.* (2017) 'Thermal energy storage materials and systems for solar energy applications', *Renewable and Sustainable Energy Reviews*, 68, pp. 693–706. Available at: https://doi.org/10.1016/j.rser.2016.10.021.

Boussaba, L. *et al.* (2021) 'Investigation and properties of a novel composite bio-PCM to reduce summer energy consumptions in buildings of hot and dry climates', *Solar Energy*, 214, pp. 119–130.

Calabrese, L. *et al.* (2019) 'Magnesium sulphate-silicone foam composites for thermochemical energy storage: Assessment of dehydration behaviour and mechanical stability', *Solar Energy Materials and Solar Cells*, 200, p. 109992. Available at: https://doi.org/10.1016/j.solmat.2019.109992.

Cárdenas, B. and León, N. (2013) 'High temperature latent heat thermal energy storage: Phase change materials, design considerations and performance enhancement techniques', *Renewable and Sustainable Energy Reviews*, 27, pp. 724–737. Available at: https://doi.org/10.1016/j.rser.2013.07.028.

Demirbas, M.F. (2006) 'Thermal energy storage and phase change materials: An overview', *Energy Sources, Part B: Economics, Planning, and Policy*, 1(1), pp. 85–95.

Dixit, P. *et al.* (2022) 'Salt hydrate phase change materials: Current state of art and the road ahead', *Journal of Energy Storage*, 51, p. 104360. Available at: https://doi.org/10.1016/J.EST.2022.104360.

Fabiani, C. *et al.* (2020) 'Palm oil-based bio-PCM for energy efficient building applications: Multipurpose thermal investigation and life cycle assessment', *Journal of Energy Storage*, 28, p. 101129.

Farid, M.M. *et al.* (2004) 'A review on phase change energy storage: Materials and applications', *Energy Conversion and Management*, 45(9–10), pp. 1597–1615. Available at: https://doi.org/10.1016/j.enconman.2003.09.015.

Fernandes, D. *et al.* (2012) 'Thermal energy storage: "How previous findings determine current research priorities"', *Energy*, 39(1), pp. 246–257.

Fu, X. *et al.* (2017) 'Synthesis and properties of bulk-biodegradable phase change materials based on polyethylene glycol for thermal energy storage', *Journal of Thermal Analysis and Calorimetry*, 128(2), pp. 643–651. Available at: https://doi.org/10.1007/S10973-016-5959-8/FIGURES/8.

Garg, H.P., Mullick, S.C. and Bhargava, V.K. (2012) *Solar thermal energy storage*. D. Reidel Publishing Company.

Hänchen, M., Brückner, S. and Steinfeld, A. (2011) 'High-temperature thermal storage using a packed bed of rocks–heat transfer analysis and experimental validation', *Applied Thermal Engineering*, 31(10), pp. 1798–1806.

Kateshia, J. and Lakhera, V. (2022) 'A comparative study of various fatty acids as phase change material to enhance the freshwater productivity of solar still', *Journal of Energy Storage*, 48, p. 103947. Available at: https://doi.org/10.1016/J.EST.2021.103947.

Khodadadi, J.M.M., Fan, L. and Babaei, H. (2013) 'Thermal conductivity enhancement of nanostructure-based colloidal suspensions utilized as phase change materials for thermal energy storage: A review', *Renewable and Sustainable Energy Reviews*, 24, pp. 418–444. Available at: https://doi.org/10.1016/j.rser.2013.03.031.

Kibria, M.A. *et al.* (2015) 'A review on thermophysical properties of nanoparticle dispersed phase change materials', *Energy Conversion and Management*, 95, pp. 69–89. Available at: https://doi.org/10.1016/j.enconman.2015.02.028.

Liang, C. *et al.* (2009) 'Microencapsulation of butyl stearate as a phase change material by interfacial polycondensation in a polyurea system', *Energy Conversion and Management*, 50(3), pp. 723–729. Available at: https://doi.org/10.1016/j.enconman.2008.09.044.

Liu, C. *et al.* (2022) 'Synthesis and characterization of microencapsulated phase change material with phenol-formaldehyde resin shell for thermal energy storage', *Solar Energy Materials and Solar Cells*, 243, p. 111789. Available at: https://doi.org/10.1016/J.SOLMAT.2022.111789.

Liu, H., Wang, X. and Wu, D. (2017) 'Fabrication of graphene/TiO$_2$/paraffin composite phase change materials for enhancement of solar energy efficiency in photocatalysis and latent heat storage', *ACS Sustainable Chemistry & Engineering*, 5(6), pp. 4906–4915. Available at: https://doi.org/10.1021/acssuschemeng.7b00321.

Maithya, O.M. *et al.* (2021) 'High-energy storage graphene oxide modified phase change microcapsules from regenerated chitin Pickering Emulsion for photothermal conversion', *Solar Energy Materials and Solar Cells*, 222, p. 110924. Available at: https://doi.org/https://doi.org/10.1016/j.solmat.2020.110924.

Miao, Q. *et al.* (2021) 'MgSO$_4$-expanded graphite composites for mass and heat transfer enhancement of thermochemical energy storage', *Solar Energy*, 220, pp. 432–439. Available at: https://doi.org/10.1016/j.solener.2021.03.008.

Nguyen, M.H. *et al.* (2022) 'Toward new low-temperature thermochemical heat storage materials: Investigation of hydration/dehydration behaviors of MgSO$_4$/Hydroxyapatite composite', *Solar Energy Materials and Solar Cells*, 240, p. 111696. Available at: https://doi.org/10.1016/j.solmat.2022.111696.

Pardo, P. *et al.* (2014) 'A review on high temperature thermochemical heat energy storage', *Renewable and Sustainable Energy Reviews*, 32, pp. 591–610.

Pielichowska, K. and Pielichowski, K. (2014) 'Phase change materials for thermal energy storage', *Progress in Materials Science*, 65, pp. 67–123. Available at: https://doi.org/10.1016/j.pmatsci.2014.03.005.

Pincemin, S. *et al.* (2008) 'Highly conductive composites made of phase change materials and graphite for thermal storage', *Solar Energy Materials and Solar Cells*, 92(6), pp. 603–613. Available at: https://doi.org/10.1016/j.solmat.2007.11.010.

Sarı, A. (2004) 'Form-stable paraffin/high density polyethylene composites as solid–liquid phase change material for thermal energy storage: Preparation and thermal properties', *Energy Conversion and Management*, 45(13), pp. 2033–2042. Available at: https://doi.org/10.1016/j.enconman.2003.10.022.

Sarı, A. and Karaipekli, A. (2007) 'Thermal conductivity and latent heat thermal energy storage characteristics of paraffin/expanded graphite composite as phase change material', *Applied Thermal Engineering*, 27(8–9), pp. 1271–1277. Available at: https://doi.org/10.1016/j.applthermaleng.2006.11.004.

Sarı, A. and Karaipekli, A. (2012) 'Fatty acid esters-based composite phase change materials for thermal energy storage in buildings', *Applied Thermal Engineering*, 37, pp. 208–216. Available at: https://doi.org/10.1016/j.applthermaleng.2011.11.017.

Sarier, N. and Onder, E. (2012) 'Organic phase change materials and their textile applications: An overview', *Thermochimica Acta*, 540, pp. 7–60. Available at: https://doi.org/10.1016/j.tca.2012.04.013.

Sarier, N. *et al.* (2011) 'Preparation of phase change material–montmorillonite composites suitable for thermal energy storage', *Thermochimica Acta*, 524(1), pp. 39–46. Available at: https://doi.org/10.1016/j.tca.2011.06.009.

Serrano, S. *et al.* (2015) 'Composite gypsum containing fatty-ester PCM to be used as constructive system: Thermophysical characterization of two shape-stabilized formulations', *Energy and Buildings*, 86, pp. 190–193.

Sharma, R.K. *et al.* (2015) 'Developments in organic solid–liquid phase change materials and their applications in thermal energy storage', *Energy Conversion and Management*, 95, pp. 193–228. Available at: https://doi.org/10.1016/j.enconman.2015.01.084.

Su, J.-F. *et al.* (2017) 'Preparation and physicochemical properties of microcapsules containing phase-change material with graphene/organic hybrid structure shells', *Journal of Materials Chemistry A*, 5(45), pp. 23937–23951. Available at: https://doi.org/10.1039/C7TA06980D.

Xu, B., Li, P. and Chan, C. (2015) 'Application of phase change materials for thermal energy storage in concentrated solar thermal power plants: A review to recent developments', *Applied Energy*, 160, pp. 286–307.

Zhang, H. and Wang, X. (2009) 'Synthesis and properties of microencapsulated n-octadecane with polyurea shells containing different soft segments for heat energy storage and thermal regulation', *Solar Energy Materials and Solar Cells*, 93(8), pp. 1366–1376. Available at: https://doi.org/10.1016/j.solmat.2009.02.021.

Zhu, Y. *et al.* (2019) 'Graphene/SiO_2/n-octadecane nanoencapsulated phase change material with flower like morphology, high thermal conductivity, and suppressed supercooling', *Applied Energy*, 250, pp. 98–108. Available at: https://doi.org/10.1016/j.apenergy.2019.05.021.

2 Latent Heat Storage
Materials and Applications

Abhijith M T and A. Sreekumar

2.1 INTRODUCTION

The world is transitioning to a decarbonized economy after the emergence of the industrial revolution. Energy is vital to the nation's economic growth and technological competitiveness. The increase of resources and energy supply has failed to satisfy the ever-increasing demands imposed by a growing population, fast urbanization, and an expanding economy. Energy storage device development that is efficient and cost-effective is critical for preserving energy, diminishing reliance on fossil fuels, and reducing greenhouse gas emissions. Energy storage systems bridge the supply-demand gap while also improving the performance and dependability of energy networks. Energy storage, when paired with viable energy sources such as solar and waste heat management, has great potential to reduce overall energy use. The type of energy source, length of storage, operational environment, economic feasibility, and other considerations all influence thermal energy storage (TES) selection. TES is the storage of heat in sensible, latent, and thermochemical forms. Because of its low energy density, sensible heat storage requires large volumes and proper architecture to release thermal energy. Latent heat storage devices are easier and more affordable to operate than are thermochemical heat storage systems.

This chapter's goal is to help the readers by giving them a thorough review of the characteristics of phase change materials (PCMs) with an emphasis on their technological applications. The uses of the PCM characteristics in storage applications are reviewed extensively, along with potential material developed by the research group of the Department of Green Energy Technology, Pondicherry University. PCMs' current state and prospects for the future are examined comprehensively.

2.2 NEED AND SIGNIFICANCE OF ENERGY STORAGE

Energy storage systems (ESS) have the potential to significantly increase the efficiency of energy conversion devices usage while also providing a wide range of fuel alternatives in the global economy. ESS are designed to store excess energy and make it available to customers on demand. They can balance energy supply and demand by using fluctuating production of reusable energy sources such as solar, wind, and bio-energy and by improving overall energy system efficiency and lowering CO_2 emissions. Energy storage cannot be adequately evaluated as to its

DOI: 10.1201/9781003345558-3

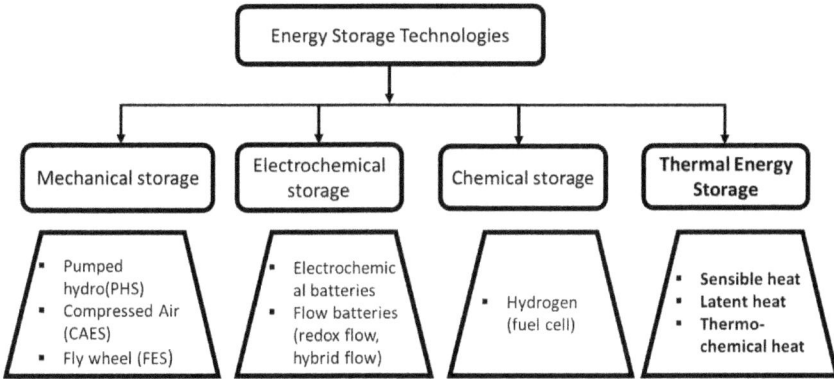

FIGURE 2.1 Energy storage classifications.

complexity without a good grasp of energy sources and end-use considerations. The following qualities can be utilized to characterize an ESS:

- **Capacity:** The stored amount of energy is determined by the storage mechanism, medium, and size.
- **Power:** Defines the rate of energy stored in the system that may be discharged or charged.
- **Efficiency:** Defines the proportion of energy given to users to the amount of energy required to charge the energy storage system. It calculates the amount of energy wasted throughout the charging/discharging cycle as well as during storage.
- **Storage duration:** Elucidates how long the energy is stored.
- **Charge-discharge time:** Explains how much time is required to charge or discharge the system.
- **Cost:** Relies on the lifespan and capital costs of the system as well as the capacity or power of the storage system.

Several different energy storage approaches are being explored. The methods of energy storage, such as mechanical, thermal, chemical, and electrochemical, are grouped in Figure 2.1.

Among all storage approaches, thermal energy storage (TES) is one of the most widely used technologies that is affordable in practical applications. It has also been demonstrated that the use of TES may enhance the overall performance of the system, increase fuel savings, minimize investment and operating costs, boost energy supply security, and reduce pollution to the environment if properly built. Capacity and power may be dependent on one another in some storage systems, making capacity, power, and discharge time non-independent variables. For instance, increased power entails improved heat transfer in TES systems (for instance, extra fins in the heat exchanger), which reduces the amount of active storage material and subsequently the capacity.

2.3 THERMAL ENERGY STORAGE (TES)

TES gains importance when heat demand and production mismatches. Thermal energy storage refers to the temporary storage technique that stores heat energy and can be retrieved/used for low or high temperatures. The charging and discharging process flow in TES medium adjusts for the imbalance between energy demand and supply while also improving system energy efficiency. Thermal energy storage works on a cyclic basis which involves three steps followed by the storage medium, as shown in Figure 2.2.

TES is becoming a noteworthy energy scenario and is acquiring a great deal of relevance in a variety of engineering fields and general applications. TES has a huge potential for energy conservation, which can reduce its carbon footprint on the environment. The major benefits that can be obtained when integrating TES into an ESS are:

- **Economically feasible:** reducing capital investment and cost of operation
- **Efficiency:** achieving efficient energy use with minimal loss
- **Pollution free:** less CO_2 emission and environmentally friendly
- Better **reliability** and **performance** of the system

2.3.1 SIGNIFICANCE OF INTEGRATING TES INTO SOLAR THERMAL SYSTEMS

The successful and expanded use of solar energy in the domains of cooling and heating, process heat, and power production relies on TES. Seasonal, daily (day-night), and hourly (cloud) flux changes in solar thermal energy prevent solar systems from producing heat or thermal electricity following the demand profiles of customer segments. Utilizing TES and/or operating in hybrid mode (solar + fossil fuels) are two solutions to this issue. Significant advantages of TES integration in solar thermal systems include:

- TES can be utilized to cover peak demands.
- A surplus amount of energy can be utilized without wasting/dumping.

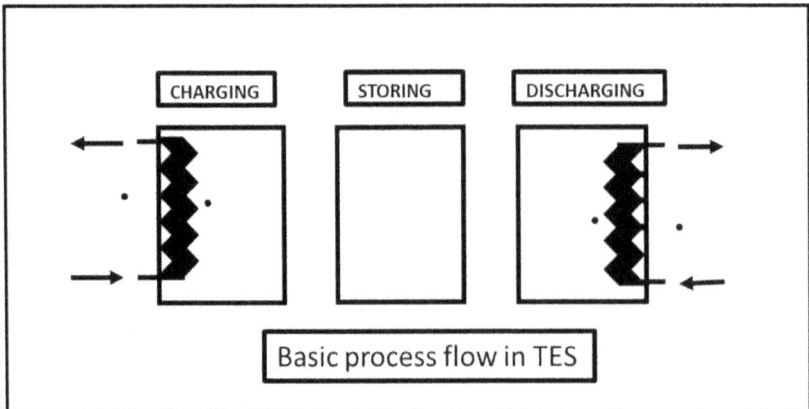

FIGURE 2.2   Principle of operation in TES.

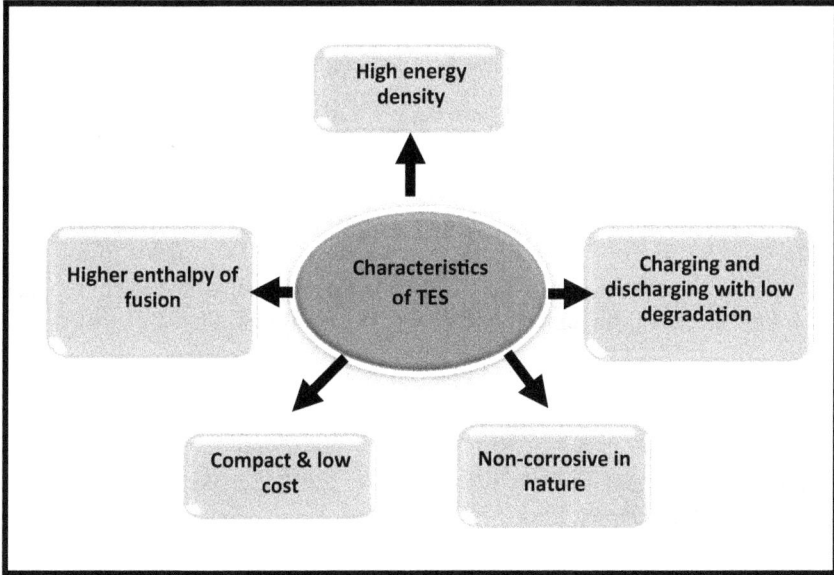

FIGURE 2.3   Characteristics of TES.

- Rapid solar flux variation may be compensated for by suitable pipework and heat exchangers, increasing system efficiency and performance.
- TES allow operation during off-peak hours (time of no solar radiation).
- The size of subsequent components can be minimized.
- TES allow better thermal management of the solar system.

Various characteristics of TES in energy storage systems are depicted in Figure 2.3.

2.4 BASIC MODES OF HEAT TRANSFER IN TES

There are three physical methods of storage in the TES system, which include sensible storage, thermochemical storage, and latent heat storage. The sensible component explains the atomic and molecular movements of translation, rotation, and vibration. The material's intermolecular forces, which affect the transition between the solid, liquid, and gaseous stages, result in the latent component. The energy held in the chemical bonds between atoms is explained by the chemical component.

Based on the properties of storage material and need, TES can be divided into three major types:

1. Sensible storage material
2. Thermochemical storage material
3. Latent heat storage material

2.4.1 Sensible Heat Storage

The process of storing thermal energy by changing the temperature of the storage material is known as sensible heat storage (SHS). The heat energy is stored in sensible storage material based on their density, specific heat capacity, and temperature swing. The temperature of the material changes during the charging or discharging cycle but does not undergo any phase transition. SHS is most typically employed for small-scale energy applications (Philip, Raam Dheep, et al., 2020).

The thermal energy stored in SHS is directly proportional to the specific capacity, mass of the material, and temperature change, which is represented by the equation:

$$Q = mC_P(T_H - T_L)$$
$$= mC_P\Delta T$$

T_H and T_L are the high and low temperatures of the SHS material in $°C$
Q = quantity of heat stored in kJ
m = mass of material in kg
C_P = specific heat capacity of the material in kJ/kg K

SHS is comparatively simpler in design and working as compared to latent heat storage. However, it has the disadvantage of being larger volume as compared to thermochemical and latent heat storage. Energy density of the storage material is considered as an important criterion for selection of material. The stratification process in tanks is currently being explored to improve the performance of SHS due to its inability to maintain a constant temperature during operation.

2.4.2 Thermochemical Storage

In the thermochemical storage (TCS) method, the difference in enthalpy of reactants and products in a chemical reaction is used to store the thermal energy. The available stored heat is directly proportional to the volume of material used to store energy, mass of the material, and heat of the reaction (endothermic), which is shown by the equation:

$$Q = a_r \, m\Delta h_r$$

Q = heat stored on kJ
m = mass in kg
a_r = fraction reacted in kg
Δh_r = endothermic heat fusion in kJ/kg

Thermal insulation in not required in this energy storage. TCS systems and their use in power-to-heat application is a pioneering technology and highlights advantages of flexibility, load management, and continuous operation to increase profitability in the industrial sector. However, this storage method is quite expensive and still needs advancement.

2.4.3 Latent Heat Storage

Because of its near isotherm functioning and high storage density, latent heat storage (LHS) is widely known for its advantages over traditional SHS. LHS is determined by a material's phase transition. Melting and solidification of a substance are employed in the solid-liquid phase transition. When sufficient heat energy is given, the latent heat material melts and stores huge amounts of thermal energy at a steady temperature and is dissipated when the material solidifies. LHS storing mediums are known as phase change materials (PCMs).

2.5 THEORY OF STORED ENERGY IN LHS MATERIALS

Unlike SHS, latent heat storage materials provide higher storage density at a narrower temperature range. In addition, LHS materials such as PCMs are able to store 5–14 times more heat for the same volume as SHS materials (Sharma et al., 2009). The amount of latent heat stored can be calculated as:

$$Q = m\Delta h$$

Where Q is the amount of heat stored in the material (J), m represents the mass of the storage medium in kg, and Δh denotes the phase transition enthalpy in J/kg.

Also, the heat storage ability of LHS with a PCM is given by the mathematical expression:

$$Q = sensible\,heat_{(melting)} + latent\,heat + sensible\,heat_{(solidifying)}$$

$$Q = m\,.C_{ps}\left(T_m - T_i\right) + m.a_m H + m.C_{pl}\left(T_f - T_m\right)$$

$$Q = m\left[C_{ps}\left(T_m - T_i\right) + a_m H + C_{pl}\left(T_f - T_m\right)\right]$$

Q = Heat stored on kJ
m = Mass in kg
a_m = Fraction reacted in kg.
H = Phase change enthalpy
T_i = Initial temperature in K
T_f = Final temperature in K
T_m = Temperature at which melting happens in K
C_{ps} =Specific capacity of solid in kJ/kg K
C_{pl} = Specific capacity of liquid in kJ/kg K

2.6 CHARACTERISTICS OF LHS MATERIALS DURING CHARGING AND DISCHARGING

Figure 2.4 explains that the solid material temperature rises in proportion to the energy it receives when it is heated until it reaches the melting point. The chemical

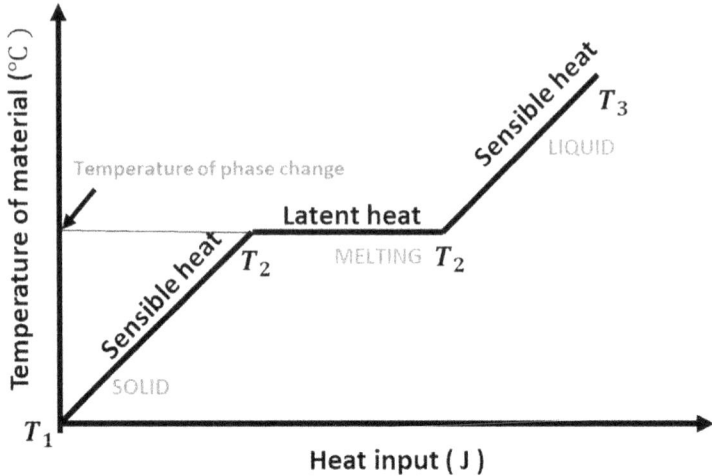

FIGURE 2.4 Temperature profile of LHS materials.

linkage in latent heat materials breaks down as the temperature rises, causing the substance to modify its state from solid to liquid. Since the phase transition is a heat-seeking process, PCM absorbs heat. During charging, when the phase transition temperature is reached, the material begins to melt. The temperature remains constant until the melting process is completed. Because the heating and cooling processes are comparable, it is possible to recover the stored thermal energy as latent heat at a constant temperature.

2.7 PHASE CHANGE MATERIALS: CHARACTERISTICS AND CLASSIFICATION

LHS materials are generally known as phase change materials. Thermal energy is stored and retrieved by undergoing a change in phase of the materials. Phase transition can take place in the following forms:

- Solid: Solid
- Solid: Liquid
- Solid: Gas
- Liquid: Gas

As previously stated, PCMs may store latent heat through solid-solid, solid-liquid, solid-gas, and liquid-gas phase shifts. However, the solid-liquid phase shift is the sole one employed for PCMs. Liquid-gas PCMs are not suitable for thermal storage due to the large volume or high pressure required to keep the components in their gas phase. The heat of transformation for liquid-gas phase transitions is greater than for solid-liquid phase transitions. Solid-solid PCMs are usually sluggish with a low heat of transition.

2.7.1 Selection of a PCM: Thermophysical Properties

There has been some general selection criteria for choosing a PCM as thermal energy storage material. PCMs are selected based on their uses in the required application (Oró et al., 2012). Thermophysical properties are initially classified as:

- **Thermodynamic properties**
  The PCM should possess
  1. The melting point in the specified working temperature range
  2. High fusion latent heat per unit volume
  3. High specific heat
  4. High density
  5. High thermal conductivity
  6. Congruent melting
  7. Minimal volume changes during the phase transformation and low vapor pressure at operating temperature to reduce the confinement issue
- **Kinetic properties**
  1. Rapid crystal formation, such that the system can fulfill heat recovery demands from the storage system
  2. High nucleation to avoid liquid phase supercooling
- **Chemical properties**
  1. Chemical endurance
  2. Charge-discharge cycle that is completely reversible
  3. No deterioration after repeated heat cycling
  4. Non-corrosive
  5. Non-toxic
  6. Non-flammable
  7. Non-explosive
- **Economic characteristics**
  1. Low cost
  2. Widespread availability of materials

2.7.2 PCM: Classification

PCMs are classified based on melting temperature and material composition, as depicted in Figure 2.5.

Meanwhile, PCMs are classified into three types, and they are further divided based on the properties of the material used, as presented in Figure 2.6 (Raam Dheep & Sreekumar, 2019).

- **Organic:** Organic PCM refers to the material that is carbon-based compounds. They are non-corrosive and can freeze and melt repeatedly without phase separation. Organic materials are classified into two types: paraffins and non-paraffins. These materials' features include high latent heat of fusion, limited thermal conductivity, instability at high temperatures, and self-nucleation with little or no supercooling.

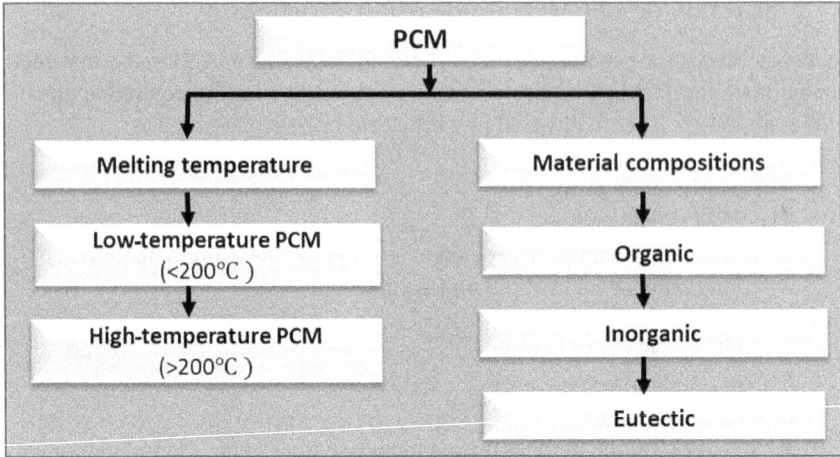

FIGURE 2.5   PCMs in terms of temperature and material composition.

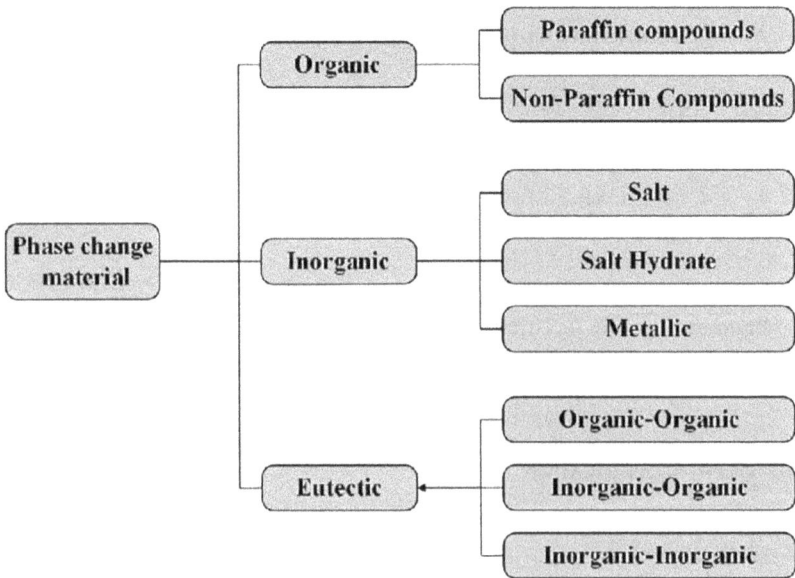

FIGURE 2.6   Classification of PCMs.

- **Inorganic:** Inorganic PCM comprises salt hydrates, nitrates, and metal-lic compounds. They can withstand temperatures up to 1500°C. Inorganic PCMs are superior in terms of being economical, having sharp melting points, high enthalpy, and very low volume change. They have drawbacks such as supercooling, segregation, material degradation and corrosion of containment, low specific heat, and reductions in enthalpy after a few ther-mal cycles owing to incongruent melting.

TABLE 2.1

Classification of Latent Heat Storage Materials Based on Their Merits and Demerits

	Merits	Demerits	Examples
ORGANIC	• Possess good thermal and chemical stability • The high heat values • Low cost and readily available • Non-flammable • Non-corrosive in nature	• Lower thermal conductivity • Low enthalpy • Large volume change during phase transition	• Paraffin • Fatty acid • Alcohol • Esters
INORGANIC	• Non-flammable • Low cost • The high heat of fusion • Good thermal conductivity	• Chances of corrosion • Shows little phase decomposition • High supercooling • Insufficient thermal stability	• Salt hydrates • Metallics
EUTECTIC	• Sharp melting point • Good structural and chemical stability • High latent heat of fusion • Little or no supercooling	• Costly • Lower thermal conductivity ranging 0.1–0.2 W $m^{-1}K^{-1}$	• Inorganic-organic • Inorganic-inorganic • Organic-organic

- **Eutectic:** Eutectics are formed by the combination of two or more components. Due to their crystallization, they form a combination of components with a limited probability of separation back into components. Eutectics are classified into organic and organic, inorganic and inorganic, and inorganic and organic combinations based on their composition. One of the distinguishing characteristics of eutectic substances is that they melt and freeze congruently with no segregation (Oró et al., 2012).

The advantages and disadvantages of phase change materials are listed in Table 2.1, showing the feasibility of choosing the appropriate material.

2.8 THERMAL ANALYSIS TO DETERMINE THE POTENCY OF LATENT HEAT STORAGE MATERIALS

The phase transition temperature and enthalpy of fusion are the major concerns of LHS materials. Generally, thermophysical properties are calculated to determine the performance parameter of the TES system. The potency of the LHS materials can be identified and measured using the following techniques:

- **Differential Scanning Calorimetry (DSC).** This is used to determine the thermal properties such as phase transition temperature and latent heat of fusion of PCM for the required application. A typical DSC run involves heating/

cooling the sample at a controlled steady rate by monitoring the heat flow
rates.
- **Differential Thermal Analysis (DTA).** The thermal stability of the PCM is
  analyzed through this technique. This characterization is a basic study for
  LHS material to evaluate how applied temperature affects the material property.

2.9 THERMAL ENERGY STORAGE APPLICATIONS OF LATENT HEAT STORAGE (LHS) MATERIALS

The design and integration of latent heat thermal energy storage materials for a
specific application is crucial in their development. Currently, researchers are conducting experimental and modeling research on PCMs using multiple possible
understandings of thermal storage and operating strategies that are both economically and sustainably viable. The PCMs' environmental cost should be assessed. For
example, if the embedded energy of a material has a long payback period, it may not
be recovered during usage. In this method, the production of PCMs from renewable,
biological sources is particularly desirable. TES can be made from a variety of fatty
molecules generated by algae and plants, such as palm oil and rapeseed. Because
these sources reproduce fast and rely heavily on solar energy, organic PCMs may be
extracted from them in a sustainable and low-energy way. Thermal lifetime analysis
is a useful method for determining the long-term benefits of a PCM option (Arkar &
Medved, 2007).

PCMs are commonly employed in TES systems nowadays and have a wide range
of applications. Here are a few examples of the various applications that are documented in the literature (Zalba et al., 2004):

- Effective storage technology for solar thermal systems
- PCM integrated passive and active storage system in the building environ-
  ment
- Cooling requirements: applicable on off-peak hours and reduction of the
  conventional electrical source of power
- Heating requirements: water heating using off-peak rates and adapting
  unloading curves
- Human thermal comfort: maintaining thermal comfort with PCM in the liv-
  ing room environment
- Thermal protection of food: ice cream storage, hotel trade, etc.
- Thermal management in electronic devices, batteries, and e-vehicles
- Engine cooling requirements
- Medical field application in the transportation of blood, vaccine storage, etc.
- Reducing exothermic temperature peaks in chemical reactions
- Solar power plant technology
- Spacecraft mission
- Food, agro-industries
- Milk dairy application: keeping milk temperature at 4°C for increasing
  usage duration

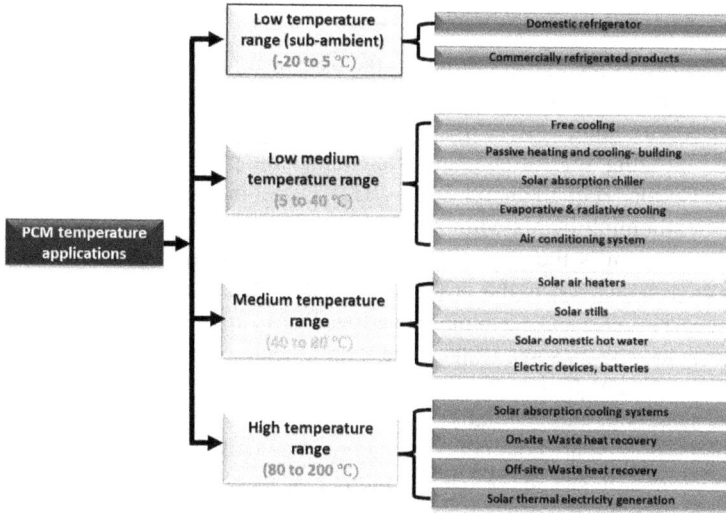

FIGURE 2.7 Application of LHS material with temperature ranges.

LHS materials are reviewed in terms of various working temperature ranges. There are four different temperature ranges from the freezing temperature range to the high-temperature range, as shown in Figure 2.7.

2.9.1 PCM APPLICATION IN SUB-AMBIENT TEMPERATURES

The notion of thermal energy storage-based cooling is gaining popularity across the world, and many environmentally conscious entities have already implemented it. In cooling applications, various materials that constitute appropriate PCMs are often used at phase transition temperatures below the ambient temperature (–20 to 5°C). Ice is the most widely used and least expensive of these PCMs. It utilizes approximately 11 kilograms of ice to store 1 kW h energy at 0°C due to its high enthalpy of fusion of 334 J g^{-1} (Dutt Sharma et al., 2004). District cooling, which is used to cool both residential and commercial structures, centralizes the production and storage of ice. In addition to ice, several of Cristopia Energy Systems' worldwide district cooling storage tanks are integrated with encapsulated PCMs composed of fatty alcohols, fatty acids, paraffin, and salt hydrates with phase transition temperature ranging from –33 to 27°C.

On a small scale, major commercial buildings may be individually cooled using devices like the Ice Bank created by CALMAC (Sharma et al., 2013). During nighttime, when power rates are lowest off-peak, this system freezes water held in huge tanks employing a cold absorption refrigeration system with an antifreeze solution comprised of water and glycol as a heat transfer fluid. The storage containers are occupied with paraffin, glycols as PCMs to retain perishable goods at or near the PCMs' cold phase transition temperature. Containers made with integrated PCM packs provide a straightforward way to keep sensitive medicinal goods at the cold temperature throughout transportation (Noël et al., 2016).

2.9.2 PCM Application in Ambient Temperatures

When power prices are higher during the daytime peak, PCMs in building materials with phase transition temperatures close to ambient (5–40°C) can lessen the burden on the air conditioning system. PCM integration into building materials has been researched utilizing several strategies, including immersion, direct incorporation, and encapsulation (Hajba-Horváth et al., 2022). Direct incorporation is the technique of directly combining PCMs with building materials (such as gypsum, concrete, or mortar); immersion is the process of impregnating building materials with liquid PCM; and encapsulation is the process of employing either tiny (1–1000 mm) or bigger particles. Cabeza et al. compiled various PCMs often used in building cooling applications, as well as some of their thermophysical properties. These PCMs may be passively implanted into a variety of building components, including wallboards, concrete, insulation, and ventilation systems. PCMs with a melting temperature in the range of 21–28°C are commonly used to provide temperature-regulated clothing for use in warm climates and provides cooling ease by lowering diurnal temperature swings from 27°C to 20–22°C (Noël et al., 2016).

Several commercial products are available, and much research has been undertaken on the basis of the thermophysical properties of PCMs used in building construction materials in a variable climatic situation. To boost the heat storage capacity of wallboards, paraffin PCM is usually used to permeate them and Micron® wallboards and metal ceiling tiles are two commercial products that use paraffin waxes. Integration of fatty acid combinations such as decanoic and dodecanoic, as well as esters such as butyl stearate and propyl palmitate, has also received a lot of attention.

2.9.3 PCM Application in Moderate Temperatures

Many PCMs including salt hydrates undergo phase shifts at temperatures ranging from 40°C to slightly more than 80°C. One or more hydrating liquids and inorganic or organic salts are combined to form salt hydrate materials. Salt hydrates were mostly studied as materials for developing stable PCMs (Nada et al., 2020). Salt hydrates have higher volumetric fusion temperatures and densities in the moderate-temperature domain than do other materials. This study looks at a variety of salt hydrate PCMs as examples. The majority of salt hydrate PCMs have moderate to high latent heat of fusion ranging from 100–300 Jg^{-1} (Saqib & Andrzejczyk, 2023). A vast number of organic compounds undergo phase shifts at moderate temperatures, including paraffin $CH_3(CH_2)n\,CH_2)$ (Bhagwat et al., 2021), fatty acids ($CH_3\,(CH_3)n$ COOH) (Hasan, 1994), and sugar alcohols (Mazman et al., 2008). The enticing feature of being obtained from renewable, plant-based sources applies to both fatty acids and sugar alcohols. According to Kürklü et al. (2002) and Shukla et al. (2009), fatty acids cycle well and retain their thermodynamic properties through a large number of cycles, making them ideal for many TES applications. Solar domestic water heating has become more envisaging technology in mild temperature range storage systems, among other thermal energy storage applications.

2.9.3.1 Solar Water Heating System

In the moderate temperature range, PCMs are extensively used in household solar hot water applications. The size and weight of this sort of practical storage prevent its usage in various residential or business situations, notably for retrofitting in limited places (Sharma et al. 2009; Atkin & Farid, 2015). In this design, solar energy may be charged into a PCM of definite volume, and chilled water may subsequently be heated up by flowing through a heat exchanger inside the PCM tank. Water acts as a heat transfer medium, which is freely allowed to pass through an evacuated tube or flat-plate solar collector to convert available solar thermal radiation to heat. This heat is then used to warm water for domestic use. Many concepts for PCM-based solar water heaters are investigated and studied for this moderate temperature range to meet the green needs of the environment and effective energy recovery.

2.9.4 PCM Application in High Temperatures

Metals and anhydrous salts are among the PCMs utilized for high-temperature heat storage, with temperatures ranging from several hundred to well over 1000°C. Various chlorides, carbonates, sulphates, nitrites, and nitrates, as well as their eutectic mixtures, are widely utilized for the latter. Despite melting at temperatures ranging from 300–550°C, nitrates and nitrites exhibit a low latent heat of (100–175) J g^{-1}. Despite having a greater latent heat, which typically exceeds 200 J g^{-1}, chlorides and carbonates do not melt until they reach 700°C (Raj & Velraj, 2010). Because molten salts may corrode many types of steel and their vapors are often reactive, costly alloys and coatings are necessary to provide the system with an acceptable lifespan (Zhou et al., 2012). Metals and their eutectic alloys show good thermal properties as PCMs such as better thermal conductivity, less corrosivity, and high latent heat of fusion, so their use is occasionally competitive with salts, despite being more expensive.

Because of the direct generation of steam in the absorber tubes, latent heat storage devices have gained interest in the development of CSP technology. The latent heat system serves as a steam storage system for the two-phase heat transporter water/steam. The melting points of possible PCMs range from 120--340°C. In this circumstance, it is often not cost effective to adopt storage methods that involve sensible heat storage. Sensible heat storage systems for two-phase heat carriers have low energy efficiency or considerable exergy due to significant temperature variations between storage medium and HTFs. A phase change in the PCM and the heat carrier, on the other hand, occurs at nearly the same temperature in a latent heat storage system. Consequently, it is feasible to further prevent greater temperature variations.

2.10 ENHANCEMENT TECHNIQUES TO IMPROVE THE PERFORMANCE OF LHS MATERIALS

Thermal conductivity and heat transfer mechanism of PCMs can be improved by carrying out some enhancement methods, which are classified as (Figure 2.8):

- Conductivity can be enhanced by adding nanomaterials, porous media.
- The rate of heat transfer can be enhanced by encapsulation methods, finned tube set-up.
- Process uniformity can be improved by using multiple PCMs.

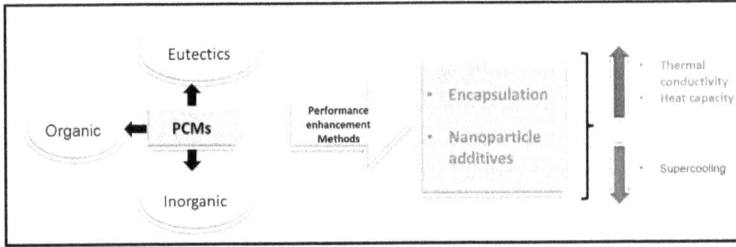

FIGURE 2.8 Enhancement techniques on PCMs.

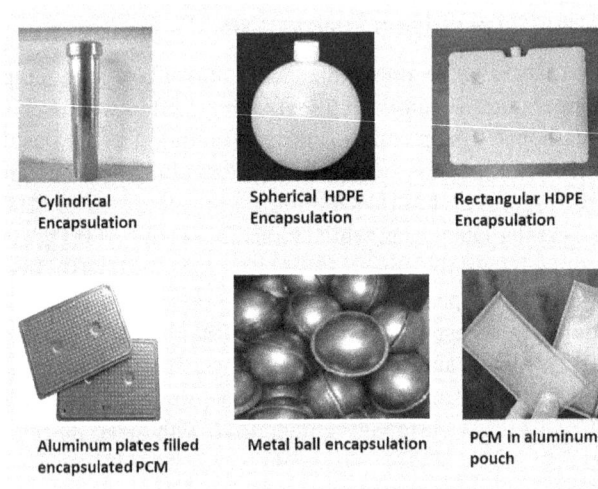

FIGURE 2.9 Different encapsulation methods.

To prevent direct contact with the environment, the PCM must be encapsulated and cannot be employed directly in the application. Additionally, during energy storage and retrieval, the PCM undergoes a physical phase transition. It should not leak while going through the phase transition. In an encapsulating environment, the PCM should have properties including flexibility, strength, corrosion resistance, and thermal stability. Furthermore, it should give the requisite surface area for heat transmission and provide greater structural stability. As a result, the PCM is enclosed following the specifications. It can be classified as nano, micro, or macro encapsulation depending on its size, which is chosen according to the required applications. Metals, polymers, and other materials are frequently employed as encapsulating materials. Encapsulation of PCM can be easily done by two generally accepted techniques such as macroencapsulation and microencapsulation. Additionally, macroencapsulation is widely used in the case of TES with a heat exchanger system (Chinnasamy & Appukuttan, 2019).

The PCM encapsulated in spherical capsules, macroscale metallic or polymeric sheet, SS316 cylinder, and pouches are referred to as macroencapsulation, which is highlighted in Figure 2.9 (Philip, Veerakumar, et al., 2020).

2.11 FUTURE ADVANCEMENT AND CHALLENGES IN PCM APPLICATIONS

To accommodate the requirements of various applications, PCMs are available in a range of melting temperatures and enthalpies. To utilize energy efficiently, TES systems are used in applications that absorb and dissipate heat utilizing PCMs. For optimal performance and to maintain systems in an ideal temperature range, PCMs are widely used in solar-based thermal collectors, building cooling technologies, glass, clothing fabrics, drug storage, thermal batteries, vegetable drying, etc. The thermally advanced PCMs draw continuously growing attention in the following applications (Mehling et al., 2022):

- Solar stills
- Thermoelectric generators
- Food storage
- Solar heating
- Advanced batteries with green environment
- Building and HVAC
- PV/T systems
- Smart textiles
- Glazing
- Heat exchangers

PCMs employed in many green storage applications have drawn a lot of benefits in TES, empowering sustainable development of energy conservation and audit. The major challenges and suggestions for cautious evaluation based on latent heat storage technology are listed as follows (Magendran et al., 2019; Mehling et al., 2022):

- Major complications are there for PCM technologies in universal standards of testing and analysis, even though a lot of research is being done towards material advancement properties of PCMs for their suitability in vivid applications.
- The cost of nanoparticles, structural instability, and thermal conductivities are all major barriers to the mainstream use of PCM materials.
- The best thermal conductivity materials, such as paraffin wax or fatty acids, are extremely combustible. Organic PCMs have a significant problem of low thermal conductivity, which adversely affects their applicability in TES systems.
- PCM encapsulation is expensive, time-consuming, and direct to supercooling properties, disturbing thermal balance. To optimize the PCM encapsulation technique, more research is required.
- Although the majority of PCMs and nano-encapsulated PCMs (NPCMs) have well-defined melting points, several investigations have shown different thermophysical characteristics of liquids and solids.

- Nanoparticles have proven to be effective as a PCM addition in recent experiments. However, there isn't much research analyzing how to use hybrid nanomaterials. Therefore, more investigation focused on the thermophysical characterization and performance evaluation of NPCM is required.
- Future steps in research will be more relevant in terms of phase segregation and supercooling of PCMs to ensure the energy-efficient performance of the storage systems.

2.12 CONCLUSION

In this chapter, several recent advancements in thermal energy storage systems were discussed. Among various TES technologies, latent heat storage materials such as PCMs contribute the best part in eradicating the imbalance between energy supply and demand through various operating principles. Mechanism of storage techniques in LHS material with energy analysis calculation shows the effectiveness of thermal energy storage system. The classifications of various PCMs and their suitability for potential applications in various operational temperature ranges are examined and reported in this study. Breakthroughs in the areas of performance enhancement of latent storage materials will increase thermal stability by various approaches also mentioned in this study. Furthermore, the possible integration of PCMs with solar thermal devices and conventional sources will cause a leap in the energy storage duration and hence achieve a sustainable energy economy.

REFERENCES

Arkar, C., & Medved, S. (2007). Free cooling of a building using PCM heat storage integrated into the ventilation system. *Solar Energy*, *81*(9), 1078–1087. https://doi.org/10.1016/j.solener.2007.01.010

Atkin, P., & Farid, M. M. (2015). Improving the efficiency of photovoltaic cells using PCM infused graphite and aluminium fins. *Solar Energy*, *114*, 217–228. https://doi.org/10.1016/j.solener.2015.01.037

Bhagwat, V. V., Roy, S., Das, B., Shah, N., & Chowdhury, A. (2021). Performance of finned heat pipe assisted parabolic trough solar collector system under the climatic condition of North East India. *Sustainable Energy Technologies and Assessments*, *45*. https://doi.org/10.1016/j.seta.2021.101171

Chinnasamy, V., & Appukuttan, S. (2019). A real-time experimental investigation of building integrated thermal energy storage with air-conditioning system for indoor temperature regulation. *Energy Storage*, *1*(3). https://doi.org/10.1002/est2.43

Hajba-Horváth, E., Németh, B., Trif, L., May, M., Jakab, M., Fodor-Kardos, A., & Feczkó, T. (2022). Low temperature energy storage by bio-originated calcium alginate-octyl laurate microcapsules. *Journal of Thermal Analysis and Calorimetry*, *147*(23), 13151–13160. https://doi.org/10.1007/s10973-022-11678-w.

Hasan, A. (1994). Thermal energy storage system with stearic acid as phase change material. *Energy Conversion and Management*, *35*(10).

Kürklü, A., Özmerzi, A., & Bilgin, S. (2002). Thermal performance of a water-phase change material solar collector. *Renewable Energy, 26,* 391–399. https://doi.org/10.1016/ S0960-1481(01)00130-6

Magendran, S. S., Khan, F. S. A., Mubarak, N. M., Vaka, M., Walvekar, R., Khalid, M., Abdullah, E. C., Nizamuddin, S., & Karri, R. R. (2019). Synthesis of organic phase change materials (PCM) for energy storage applications: A review. In *Nanostructures and nano-objects* (Vol. 20). Elsevier B.V. https://doi.org/10.1016/j. nanoso.2019.100399

Mazman, M., Cabeza, L. F., Mehling, H., Paksoy, H. Ö., & Evliya, H. (2008). Heat transfer enhancement of fatty acids when used as PCMs in thermal energy storage. *International Journal of Energy Research, 32*(2), 135–143. https://doi.org/10.1002/ er.1348

Mehling, H., Brütting, M., & Haussmann, T. (2022). PCM products and their fields of application – An overview of the state in 2020/2021. *Journal of Energy Storage, 51.* https://doi. org/10.1016/j.est.2022.104354

Nada, S. A., Alshaer, W. G., & Saleh, R. M. (2020). Effect of phase change material plates' arrangements on charging and discharging of energy storage in building air free cooling. *Energy Storage, 2*(4). https://doi.org/10.1002/est2.142

Noël, J. A., Kahwaji, S., Desgrosseilliers, L., Groulx, D., & White, M. A. (2016). Phase change materials. In *Storing energy: With special reference to renewable energy sources* (pp. 249–272). Elsevier. https://doi.org/10.1016/B978-0-12-803440-8.00013-0

Oró, E., de Gracia, A., Castell, A., Farid, M. M., & Cabeza, L. F. (2012). Review on phase change materials (PCMs) for cold thermal energy storage applications. *Applied Energy, 99*, 513–533). https://doi.org/10.1016/j.apenergy.2012.03.058

Philip, N., Raam Dheep, G., & Sreekumar, A. (2020). Cold thermal energy storage with lauryl alcohol and cetyl alcohol eutectic mixture: Thermophysical studies and experimental investigation. *Journal of Energy Storage, 27.* https://doi.org/10.1016/j. est.2019.101060

Philip, N., Veerakumar, C., & Sreekumar, A. (2020). Lauryl alcohol and stearyl alcohol eutectic for cold thermal energy storage in buildings: Preparation, thermophysical studies and performance analysis. *Journal of Energy Storage, 31.* https://doi.org/10.1016/j. est.2020.101600

Raam Dheep, G., & Sreekumar, A. (2019). Thermal reliability and corrosion characteristics of an organic phase change materials for solar space heating applications. *Journal of Energy Storage, 23*, 98–105. https://doi.org/10.1016/j.est.2019.03.009

Raj, V. A. A., & Velraj, R. (2010). Review on free cooling of buildings using phase change materials. *Renewable and Sustainable Energy Reviews, 14*(9). https://doi.org/10.1016/j. rser.2010.07.004

Saqib, M., & Andrzejczyk, R. (2023). A review of phase change materials and heat enhancement methodologies. *Wiley Interdisciplinary Reviews: Energy and Environment, 12*(3). https://doi.org/10.1002/wene.467

Sharma, A., Shukla, A., Chen, C. R., & Dwivedi, S. (2013). Development of phase change materials for building applications. *Energy and Buildings, 64*, 403–407. https://doi. org/10.1016/j.enbuild.2013.05.029

Sharma, A., Tyagi, V. V., Chen, C. R., & Buddhi, D. (2009). Review on thermal energy storage with phase change materials and applications. *Renewable and Sustainable Energy Reviews, 13*(2), 318–345. https://doi.org/10.1016/j.rser.2007.10.005

Sharma, S. D., Kitano, H., & Sagara, K. (2004). Phase change materials for low temperature solar thermal applications. *Research Reports of the Faculty of Engineering Mie University, 29.*

Shukla, A., Buddhi, D., & Sawhney R.L. (2009). Solar water heaters with phase change material thermal energy storage medium: A review. *Renewable and Sustainable Energy Reviews*, *13*(8), 2119–2125. https://doi.org/10.1016/j.rser.2009.01.024

Zalba, B., Marín, J. M., Cabeza, L. F., & Mehling, H. (2004). Free-cooling of buildings with phase change materials. *International Journal of Refrigeration*, *27*(8), 839–849. https://doi.org/10.1016/j.ijrefrig.2004.03.015

Zhou, D., Zhao, C. Y., & Tian, Y. (2012). Review on thermal energy storage with phase change materials (PCMs) in building applications. *Applied Energy*, *92*, 593–605. https://doi.org/10.1016/j.apenergy.2011.08.025

3 Thermochemical Heat Storage

Materials and Application

B. K. Purohit, Karunesh Kant, Amritanshu Shukla, and Atul Sharma

3.1 INTRODUCTION

Energy can be measured as a universal currency, considering that it is essential for carrying out all activities and is closely linked to the development of a country. Countries with more energy with respect to production and consumption levels are highly developed. The different types of energy include gravitational, thermal, chemical, mechanical, geothermal, electrical, etc. (Ding and Riffat, 2013). The energy sources can be classified as non-renewable and renewable. Numerous non-renewable (coal, natural gas, oil, nuclear power, etc.) and renewable (solar, water, wind, biomass, geothermal) sources are utilized to produce the required energy. These produced energies are supplied to a system to accomplish the work needed. However, due to the energy efficiency factors, the utilization of these supplied energies is observed to be significantly less, and most supplied energy is wasted on the environment. Therefore, new energy storage techniques were introduced for efficient and economical utilization of produced/available energy. The objective is to meet the peak demand to ensure a steady supply of energy during the demand period when primary energy sources are unavailable or with energy sources such as wind and solar energy (Dincer, 2002; Sharma *et al.*, 2009; Rammelberg, Schmidt and Ruck, 2012; Sutjahja *et al.*, 2016; Kant, Shukla and Sharma, 2017). This book is limited to the discussion on available thermal energy storage (TES) techniques.

High storage density (expressed in energy units per unit volume) and superior thermal energy charging and discharging capabilities are desirable qualities for any TES system (Prim, 2013; Purohit and Sistla, 2021). The techniques for TES systems that are generally recognized include sensible, latent, and chemical heat storage techniques. Sensible heat storage (SHS) systems typically store thermal energy by raising or lowering the temperature of a solid or liquid medium (within the interval where no phase change involves). Among the other methods for thermal heat storage, latent heat thermal energy storage systems (LHS) were attractive and frequently used. Within the applicable temperature range, the storage medium for latent heat undergoes a phase transition from solid to liquid, liquid to gas, or vice versa. This material

DOI: 10.1201/9781003345558-4

is referred to as a phase change material (PCM). Chemical heat storage (CHS) systems are further classified as sorption and thermochemical storage systems (Sharma *et al.*, 2009; Abedin, 2011; Sunku Prasad *et al.*, 2019; Kant and Pitchumani, 2022a). SHS and LHS were generally used for TES applications; however, the thermochemical heat storage (TCHS) system has a high energy density and lower energy losses throughout the storage period (Kant *et al.*, 2016). This chapter will focus solely on various TCHS systems, including their mechanisms, advantages, drawbacks, applications, and recent advancements.

3.2 THERMOCHEMICAL HEAT STORAGE SYSTEMS

The intermolecular connection between specific chemical material pairs can be broken by applying heat energy to them, and they can be divided into distinct reactive components. This material would eventually be able to store heat energy. In reverse, the different reactive components can again be reacted to obtain the material by releasing heat energy. Sorption and chemical storage are the procedures that address chemical heat storage (Farulla *et al.*, 2020; Bao and Ma, 2022). Sorption is a process by which a gas or a vapor (sorbate) is captured selectively by a sorbent media (either a solid or liquid substance). This sorption storage phenomenon is carried out through the processes known as adsorption and absorption. According to Poppi *et al.* (2018), *absorption* is defined as "the process of one material (absorbate) being retained by another (absorbent)." On the one hand, the process of capturing a gas or a vapor by a liquid (absorbent) media is known as absorption.

The capturing process may be either physical or chemical. On the other hand, the process of capturing a gas or a vapor on the surface of a solid or a porous (adsorbent) media is known as adsorption. *Adsorption* is defined by Poppi *et al.* (2018) as "a phenomenon occurring at the interface between two phases, in which cohesive forces act between the molecules of all substances irrespective of their state of aggregation." Additionally, in the adsorption process, physical adsorption (or physisorption) is used to describe binding that happens as a result of van der Waals forces. In contrast, chemical adsorption is used to describe binding that takes place as a result of valency forces (or chemisorption) (Ding and Riffat, 2013). Chemical storage systems allow chemical energy conversion into other kinds of energy through chemical reactions. These systems use the breaking and reformation of molecular bonds in reversible chemical or physical reactions to collect and store thermal energy and release it. The processes such as electrochemical, electromagnetic (photochemical or photosynthesis), thermochemical without sorption, chemical adsorption (chemisorption), and chemical absorption are used to store chemical energy effectively. The required thermochemical energy storage systems, in which the thermal energy utilizations or transactions occur, consist of the processes such as thermochemical without sorption, chemical adsorption (chemisorption), and chemical absorption. Table 3.1 displays some of the most promising materials and pertinent storage qualities. These chemical storage systems are classified in Figure 3.1 (Ding and Riffat, 2013; Kerskes, 2016; Bao and Ma, 2022).

TABLE 3.1
Promising Materials for Thermochemical Energy Storage (Abedin, 2011; Ding and Riffat, 2013; Sunku Prasad *et al.*, 2019)

Thermochemical Material	Solid Reactant	Working Fluid	Energy Storage Density (GJ/m³)	Charging Temperature (°C)
Calcium carbonate ($CaCO_3$)	CaO	CO_2	3.3	837
Calcium hydride (CaH_2)	Ca	H_2	7.37	1100–1400
Calcium hydroxide ($Ca(OH)_2$)	CaO	H_2O	1.9	479
Calcium sulphate ($CaSO_4 \cdot 2H_2O$)	$CaSO_4$	H_2O	1.4	89
Iron carbonate ($FeCO_3$)	FeO	CO_2	2.6	180
Iron hydroxide ($Fe(OH)_2$)	FeO	H_2O	2.2	150
Magnesium hydride (MgH_2)	Mg	H_2	3.9	300–480
Magnesium nickel hydride (Mg_2NiH_4)	Mg_2Ni	H_4	3.12	253–523
Magnesium sulphate ($MgSO_4 \cdot 7H_2O$)	$MgSO_4$	H_2O	2.8	122
Sodium hydride (NaH)	Na	H2	2.89	600
Silicon oxide (SiO_2)	Si	O_2	37.9	4065
Sodium sulphide ($Na_2S.5H_2O$)	Na_2S	H_2O	2.8	110
Strontium bromide ($SrBr_2.6H_2O$)	$SrBr_2.H_2O$	H_2O	0.22	43
Strontium carbonate ($SrCO_3$)	SrO	CO_2	1.2–1.5	900–1200

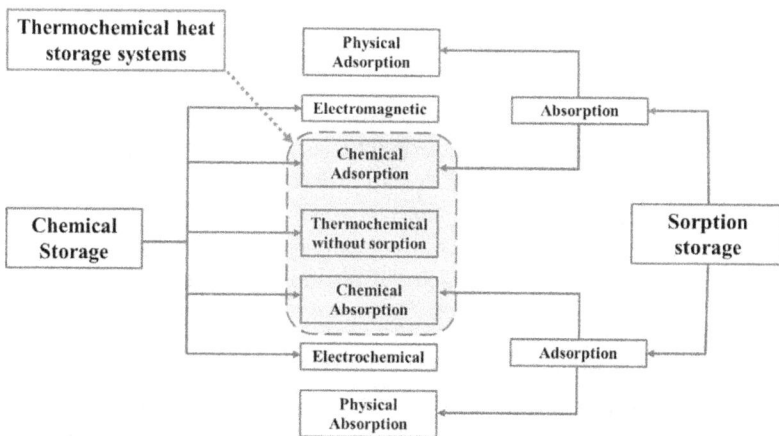

FIGURE 3.1 Classification of chemical and sorption storage systems (Ding and Riffat, 2013; Kerskes, 2016; Bao and Ma, 2022).

3.3 MECHANISM OF HEAT STORAGE FOR THERMOCHEMICAL HEAT STORAGE SYSTEMS

The mechanisms of heat transition for thermochemical energy storage systems can be classified into two types and are different for the thermochemical process without sorption and with sorption (chemisorption and chemical absorption).

3.3.1 MECHANISM OF HEAT STORAGE IN THERMOCHEMICAL PROCESS WITHOUT SORPTION

The three primary steps in any typical thermal energy storage cycle are charging, storing, and discharging. In thermochemical energy storage, capturing, storing, and then releasing thermal energy is carried out by absorbing, breaking/desorption, and reforming molecular bonds in a reversible chemical reaction (Mehling and Cabeza, 2008; Sharma *et al.*, 2009).

The amount of heat associated with this type of storage medium (Q) depends mainly on the factors such as the mass of storage materials (m), materials heat of reaction (ΔH_R), and the extent of chemical reaction (ε), that is, $Q = \in .m.\Delta H_R$ (Sharma *et al.*, 2009). The reversible chemical reactions as follows explain the basic principle for thermochemical heat storage:

$$AB \quad \underset{(Q-)}{\overset{(Q\downarrow)}{\rightleftharpoons}} \quad A+B \tag{1}$$

Initially, a required energy quantity of *Q* was supplied to dissociate compound *AB* by absorbing *Q* thermal energy (charging step). The charging process is an endothermic reaction. After the charging process, *AB* breaks into its constituent components, A and B, as present in Eq. (1). After the charging process, the components A and B can be kept/stored apart throughout the charging process to prevent a quick reaction backlash with little or no energy losses and to control utilization of energy during the peak load demand conditions (Sunliang, 2010). The materials are usually stored at ambient temperatures, leading to no thermal losses (except during the initial cooling of components "A" and "B" after charging). Any other energy losses are due to the degradation of the materials. Because the reaction is reversible, components *A* and *B* combine (discharging step) to regenerate compound AB. The stored energy Q is then released/recovered during this reversible (exothermic reaction) reaction.

3.3.2 MECHANISM OF HEAT STORAGE THROUGH SORPTION PROCESS

Both chemical absorption and chemical adsorption processes are responsible for thermal energy storage through sorption energy storage systems. Both processes will capture gas or vapor medium on the solid or liquid medium. These may be further classified into two distinct categories of open and closed systems. In an open system, during the charging of heat to the storage medium, the fluid available in the storage medium is converted directly to the gaseous state. Thermal energy and gas are released into the environment through a heat exchanger interface. In the case of a closed system, during the charging of heat to the storage medium, the fluid available in the storage medium is converted directly to the gaseous state, but this gas is not released directly; the thermal energy is released to the environment through a heat exchanger interface.

3.3.2.1 Open System

In an open system, hot and dry air is allowed to flow across the energy storage unit, and the adsorbent medium absorbs the temperature from supplied hot air. Hence, the water content in the storage material will be trapped and then move out to the environment with the supplied air (endothermic process). During this desorption reaction in the storage unit, the storage material will absorb the heat from the externally supplied hot air, making the exit-supplied air warmer and more humid. The humid, warm air flowing out of the unit can then be used for primary heating purposes and later can be stored separately as cold, moist air with moisture. The storage material is now charged with thermal energy and can be stored in an adiabatic system until further demand. The cold, humid air stream can be passed over this hot storage medium for discharging or heat recovery, and the storage unit will absorb the moisture in the cold air stream (exothermic process). Hence, the stored heat energy from the storage medium will again transfer to the supplied air stream and become dry and warmer. The operating principle of an open system is schematically depicted in Figure 3.2 (Abedin and Rosen, 2012; Ding and Riffat, 2013; Bao and Ma, 2022).

3.3.2.2 Closed System

In a closed system, hot and dry air is allowed to flow across the energy storage unit, and the adsorbent medium absorbs the temperature from supplied hot air. Hence, the

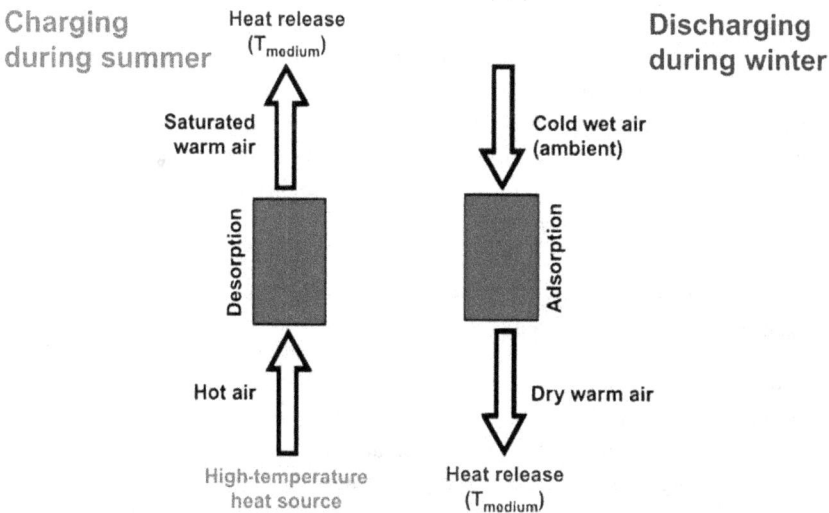

FIGURE 3.2 Operating principle of an open adsorption energy storage system (Abedin and Rosen, 2012; Ding and Riffat, 2013; Bao and Ma, 2022).

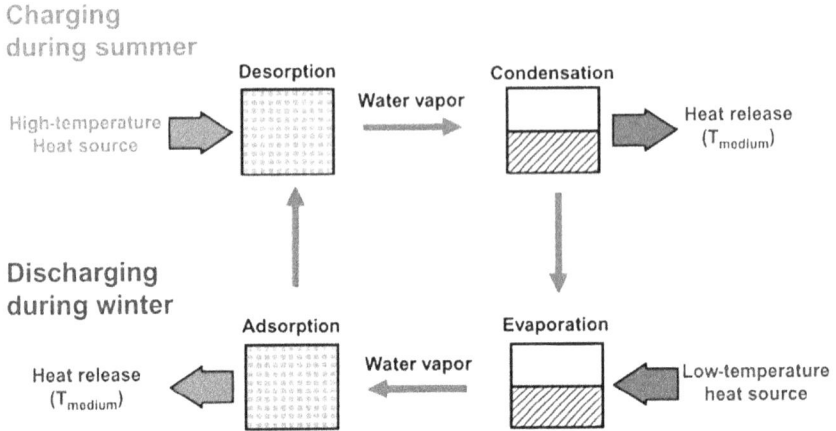

FIGURE 3.3 Operating principle of a closed adsorption energy storage system (Abedin and Rosen, 2012; Ding and Riffat, 2013; Bao and Ma, 2022).

water content in the storage material will be trapped and then moved out with the supplied air (endothermic process). During this desorption reaction in the storage unit, the storage material will absorb the heat from the externally supplied hot air, making the exit-supplied air warmer and more humid. The released humid, warm air (water vapor + air) is then passed through a condenser to utilize the available energy, and the condensed water is stored.

The storage material is now charged with thermal energy and can be stored in an adiabatic system until further demand. The low-temperature heat source evaporates the collected water in the discharging cycle depending on the energy demand. The water vapor is subjected to flow over the hot storage medium present in the adsorber unit. The storage unit will again absorb moisture in the cold air stream (exothermic process). Hence, the stored heat energy from the storage medium will again transfer to the supplied air stream and become dry and warmer. The operating principle of the close system is schematically represented in Figure 3.3 (Ding and Riffat, 2013; Bao and Ma, 2022; Abedin and Rosen, 2012).

3.4 MATERIALS USED FOR THERMOCHEMICAL HEAT STORAGE SYSTEMS

When selecting a thermochemical material, several factors should be taken into consideration since they have an impact on TES systems. Some significant parameters are energy storage density, reaction temperature, rate, heat transfer capabilities, flow characteristics, cycling behavior (reversibility and deterioration over numerous cycles), cost, availability, corrosiveness, toxicity, and safety (Abedin, 2011). The creation of materials for TCHS is currently mainly in the laboratory stage. It is still a long way from any design or material successfully implemented on a large scale for commercial use. However, compared to other thermal storage techniques, it offers exceptional

```
                                                                  ┌──────────────────┐
                                                              ┌──▶│   Ammoniates     │
                                             ┌─────────────┐  │   └──────────────────┘
                                        ┌───▶│  Sorption storage │  ┌──────────────────┐
┌──────────────────────┐               │    │   processes  ├──┼──▶│    Hydrates      │
│ Materials/ Reactants │───────────────┤    └─────────────┘  │   └──────────────────┘
│  for Thermochemical  │               │                     │   ┌──────────────────┐
│  heat storage systems│               │                     └──▶│  Metal-hydrides  │
└──────────────────────┘               │    ┌─────────────┐      └──────────────────┘
                                        │    │  Chemical reaction │  ┌──────────────────┐
                                        └───▶│ storage processes├──┬──▶│   Carbonates    │
                                             └─────────────┘      │   └──────────────────┘
                                                                  │   ┌──────────────────┐
                                                                  └──▶│   Hydroxides     │
                                                                      └──────────────────┘
```

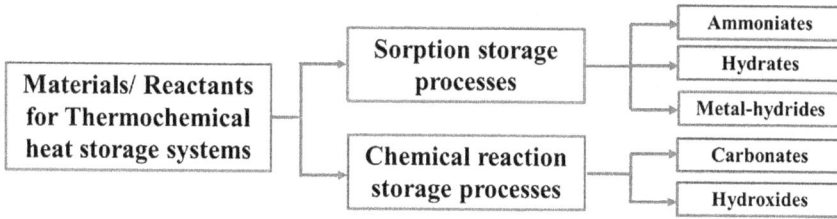

FIGURE 3.4 Overview of thermochemical materials/reactants used for energy storage (Cot-Gores, Castell and Cabeza, 2012; Bao and Ma, 2022).

volumetric energy density values (Cot-Gores, Castell and Cabeza, 2012; Prieto *et al.*, 2016). The materials used for storage can be categorized independently as chemical reactions and sorption storage (Figure 3.4). Reversible gas-solid, gas-liquid, and gas-gas reactions are beneficial working pairs for heat storage in sorption processes.

Sorption storage processes generally store low-grade heat (80–100°C) and medium-grade heat (100–400°C) for applications such as space heating, domestic hot water preparation, etc. Some of the examples of working pairs (solid, liquid, and composite sorbents) are LiBr solution/H_2O; LiCl solution/H_2O; LiCl/activated alumina; LiCl/expanded graphite; $LiCl_2$ solution/H_2O; $CaCl_2$ solution/H_2O; aluminophosphates (ALPOs)/silico-aluminophosphates; the combination of a salt hydrate with an additive with a porous structure such as expanded graphite, metal foam, carbon fiber, and activated carbon (Farulla *et al.*, 2020). Chemical reactions are used to store medium-(100–400°C) and high-grade (>400°C) heat. Examples of chemical reactions are dehydration of metal hydroxides (250–800°C); dehydration of metal hydrides (80–400°C); dehydration of salt hydrates; de-ammoniation of ammonium chlorides; decarboxylation of metal carbonates (100–950°C); methane steam reforming; catalytic dissociation; and metal oxide redox (600–1000°C) (Farulla *et al.*, 2020; Prieto *et al.*, 2016).

3.4.1 AMMONIATES

Benfer *et al.* (2004) created the ammoniates pair of $BaCl_2$/$MnCl_2$, which had previously been combined and compressed in expanded graphite, for a refrigeration prototype. At 15°C, 25°C, and 195°C, the material's heat-power conversion from 20% to 80% was examined. The prototype's reported average heating and cooling power was 1 kW and 1.3 kW, respectively. Coefficient of performance (COP) was calculated to be 0.35. In the same procedure (Benfer *et al.*, 2004), they also created ammoniates of $PbCl_2$ and $MnCl_2$ to cool and maintain an 88 L cold box at 0°C for three hours. The $MnCl_2$ reactor emitted heat into the environment, whereas the $PbCl_2$ reactor was placed within the box to collect heat from the air. The resorption refrigerator's cooling capacity was 48 kWh m^{-3} (Lépinasse, Marion and Goetz, 2001).

With the same $BaCl_2$/$MnCl_2$ ammoniates pair but previously impregnated in expanded graphite and solidified in the shape of cylindrical blocks, Li et al. (Li, Wang, Oliveira, *et al.*, 2009; Li, Wang, Kiplagat, *et al.*, 2009; Li, Wang, Chen, *et al.*, 2009) studied a double-effect sorption refrigeration system. An experimental COP of 0.52 at 60% conversion at operating temperatures of 180°C, 30°C, and 10°C was reported. An average SCP of 301 W kg^{-1} was attained during the 60 minutes (Li, Wang, Oliveira, et al., 2009). A resorption prototype for refrigeration at 0°C based on the ammoniates of $BaCl_2$ and $NiCl_2$ that had been combined and compacted in expanded graphite (Benfer et al., 2004) was studied by Goetz, Spinner, and Lepinasse (1997). At a cooling temperature of 0°C and a heat sink temperature of 40°C, the mean cooling power during the 15 minutes of the refrigeration phase (41% conversion) was 40 W (396 W kg^{-1}). The other ammoniates pairs used reported are NH_4Cl/$MnCl_2$ (Bao et al., 2010; Bao, Wang and Wang, 2011), NH_4Cl/$MnCl_2$ (Xu, Oliveira and Wang, 2011), $CaCl_2$/$MnCl_2$ (Wang, Zhang and Wang, 2010), $SrCl_2$/$CoCl_2$ (Spinner, 1993; Vasiliev et al., 1999; Llobet and Goetz, 2000), and $MgCl_2$/$LiCl$ (Haije et al., 2007; Pal et al., 2009, 2011; Cot-Gores, Castell and Cabeza, 2012).

3.4.2 HYDRATES

The hydrates of $MgSO_4$, $MgCl_2$, $CaCl_2$, and $Al_2(SO_4)_3$ were proposed for seasonal heat storage of solar energy. The authors (van Essen, Cot Gores, et al., 2009; van Essen, Zondag, et al., 2009; Cot-Gores, Castell and Cabeza, 2012) aimed to store the surplus of solar energy generated during the summer and meet the heating demand for space heating and domestic hot water during the winter when the demand exceeds the solar supply. The materials were tested under practical conditions, and the performance of the material was assessed by measuring the temperature lift of the bed during the sorption process. The system with $MgCl_2$, $CaCl_2$, $MgSO_4$, and $Al_2(SO_4)_3$ reached a maximum temperature lift of 19°C, 11°C, 4°C, and 1°C, respectively. However, the $MgCl_2$ and $CaCl_2$ tend to form a gel-like material during sorption due to their hygroscopic nature.

Zondag et al. (2010) found that HCl was formed during the dehydration of $MgCl_2.7H_2O$ at temperatures above 135°C. The formation of this gas not only degrades the storage material but is also strongly corrosive. Therefore, dehydration was limited to a temperature below 135°C. The hydration experiments were performed in an open system (atmospheric) instead of the closed system, using a moistened nitrogen flow of 20 slpm with a vapor pressure of 1.2 kPa. In addition, the overhydration of $MgCl_2$ was avoided by impregnation into a carrier material (cellulose). The measured temperature lift was 15°C, and the calculated energy density of the material after 23 h hydration was 0.14 MJ kg^{-1}. The other hydrate materials reported are Na_2S (Boer et al., 2004), $MnCl_2$ (Stitou, Mazet and Bonnissel, 2004), and $SrBr_2$ (Lahmidi, Mauran and Goetz, 2006; Mauran, Lahmidi and Goetz, 2008; Cot-Gores, Castell and Cabeza, 2012).

3.4.3 METAL-HYDRIDES

Two identical tube and coil-type reactors working with the $MmNi_{4.5}Al_{0.5}$/ZrMnFe pair were used in Ram Gopal and Srinivasa Murthy's (Gopal and Murthy, 1999) experiments for the metal hydride cooling system. The specific cooling power was found to range from 30 to 45 W per kg of total hydride mass (60 to 90 W per kilogram of desorbing hydride), while the COP ranged from 0.2 to 0.35 depending on the operating conditions.

Chernikov *et al.* (2002) investigated a resorption system for the simultaneous production of chilled water at less than 4°C and hot water at about 50°C using high-temperature and the low-temperature metal hydrides pair $MmNi_{4.15}Fe_{0.85}/LaNi_{4.6}Al_{0.4}$. The system was run for 8 half-cycles, each lasting 20 minutes. The output temperature of the chilled water was 1.5°C, with an average cooling power of 154 W. Total energy used to produce heat and cold was 12.6 MJ and 1.48 MJ, respectively. With the help of waste heat released from manufacturing and automotive facilities, Lee (1995) developed a resorption metal hydride air conditioner employing the $Zr_{0.9}Ti_{0.1}Cr_{0.6}Fe_{1.4}/Zr_{0.9}Ti_{0.1}Cr_{0.9}Fe_{1.1}$ pair in a tubular type reactor. Reactors of the tubular type were employed. The maximum reported SCP was approximately 151 W per kg of total hydride mass at an operating temperature of 220°C/30°C/18°C, 4 min heating time, and 7 min cooling time (271 W per kg of desorbing metal hydride). Kang (1996) studied a resorption system utilizing $MmNi_{4.15}Fe_{0.85}/LaNi_{4.7}Al_{0.3}$. Finned tube reactors were used. The reported COP was roughly 0.2 at the operating parameters of 150°C, 30°C, 20°C, and 20 min cycle time. Imoto (1996) developed a solar-assisted resorption refrigerator system based on a $La_{0.6}Y_{0.4}Ni_{4.8}Mn_{0.2}/LaNi_{4.6}Mn_{0.3}Al_{0.1}$ pair. The observed thermal power and COP were 1860 W and 0.42, respectively, at the operating temperatures of 140°C, 20°C, –20°C, and 19 min (80% conversion). The other metal hydrate material reported are pair of $La_{0.6}Y_{0.4}Ni_{4.8}Mn_{0.2}/LaNi_{4.61}Mn_{0.26}Al_{0.13}$ (Qin *et al.*, 2007), $La_{0.6}Y_{0.4}Ni_{4.8}Mn_{0.2}/LaNi_{4.61}Mn_{0.26}Al_{0.13}$ (Ni and Liu, 2007), and $Ti_{0.99}Zr_{0.01}V_{0.43}Fe_{0.09}Cr_{0.05}Mn_{1.5}/LmNi_{4.91}Sn_{0.15}$ (Linder, Mertz and Laurien, 2010). Authors also reported a two-stage metal hydride heat transformer using pairs of $LmNi_{4.85}Sn_{0.15}/LmNi_{4.49}Co_{0.1}Mn_{0.205}Al_{0.205}/LmNi_{4.08}Co_{0.2}Mn_{0.62}Al_{0.1}$ (Werner and Groll, 1991) and $LmNi_{4.91}Sn_{0.15}/LaNi_{4.1}Al_{0.52}Mn_{0.38}/Ti_{0.99}Zr_{0.01}V_{0.43}Fe_{0.09}Cr_{0.05}Mn_{1.5}$ (Hans-Peter, 2007; Cot-Gores, Castell and Cabeza, 2012).

3.4.3.1 Carbonates

Because of their high volumetric density, low working pressure, non-toxic and non-corrosive chemical makeup, and very high operating temperatures (usually over 800°C), carbonates are particularly appealing materials for storing thermal energy (Kyaw *et al.*, 1997). The thermal energy is stored and released using the exothermic carbonation and endothermic calcination processes. The metal carbonate (MCO_3) disintegrates into CO_2 and MO within the calcination reactor. The low partial pressure of CO_2 and 600–700°C are commonly used for this reaction. To utilize the heat produced by a high-temperature exothermic process, the byproducts of the reaction (MO and CO_2) are then combined in a carbonation reactor. The high partial pressure of CO_2 and 900°C are used for the carbonation reaction. The carbon dioxide released during the calcination reaction can either be delivered to a carbonate reactor or held in a storage tank following proper compression. The calcination and carbonation reactions might happen simultaneously in one reactor or separately in two reactors (Sunku Prasad *et al.*, 2019).

3.4.4 HYDROXIDES

At near-atmospheric pressures, reversible hydration/dehydration of metal oxides can be exploited as a high-temperature (500°C) thermal energy storage solution (Yan and Zhao, 2016). Common hydroxides are inexpensive, plentiful, and non-toxic

(Yan and Zhao, 2016). The most extensively researched common hydroxides for a high-temperature TCESS and chemical heat pumps are calcium hydroxide and magnesium hydroxide because of their high volumetric energy density (up to 600 MJ/m^3) (Rougé et al., 2017; Yan, Zhao and Pan, 2017). The pressure of water vapor supplied to the reaction bed determines the charging and discharging operations of the hydroxide-based TCES in an indirect-type, fixed-bed heat exchanger and the decomposition temperature. To enhance the reaction kinetics of the hydroxides, fluidized-bed and moving-bed reactors have also been suggested in the literature (Rougé et al., 2017). Compared to the Mg(OH)$_2$/MgO/H$_2$O system, which is inert to hydration in highly superheated steam and whose reaction rate decreases with increasing temperature, the Ca(OH)$_2$/CaO/H$_2$O system is more appealing (Ervin, 1977; Sunku Prasad et al., 2019).

3.5 APPLICATION OF THERMOCHEMICAL HEAT STORAGE SYSTEMS

TCES can help balance energy supply and demand in thermal systems daily, weekly, and even seasonally. TES can also improve the overall efficiency of energy systems while decreasing peak demand, energy consumption, CO emissions, and prices. TCES is most commonly used to combine waste heat and district heating systems, heat pumps, and heat and power (CHP) generators in district heating networks in building constructions. Moreover, it is also utilized in concentrating solar power (CSP) systems as thermal energy storage.

3.5.1 TCES in Buildings

Kant and Pitchumani (2022b) analyzed a novel constructal fin tree embedded in thermochemical energy storage for building applications. The effect of construction fin design parameters on energy storage density and levelized cost of storage was investigated to develop design envelopes that meet the requirements of the US Department of Energy Buildings Thermal Energy Storage program, which includes a round-trip thermal energy storage density of more than 80 kWh/m^3 and a storage cost of less than $15/kWh. The study presents ideal designs for constructing a tree-augmented thermochemical reactor bed based on the lowest levelized storage cost and the highest energy storage density for building-integrated thermal energy storage applications. Later authors also discuss the advances and opportunities in thermochemical heat storage systems for building applications (Kant and Pitchumani, 2022a). Clark and Farid (Clark and Farid, 2022) experimentally investigated the cascade TCESS using SrCl$_2$-cement and zeolite-13X materials. The two materials were chosen based on their respective hydration and dehydration requirements. The volumetric energy density ranged from 108–138 kWhm^{-3} with dehydration temperatures of 50–130°C and hydration conditions of 12°C, 75% RH. The cascaded system was revealed to enhance power output and temperature lift. Additionally, as compared to a standard salt-based system, the cascade system increased exergy efficiency by 6–38%. Kant et al. (2021) evaluated the performance of a K$_2$CO$_3$-based

TCESS using a honeycomb-structured heat exchanger. The results of this study provide detailed insight into the heat release processes occurring in a fixed bed of K_2CO_3. Clark *et al.* (2022) screened salt hydrates for thermochemical energy storage for building heating applications. The criteria set were volumetric energy density of >500 kWh m^{-3} with a dehydration temperature of <100°C, a material cost of < \$3.5 USD kg^{-1} (< \$15 USD kWh^{-1}), melting does not occur during dehydration, and safety. Based on the data for the selected materials, it was determined that $SrCl_2$ and $SrBr_2$ are the most promising salts. This research discusses the benefits and drawbacks of each salt and the situations for which they are best suited. Li, Luo, *et al.* (2022) predicted solar-thermal energy conversion of building envelope using thermochemical sorbent based on established reaction kinetics. Compared to earlier research, the results show that this passive building envelope can achieve more heat harvesting and utilization efficiency in a more compact space. Moreover, the effect of radiation intensity on air purification and thermal performance was studied. Several studies have been conducted on thermochemical thermal energy storage for building applications and reviewed in references (Li, Klemeš, *et al.*, 2022; Marie *et al.*, 2022).

3.5.2 TCES IN CSP

Bellan *et al.* (2022) reviewed high-temperature thermochemical heat storage: particle reactors and materials based on solid–gas reactions to produce electricity beyond insolation hours and supply to the electrical grid. The paper analyzes and summarizes several thermochemical storage systems currently under investigation, mainly TCS based on metal oxides. Several experimental, numerical, and technical research examples are provided on developing particle reactors and materials for high-temperature TCS applications. The benefits and drawbacks of various heat storage systems (sensible, latent, and thermochemical) and particle receivers (stacked, fluidized, and entrained) have been explored and documented. Gigantino, Sas Brunser, and Steinfeld (2020) investigated high-temperature thermochemical heat storage via the CuO/Cu_2O redox cycle: from material synthesis to packed-bed reactor engineering and cyclic operation. A technique was devised for producing porous CuO-based granules with yttria-stabilized zirconia (YSZ) as a sintering inhibitor. During 100 consecutive cycles in the air between 950–1050°C, the synthesized granules demonstrate high and reversible redox conversion, yielding a gravimetric energy storage density associated with endothermic/exothermic redox processes in the range of 470 to 615 kJ/kg for 50 to 65 wt% CuO-YSZ granules.

Ortiz *et al.* (2021) proposed integrating a high-temperature thermochemical energy storage cycle to boost the solar contribution in solar combined cycles. The unique solar combined cycle investigated in this paper raises the yearly solar share beyond 50%, whereas existing state-of-the-art technology is below 15%. Results from actual solar irradiation data and clustering analysis reveal total plant efficiencies of more than 45% (considering off-design performance) with extremely high dispatch ability, justifying the interest in further developing this unique cycle. Funayama *et al.* (2019) develop a suitable composite material by enhancing heat transfer through the reaction bed and mitigating problems of pure $CaO/Ca(OH)_2$ materials, such as forming centimeter-scale agglomerates and change in bulk volume during repetitive reactions. The composite material, in which $CaO/Ca(OH)_2$ samples were partitioned into pores with a mean diameter of 400 μm, preserved strong reactivity and bulk volume during cycle reactions. Carrillo *et al.* (2019) reviewed high-temperature thermochemical heat storage systems and materials for solar energy on demand. This paper provides a detailed analysis of major TCS systems, with a focus on those employed at high temperatures (500–1000°C) and based on redox processes. Numerous technical solutions are already available, and only more extensive economic analyses can determine which is best suited to a particular plant design. Thermochemical storage methods based on ammonia dissociation, hydrides (TiH1.7 and CaH_2), and carbonates ($CaCO_3$ and $SrCO_3$) provide greater energy storage densities and use affordable materials in most circumstances. However, many chemical systems have kinetic and/or reversibility constraints and necessitate gas storage, which adds complexity to the process and necessitates the installation of equipment that may have to run at high pressures, increasing the plant cost. Rodat *et al.* (2020) reviewed high-temperature solar thermochemical processes to fulfill the commitment toward day and night solar process operation since it can potentially increase product durability, quality, efficiency, and economics. The industrial processes are typically continuous, and daily start-up and shut-down severely limit the output capacity of solar-powered processes, representing a substantial barrier to scaling up. This work reviews and categorizes day and night continuous high-temperature solar processes for the first time.

3.6 SUMMARY AND CONCLUSIONS

This chapter describes a thermochemical TES system with an extra thermal upgrade of stored energy, high specific thermal powers, and the possibility of colossal storage capacity due to its scalable design. Because it provides exceptional flexibility in various dimensions, this technique has enormous potential for deployment in industrial applications: The experimental results show that the TES has a wide charging and discharging temperature range that is particularly important for industrial operations (180–280°C using strontium bromide (SrBr2) and water vapor as the reference thermochemical working pair). The inorganic salt hydration and dehydration process has a high specific energy density of 291 kJ/kg (or 81 kWh/t), equivalent to high-performance phase change materials in latent TES at similar temperatures. Thermal outputs of up to 1.2 kW for charging and discharging have been demonstrated for the storage module comprising 4.7 kg of $SrBr_2$ (0.38 kWh capacity). Because the dehydration and hydration processes follow an equilibrium line, varying the steam pressure may change the reaction temperatures. This effect may be exploited to precisely match charging and discharging temperatures to a given storage application, even if those temperatures fluctuate over time. Conversely, this degree of freedom could be leveraged to create an effectively wider temperature difference between the storage material and the heat transfer fluid, allowing for a smaller heat exchanger while maintaining the power of TES.

REFERENCES

Abedin, A.H. (2011) 'A critical review of thermochemical energy storage systems', *The Open Renewable Energy Journal*, 4(1), pp. 42–46. Available at: https://doi.org/10.2174/187 6387101004010042.

Abedin, A.H. and Rosen, M.A. (2012) 'Closed and open thermochemical energy storage: Energy- and exergy-based comparisons', *Energy*, 41(1), pp. 83–92. Available at: https://doi.org/10.1016/j.energy.2011.06.034.

Bao, H. and Ma, Z. (2022) 'Thermochemical energy storage', *Storing Energy: With Special Reference to Renewable Energy Sources*, pp. 651–683. Available at: https://doi.org/10.1016/B978-0-12-824510-1.00028-3.

Bao, H.S. *et al.* (2010) 'Choice of low temperature salt for a resorption refrigerator', *Industrial & Engineering Chemistry Research*, 49(10), pp. 4897–4903. Available at: https://doi.org/10.1021/ie901575k.

Bao, H.S., Wang, R.Z. and Wang, L.W. (2011) 'A resorption refrigerator driven by low grade thermal energy', *Energy Conversion and Management*, 52(6), pp. 2339–2344. Available at: https://doi.org/10.1016/j.enconman.2010.12.045.

Bellan, S. *et al.* (2022) 'A review on high-temperature thermochemical heat storage: Particle reactors and materials based on solid–gas reactions', *WIREs Energy and Environment*, 11(5). Available at: https://doi.org/10.1002/wene.440.

Benfer, R. *et al.* (2004) 'Reactor for gas/liquid or gas/liquid/solid reactions', United States Patent Application US20040151640A1.

Carrillo, A.J. *et al.* (2019) 'Solar energy on demand: A review on high temperature thermochemical heat storage systems and materials', *Chemical Reviews*, 119(7), pp. 4777–4816. Available at: https://doi.org/10.1021/acs.chemrev.8b00315.

Chernikov, A. *et al.* (2002) 'An installation for water cooling based on a metal hydride heat pump', *Journal of Alloys and Compounds*, 330–332, pp. 907–910. Available at: https://doi.org/10.1016/S0925-8388(01)01479-7.

Clark, R.-J. and Farid, M. (2022) 'Experimental investigation into cascade thermochemical energy storage system using SrCl2-cement and zeolite-13X materials', *Applied Energy*, 316, p. 119145. Available at: https://doi.org/10.1016/j.apenergy.2022.119145.

Clark, R.-J. *et al.* (2022) 'Experimental screening of salt hydrates for thermochemical energy storage for building heating application', *Journal of Energy Storage*, 51, p. 104415. Available at: https://doi.org/10.1016/j.est.2022.104415.

Cot-Gores, J., Castell, A. and Cabeza, L.F. (2012) 'Thermochemical energy storage and conversion: A-state-of-the-art review of the experimental research under practical conditions', *Renewable and Sustainable Energy Reviews*, 16(7), pp. 5207–5224. Available at: https://doi.org/10.1016/j.rser.2012.04.007.

de Boer, R. *et al.* (2004) *Solid-sorption cooling with integrated thermal storage: The SWEAT prototype.* Energy Research Centre of the Netherlands (ECN).

Dincer, I. (2002) 'On thermal energy storage systems and applications in buildings', *Energy and Buildings*, 34(4), pp. 377–388.

Ding, Y. and Riffat, S.B. (2013) 'Thermochemical energy storage technologies for building applications: A state-of-the-art review', *International Journal of Low-Carbon Technologies*, 8(2), pp. 106–116. Available at: https://doi.org/10.1093/ijlct/cts004.

Ervin, G. (1977) 'Solar heat storage using chemical reactions', *Journal of Solid State Chemistry*, 22(1), pp. 51–61. Available at: https://doi.org/10.1016/0022-4596(77)90188-8.

Farulla, G.A. *et al.* (2020) 'A review of thermochemical energy storage systems for power grid support', *Applied Sciences (Switzerland)*, 10(9). Available at: https://doi.org/10.3390/app10093142.

Funayama, S. *et al.* (2019) 'Composite material for high-temperature thermochemical energy storage using calcium hydroxide and ceramic foam', *Energy Storage*, 1(2), p. e53. Available at: https://doi.org/10.1002/est2.53.

Gigantino, M., Sas Brunser, S. and Steinfeld, A. (2020) 'High-temperature thermochemical heat storage via the CuO/Cu_2O redox cycle: From material synthesis to packed-bed reactor engineering and cyclic operation', *Energy & Fuels*, 34(12), pp. 16772–16782. Available at: https://doi.org/10.1021/acs.energyfuels.0c02572.

Goetz, V., Spinner, B. and Lepinasse, E. (1997) 'A solid-gas thermochemical cooling system using $BaCl_2$ and $NiCl_2$', *Energy*, 22(1), pp. 49–58. Available at: https://doi.org/10.1016/S0360-5442(96)00081-3.

Gopal, M.R. and Murthy, S.S. (1999) 'Experiments on a metal hydride cooling system working with $ZrMnFe/MmNi_{4.5}Al_{0.5}$ pair', *International Journal of Refrigeration*, 22(2), pp. 137–149. Available at: https://doi.org/10.1016/S0140-7007(98)00033-4.

Haije, W.G. *et al.* (2007) 'Solid/vapour sorption heat transformer: Design and performance', *Applied Thermal Engineering*, 27(8–9), pp. 1371–1376. Available at: https://doi.org/10.1016/j.applthermaleng.2006.10.022.

Hans-Peter, K. (2007) 'Operation behaviour of a double-stage metal hydride sorption device for cold generation', Dissertation, University of Stuttgart. Available at: https://doi.org/10.18419/opus-1744.

Imoto, T. (1996) 'Development of an F-class refrigeration system using hydrogen-absorbing alloys', *International Journal of Hydrogen Energy*, 21(6), pp. 451–455. Available at: https://doi.org/10.1016/0360-3199(95)00106-9.

Kang, B. (1996) 'Dynamic behavior of heat and hydrogen transfer in a metal hydride cooling system', *International Journal of Hydrogen Energy*, 21(9), pp. 769–774. Available at: https://doi.org/10.1016/0360-3199(96)00017-1.

Kant, K. and Pitchumani, R. (2022a) 'Advances and opportunities in thermochemical heat storage systems for buildings applications', *Applied Energy*, 321, p. 119299. Available at: https://doi.org/10.1016/j.apenergy.2022.119299.

Kant, K. and Pitchumani, R. (2022b) 'Analysis of a novel constructal fin tree embedded thermochemical energy storage for buildings applications', *Energy Conversion and Management*, 258, p. 115542. Available at: https://doi.org/10.1016/j.enconman.2022.115542.

Kant, K. *et al.* (2016) 'Thermal energy storage based solar drying systems: A review', *Innovative Food Science & Emerging Technologies*, 34, pp. 86–99. Available at: https://doi.org/10.1016/j.ifset.2016.01.007.

Kant, K. *et al.* (2021) 'Performance analysis of a K_2CO_3-based thermochemical energy storage system using a honeycomb structured heat exchanger', *Journal of Energy Storage*, 38, p. 102563. Available at: https://doi.org/10.1016/j.est.2021.102563.

Kant, K., Shukla, A. and Sharma, A. (2017) 'Advancement in phase change materials for thermal energy storage applications', *Solar Energy Materials and Solar Cells*, 172, pp. 82–92. Available at: https://doi.org/10.1016/j.solmat.2017.07.023.

Kerskes, H. (2016) 'Thermochemical energy storage', *Storing Energy*, pp. 345–372. Available at: https://doi.org/10.1016/B978-0-12-803440-8.00017-8.

Kyaw, K. *et al.* (1997) 'Applicability of zeolite for CO_2 storage in a CaO-CO_2 high temperature energy storage system', *Energy Conversion and Management*, 38(10–13), pp. 1025–1033. Available at: https://doi.org/10.1016/S0196-8904(96)00132-X.

Lahmidi, H., Mauran, S. and Goetz, V. (2006) 'Definition, test and simulation of a thermochemical storage process adapted to solar thermal systems', *Solar Energy*, 80(7), pp. 883–893. Available at: https://doi.org/10.1016/j.solener.2005.01.014.

Lee, S. (1995) 'Operating characteristics of metal hydride heat pump using Zr-based laves phases', *International Journal of Hydrogen Energy*, 20(1), pp. 77–85. Available at: https://doi.org/10.1016/0360-3199(93)E0005-6.

Lépinasse, E., Marion, M. and Goetz, V. (2001) 'Cooling storage with a resorption process. Application to a box temperature control', *Applied Thermal Engineering*, 21(12), pp. 1251–1263. Available at: https://doi.org/10.1016/S1359-4311(00)00113-7.

Li, T., Wang, R., Chen, H. *et al.* (2009) 'Performance improvement of a combined double-way thermochemical sorption refrigeration cycle with reheating process', *AIChE Journal*, 56(2), pp. 477–484. Available at: https://doi.org/10.1002/aic.11996.

Li, T.X., Wang, R.Z., Kiplagat, J.K. *et al.* (2009) 'Thermodynamic study of a combined double-way solid–gas thermochemical sorption refrigeration cycle', *International Journal of Refrigeration*, 32(7), pp. 1570–1578. Available at: https://doi.org/10.1016/j.ijrefrig.2009.05.004.

Li, T.X., Wang, R.Z., Oliveira, R.G. *et al.* (2009) 'A combined double-way chemisorption refrigeration cycle based on adsorption and resorption processes', *International Journal of Refrigeration*, 32(1), pp. 47–57. Available at: https://doi.org/10.1016/j.ijrefrig.2008.07.007.

Li, W., Klemeš, J.J. *et al.* (2022) 'Salt hydrate–based gas-solid thermochemical energy storage: Current progress, challenges, and perspectives', *Renewable and Sustainable Energy Reviews*, 154, p. 111846. Available at: https://doi.org/10.1016/j.rser.2021.111846.

Li, W., Luo, X. *et al.* (2022) 'Solar-thermal energy conversion prediction of building envelope using thermochemical sorbent based on established reaction kinetics', *Energy Conversion and Management*, 252, p. 115117. Available at: https://doi.org/10.1016/j.enconman.2021.115117.

Linder, M., Mertz, R. and Laurien, E. (2010) 'Experimental results of a compact thermally driven cooling system based on metal hydrides', *International Journal of Hydrogen Energy*, 35(14), pp. 7623–7632. Available at: https://doi.org/10.1016/j.ijhydene.2010.04.184.

Llobet, J. and Goetz, V. (2000) 'Production de froid par transformation thermochimique : expérimentation d'un nouveau système à double effet à contact', *International Journal of Refrigeration*, 23(4), pp. 312–329. Available at: https://doi.org/10.1016/S0140-7007(99)00052-3.

Marie, L.F. *et al.* (2022) 'Advances in thermochemical energy storage and fluidised beds for domestic heat', *Journal of Energy Storage*, 53, p. 105242. Available at: https://doi.org/10.1016/j.est.2022.105242.

Mauran, S., Lahmidi, H. and Goetz, V. (2008) 'Solar heating and cooling by a thermochemical process. First experiments of a prototype storing 60kWh by a solid/gas reaction', *Solar Energy*, 82(7), pp. 623–636. Available at: https://doi.org/10.1016/j.solener.2008.01.002.

Mehling, H. and Cabeza, L.F. (2008) *Heat and cold storage with PCM: An up to date introduction into basics and applications.* 1st edn. Edited by H. Mehling and L.F. Cabeza. Springer. Available at: https://doi.org/10.1007/978-3-540-68557-9.

Ni, J. and Liu, H. (2007) 'Experimental research on refrigeration characteristics of a metal hydride heat pump in auto air-conditioning', *International Journal of Hydrogen Energy*, 32(13), pp. 2567–2572. Available at: https://doi.org/10.1016/j.ijhydene.2006.09.038.

Ortiz, C. *et al.* (2021) 'Solar combined cycle with high-temperature thermochemical energy storage', *Energy Conversion and Management*, 241, p. 114274. Available at: https://doi.org/10.1016/j.enconman.2021.114274.

Pal, M. *et al.* (2009) 'Thermally driven ammonia-salt type II heat pump: Development and test of a prototype', *Proceedings of the Heat Powered Cycles Conference.* https://publications.tno.nl/publication/34631115/37c2FC/m09059.pdf

Pal, M. *et al.* (2011) 'Experimental setup for determining ammoniasalt adsorption and desorption behavior under typical heat pump conditions: A description of the setup and experimental results', *ISHPC Conference 2011*, TNO, Padua.

Poppi, S. *et al.* (2018) 'Techno-economic review of solar heat pump systems for residential heating applications', *Renewable and Sustainable Energy Reviews*, 81, pp. 22–32. Available at: https://doi.org/10.1016/j.rser.2017.07.041.

Prieto, C. *et al.* (2016) 'Review of technology: Thermochemical energy storage for concentrated solar power plants', *Renewable and Sustainable Energy Reviews*, 60, pp. 909–929. Available at: https://doi.org/10.1016/j.rser.2015.12.364.

Prim, E.O. (2013) 'Thermal energy storage (TES) using phase change materials (PCM) for cold applications', Doctoral dissertation, University of Lleida, Spain.

Purohit, B.K. and Sistla, V.S. (2021) 'Inorganic salt hydrate for thermal energy storage application: A review', *Energy Storage*, 3(2). Available at: https://doi.org/10.1002/est2.212.

Qin, F. *et al.* (2007) 'Development of a metal hydride refrigeration system as an exhaust gas-driven automobile air conditioner', *Renewable Energy*, 32(12), pp. 2034–2052. Available at: https://doi.org/10.1016/j.renene.2006.10.014.

Rammelberg, H.U., Schmidt, T. and Ruck, W. (2012) 'Hydration and dehydration of salt hydrates and hydroxides for thermal energy storage – Kinetics and energy release', *Energy Procedia*, 30, pp. 362–369. Available at: https://doi.org/10.1016/j.egypro.2012.11.043.

Rodat, S. *et al.* (2020) 'On the path toward day and night continuous solar high temperature thermochemical processes: A review', *Renewable and Sustainable Energy Reviews*, 132, p. 110061. Available at: https://doi.org/10.1016/j.rser.2020.110061.

Rougé, S. *et al.* (2017) 'Continuous $CaO/Ca(OH)_2$ fluidized bed reactor for energy storage: First experimental results and reactor model validation', *Industrial & Engineering Chemistry Research*, 56(4), pp. 844–852. Available at: https://doi.org/10.1021/acs.iecr.6b04105.

Sharma, A. *et al.* (2009) 'Review on thermal energy storage with phase change materials and applications', *Renewable and Sustainable Energy Reviews*, 13(2), pp. 318–345. Available at: https://doi.org/10.1016/j.rser.2007.10.005.

Spinner, B. (1993) 'Ammonia-based thermochemical transformers', *Heat Recovery Systems and CHP*, 13(4), pp. 301–307. Available at: https://doi.org/10.1016/0890-4332(93)90053-X.

Stitou, D., Mazet, N. and Bonnissel, M. (2004) 'Performance of a high temperature hydrate solid/gas sorption heat pump used as topping cycle for cascaded sorption chillers', *Energy*, 29(2), pp. 267–285. Available at: https://doi.org/10.1016/j.energy.2003.08.011.

Sunku Prasad, J. *et al.* (2019) 'A critical review of high-temperature reversible thermochemical energy storage systems', *Applied Energy*, 254, p. 113733. Available at: https://doi.org/10.1016/j.apenergy.2019.113733.

Sunliang, C. (2010) 'State of the art thermal energy storage solutions for high performance buildings', Master's thesis. University of Jyväskylä, Norway.

Sutjahja, I.M. *et al.* (2016) 'The role of chemical additives to the phase change process of $CaCl_2.6H_2O$ to optimize its performance as latent heat energy storage system', *Journal of Physics: Conference Series*, 739(1).

van Essen, V.M., Cot Gores, J. *et al.* (2009) 'Characterization of salt hydrates for compact seasonal thermochemical storage', in *ASME 2009 3rd International Conference on Energy Sustainability, Volume 2*. ASMEDC, pp. 825–830. Available at: https://doi.org/10.1115/ES2009-90289.

van Essen, V.M., Zondag, H.A. *et al.* (2009) 'Characterization of $MgSO_4$ hydrate for thermochemical seasonal heat storage', *Journal of Solar Energy Engineering*, 131(4). Available at: https://doi.org/10.1115/1.4000275.

Vasiliev, Leonard Leonardovich *et al.* (1999) 'A solar and electrical solid sorption refrigerator', *International Journal of Thermal Sciences*, 38(3), pp. 220–227. Available at: https://doi.org/10.1016/S1290-0729(99)80085-4.

Wang, C., Zhang, P. and Wang, R.Z. (2010) 'Performance of solid–gas reaction heat transformer system with gas valve control', *Chemical Engineering Science*, 65(10), pp. 2910–2920. Available at: https://doi.org/10.1016/j.ces.2010.01.011.

Werner, R. and Groll, M. (1991) 'Two-stage metal hydride heat transformer laboratory model: Results of reaction bed tests', *Journal of the Less Common Metals*, 172–174, pp. 1122–1129. Available at: https://doi.org/10.1016/S0022-5088(06)80019-6.

Xu, J., Oliveira, R.G. and Wang, R.Z. (2011) 'Resorption system with simultaneous heat and cold production', *International Journal of Refrigeration*, 34(5), pp. 1262–1267. Available at: https://doi.org/10.1016/j.ijrefrig.2011.03.012.

Yan, J. and Zhao, C.Y. (2016) 'Experimental study of $CaO/Ca(OH)_2$ in a fixed-bed reactor for thermochemical heat storage', *Applied Energy*, 175, pp. 277–284. Available at: https://doi.org/10.1016/j.apenergy.2016.05.038.

Yan, J., Zhao, C.Y. and Pan, Z.H. (2017) 'The effect of CO_2 on $Ca(OH)_2$ and $Mg(OH)_2$ thermochemical heat storage systems', *Energy*, 124, pp. 114–123. Available at: https://doi.org/10.1016/j.energy.2017.02.034.

Zondag, H.A. *et al.* (2010) 'An evaluation of the economical feasibility of seasonal sorption heat storage', in *Proceedings of the 5th International Renewable Energy Storage Conference IRES 2010*, Berlin.

4 Underground Thermal Energy Storage Systems and Their Applications

*Hamed Mokhtarzadeh, Shiva Gorjian,
Yaghuob Molaie, Kamran Soleimani, and
Alireza Gorjian*

4.1 INTRODUCTION

It is predicted that energy consumption will face significant problems in the future decades due to the rapidly growing global population and the adoption of new technology. How to extract energy from nature is another difficulty of energy consumption; therefore, environmental protection has become a global imperative (Gorjian *et al.*, 2022). The future perspective should be created to take into account the increased use of clean alternative energy sources, improving energy efficiency, and considering the finite nature of fossil fuel resources, the rise in fuel prices, and the global warming phenomenon as a result of the release of greenhouse gases (GHGs), primarily caused by the combustion of fossil fuels to meet the energy needs of the growing population (Xu, Wang and Li, 2014; *2030 Targets | European Commission*, 2022). As a result, it is essential to create and deploy low-carbon, long-lasting, and reasonably priced energy systems (Kalaiselvam and Parameshwaran, 2014). The creation of energy storage technology is one possibility in this area, which is just as crucial as the creation of new energy sources. Energy storage is mostly utilized to lessen the imbalance between supply and demand, but it may also boost the efficiency and dependability of energy systems, contributing significantly to energy conservation (Al Shaqsi, Sopian and Al-Hinai, 2020; De Rosa *et al.*, 2021; Olabi and Abdelkareem, 2021). The capture and storage of thermal energy within the framework of concepts for renewable energies and waste heat use is one of the most crucial areas of the energy strategy. About 50% of the final energy consumed globally comes from thermal energy, which results in 40% of the CO_2 emissions (*Heat – Renewables 2019 – Analysis – IEA*, 2022). The majority of thermal demand for cooling will expand drastically as a result of the vast scale of human thermal energy consumption, and any improvement in thermal energy management methods may significantly boost society's financial gain (*Heating – Analysis –IEA*, 2022; *The Future of Cooling: Opportunities for Energy-Efficient Air Conditioning | U.S. Climate Resilience Toolkit*, 2022). By increasing efficiency and lowering the ratio of energy consumption, thermal energy storage (TES) technology will close the gap between energy demand and supply. It is a viable solution to address the problems related to energy security, which are caused by GHG emissions, and climate change (Finck *et al.*, 2018).

DOI: 10.1201/9781003345558-5

TES technology uses a storage medium to keep heat or cold in reserve. In this method, the stored energy may be utilized later to generate power as well as for heating and cooling purposes. TES technologies assist in balancing energy supply and demand across daily, weekly, monthly, or seasonal timescales, depending on the availability of resources (Sarbu and Sebarchievici, 2016; Gorjian, Calise, *et al.*, 2021). The noted difference might be due to variations in power, location, time, or temperature. With three primary phases of charging, discharging, and storage that together create a whole cycle, energy is added to the storage system in TES technologies so that it may be utilized again at a later time. As a result, TES systems should take into account the cycles described, including the storage's short-, medium-, or long-term (seasonal) capacity (Enescu *et al.*, 2020; Cabeza *et al.*, 2021).

Buildings (for space heating and cooling, and water heating), solar power production systems, and greenhouses are just a few of the thermal applications that may be connected with TES technologies (for heating or cooling) (Dinçer and Rosen, 2010). People have utilized stored energy derived from natural resources throughout history. Stones, water, soil, and phase change materials (PCMs) are examples of natural resources that have been employed for energy storage, together with the earth, sky, ambient air, and water evaporation (Morofsky, 2007; Jalili Jamshidian, Gorjian and Far, 2018). Extracting natural ice or snow from mountains, lakes, and rivers to keep food and beverages cold and to chill the area is one of the most traditional ways of storing energy. Living in natural caves or excavating in rocks and dirt that were chilly in the summer and warm in the winter is another method of storing energy since seasonal temperature fluctuations do not penetrate the depths of the ground. Ice was utilized for air conditioning before mechanical cooling systems were invented. When put within the air ducts in the early 19th century, the ice served as a cooling medium, chilling and humidifying the heated air blasted into it (Morofsky, 2007; Drijver, Dinkla and Grotenhuis, 2010). Large water storage tanks were one of the first kinds of constructed energy storage that helped lower peak energy demand and were the most popular in solar and district heating applications. Storage devices are also required for solar thermal applications because of the daily changes in solar energy. In this manner, the heat produced by the sun may be utilized after dusk. Additionally, variations in sun intensity lead to a requirement for weekly and seasonal storage (Lee, 2013). With the emergence of the energy crisis in the early 1970s, interest in TES technologies started to pick up steam. The long-term objective was to store solar heat primarily for space heating from summer to winter at the outset of large-scale seasonal storage research and development. Industrial waste heat, which had a high potential as an energy source, was another one exploited in this context (Xu, Wang and Li, 2014). In recent years, cooling has also grown to be a significant problem, leading to a rise in the usage of district cooling systems. These systems have up to now relied on passive cold storage, although interest in large-scale seasonal cold storage systems is developing. The three key forces driving the development of large-scale seasonal energy storage systems are (Morofsky, 2007; Xu, Wang and Li, 2014):

- Separation of the production of heat and electricity in cogeneration facilities employing heat storage to improve the proportion of the yearly heat demand that can be satisfied by cogeneration

- Solar energy may be used to cover winter heating demands with the help of solar central heating systems with seasonal storage
- Wintertime air temperature storage for summertime cooling

Thus, thermal energy storage may be used for the following primary applications (Liu *et al.*, 2020):

- Energy conservation through the use of new alternative energy sources
- Peak shaving in grid networks and district heating systems
- Energy conservation involves running energy conversion machinery at full (optimal) load rather than partial load, such as heat pumps and co-generating plants. This lowers the need for power and boosts efficiency.
- Lowering the emissions of greenhouse gases
- Release of high-quality electrical energy for commercial applications with additional value

TES's role and significance will be given a lot of emphasis for the following reasons:

- Approximately 90% of the global energy budget is utilized for thermal energy, which is used for a variety of functions, including the storage, conversion, and transfer of heat. Figure 4.1 demonstrates the significance of thermal energy as a bridge between main and secondary energy sources. Therefore, by

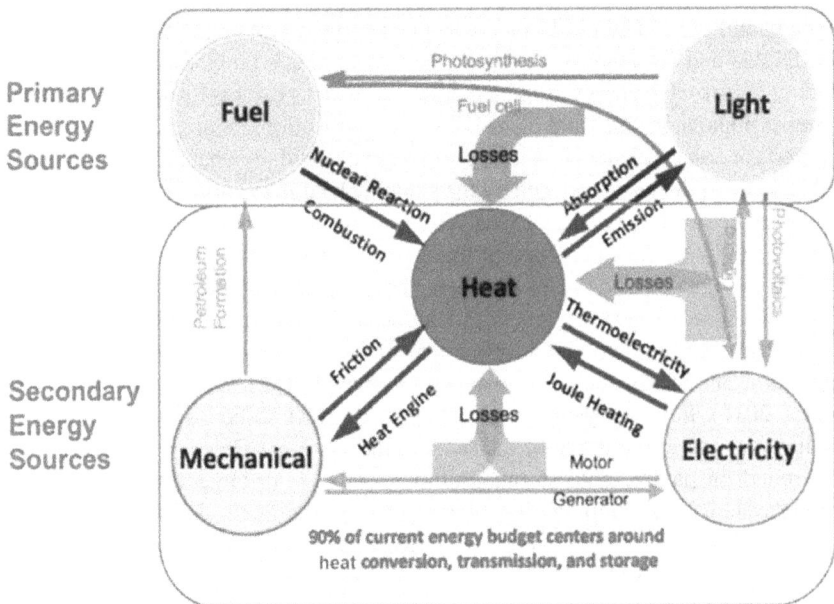

FIGURE 4.1 Thermal energy as the heart of the whole energy chain (Sadeghi, 2022).

bridging the gap between heat demand and supply, TES technologies can help contribute to a more suitable generation and use of thermal energy (Sadeghi, 2022).

- A sizable portion of energy consumption on the demand side is used as thermal energy (approximately 50% in the UK) (Taylor *et al.*, 2013).
- Thermal energy must be a byproduct or intermediary in many industrial operations due to thermodynamic restrictions. The use of TES offers a significant potential to increase process energy productivity, but conventional energy storage technologies are scarcely capable of filling this gap (Taylor *et al.*, 2013; Sadeghi, 2022).
- TES is essential for the storage of compressed air, the creation of solar thermal power, cold energy, and pumped thermal energy (Hartmann *et al.*, 2012).
- With an existing range of fewer than 200 km, TES can significantly increase the effectiveness and performance of electric cars. However, the usage of air conditioning will result in a range decrease of between 30–40%.
- Adding TES to power plants can dramatically improve their capacity to meet peak demand and lower the cost of carbon capture (Li *et al.*, 2011).

According to various criteria, there are many classes for thermal energy storage methods, as illustrated in Figure 4.2. Thermal storage methods can be divided into 'heat storage' and 'cold storage' depending on the temperature level of the thermal energy that is being stored. While 'short-term' and 'long-term' thermal energy storage may be obtained if the amount of time the thermal energy is held is the criterion. Finally, 'sensible heat storage', 'latent heat storage', and 'thermochemical heat storage' can all be taken into consideration if the material's state (a mechanism) is the criteria (Hasnain, 1998a, 1998b; Alva, Lin and Fang, 2015).

Based on the properties of the materials used for energy storage, Figure 4.3 presents a very clear categorization scheme for the types of heat storage.

Heat- and cold-storage thermal energy technologies are categorized as (Liu *et al.*, 2020):

- Underground thermal energy storage (UTES)
- Above-ground water tanks

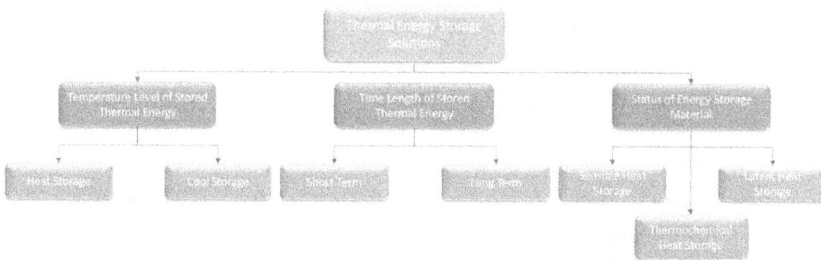

FIGURE 4.2 Solutions for storing thermal energy and their categorization (Sharma *et al.*, 2009).

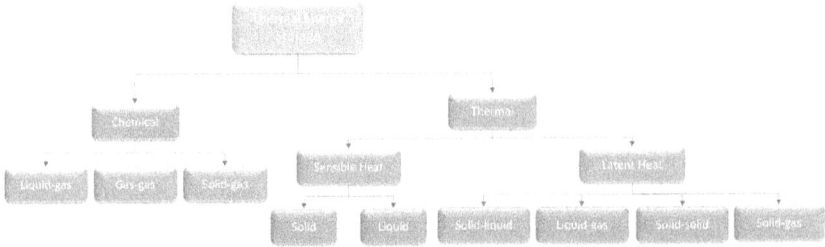

FIGURE 4.3 TES technologies categorized using the material state criterion (Sharma *et al.*, 2009; Alva, Lin and Fang, 2018).

- Rock-filled reservoirs with air circulation
- Phase change materials (PCM)
- Thermochemical storage

Nature is the driving force behind the development of storage systems between seasons of the year because seasonal climatic variations result in the passive storage of thermal energy in subterranean waters. Underground water is an appropriate source for heat extraction in the winter since the average ground temperature is greater in the winter than it is in the summer. In other words, the heat extracted in the summer is collected and employed as a storage system. These extraction systems are mostly used to deliver heat in the winter and cool in the summer. It should be noted that additional energy storage will be required if the system's heating demand is lower or greater than its cooling demand. Subsurface thermal energy storage (UTES) systems employ naturally occurring underground places to store thermal energy (Lee, 2013). The introduction of TES technologies, including the several forms of thermal storage, their benefits, and their significance, is the first topic covered in this chapter. The chapter then covers UTES technologies, including aquifer thermal energy storage, borehole thermal energy storage, cave thermal energy storage, pit thermal energy storage, and water reservoirs, and discusses their performance and uses.

4.2 UNDERGROUND THERMAL ENERGY STORAGE TECHNOLOGIES

Sensible heat is the primary kind of energy stored during long-term (seasonal) thermal energy storage. However, energy losses happen while the sensible heat is being stored. TES is now the most researched and applied method in sensible heat storage. Due to their ability to retain heat, availability of space, and affordability in comparison to water, storage materials have received interest (Gorjian, Ebadi, *et al.*, 2021). A method called underground thermal energy storage (UTES) makes it possible to store thermal energy on a massive scale. With the help of this technology, it is possible to store heat in the soil and subterranean water during the summer and recover it during the winter, similar to how cold may be stored in the winter and utilized for cooling in the summer. The data show that the temperature of the soil at a given depth is basically consistent, greater in winter than the ambient air temperature, but lower in summer than the air temperature. Up to a depth of about 10 meters, the shallow soil and groundwater are affected by changes in ambient air temperature. The geothermal gradient causes the sun's surface temperature to rise by an average of 3°C every 100 m as depth rises. Consequently, the earth and subterranean water are appropriate spaces for storing and periodically extracting heat and cold (Lee, 2013).

UTES systems are divided into two main groups (Novo *et al.*, 2010): (1) open loop and (2) closed loop. The term 'closed' systems refer to those in which a technical fluid (in most instances, water) is pumped via heat exchangers in the ground as opposed to those in which groundwater is pumped from the earth and injected into the ground through wells or in subterranean caves. Aquifer thermal energy storage (ATES), pit thermal energy storage (PTES), and reservoir/tank thermal energy storage (TTES) are examples of open-loop UTES systems. The opposite is true for closed-loop UTES systems, which have borehole thermal energy storage (BTES). Other UTES methods exist, such as pit and cave TES, although they are rarely used in the industrial setting. While BTES, TTES, and PTES systems are less likely to use cooling, ATES actively benefits from a balanced heating and cooling load. With or without a heat pump, UTES systems may be utilized for both cooling and heating (Li *et al.*, 2011; Hartmann *et al.*, 2012). As a result, several UTES systems have been developed since the 1970s (Paksoy and Beyhan, 2015):

- Aquifer thermal energy storage (ATES)
- Borehole thermal energy storage (BTES)
- Cavern thermal energy storage (CTES)
- Pit thermal energy storage (PTES)
- Water tank thermal energy storage (TTES)

4.2.1 AQUIFER THERMAL ENERGY STORAGE (ATES)

Minerals and subsurface water in the earth's crust contain thermal energy. The stored energy is transferred through one or more wells. The aquifer, a saturated and porous subterranean layer, is used as a storage medium for thermal energy in aquifers. Thermal energy is transferred by withdrawing groundwater from the aquifer and reinjecting it at a modified temperature into a separate well nearby. The ATES is the most cost-effective natural UTES system. ATES is expanding quickly in recent years as a practical solution for seasonal TES, especially in European countries, and several ATES studies have been conducted, where the typical modeling techniques and thermal performance indicators are provided (Schmidt *et al.*, 2015). Figure 4.4 is a schematic illustration of this kind of TES system. Since the selection of the proper aquifer and well spacing heavily influences the performance of the ATES system, research is currently primarily focused on the variables influencing the subsurface component's performance. Thermal recovery effectiveness and thermal disturbance intensity are the two most often utilized indicators for analyzing thermal performance. Thermal interference should be avoided in a doublet system to improve thermal recovery efficiency, whereas in a multi-well or multi-system, an adequate intensity of thermal

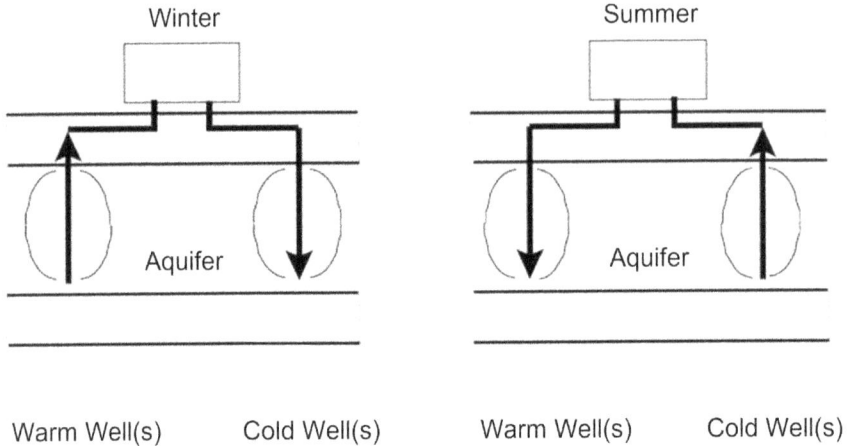

Warm Well(s)   Cold Well(s)   Warm Well(s)   Cold Well(s)

FIGURE 4.4 ATES system schematic; flow direction is reversed in the summer to maximize efficiency (Schmidt *et al.*, 2015).

interference should be chosen to efficiently boost the energy supply in a particular location (Gao *et al.*, 2017).

4.2.2 BOREHOLE THERMAL ENERGY STORAGE (BTES)

In general, the most practical UTES are BTES systems. To facilitate the passage of thermal energy into and out of the earth, BTES systems comprise vertical heat exchangers buried 20 to 300 m under the surface (clay, sand, rock, etc.). The storage media in these systems is bedrock. A heat transfer medium is injected through the boreholes that pierce the storage volume to transmit heat. A heat transfer medium can be circulated through a borehole thanks to a pipe system that has been placed there. The temperature of the storage media rises when thermal energy is introduced (Schmidt *et al.*, 2015). Figure 4.5a depicts a BTES system in schematic form. For the purpose of heating houses or businesses in the summer, many projects incorporate the storage of solar heat. Additionally, geothermal heat exchangers (also known as geothermal heat pumps) are frequently used in conjunction with heat pumps to collect or transmit low-temperature heat from or to the earth. BTES systems are among the most widely used types of UTES because of their versatility in virtually all soil situations (Gehlin, 2016).

4.2.3 ROCK CAVERN THERMAL ENERGY STORAGE (CTES)

In a cavern thermal energy storage system (CTES), energy is stored in the form of hot water in an underground cavern (Figure 4.5 a, b). In this system, with a large volume of water, maintaining the characteristics of the classified temperature in the cave is of great importance. In this system, during injection, hot water is injected into the tank from above, while colder water is withdrawn from below (Schmidt *et al.*,

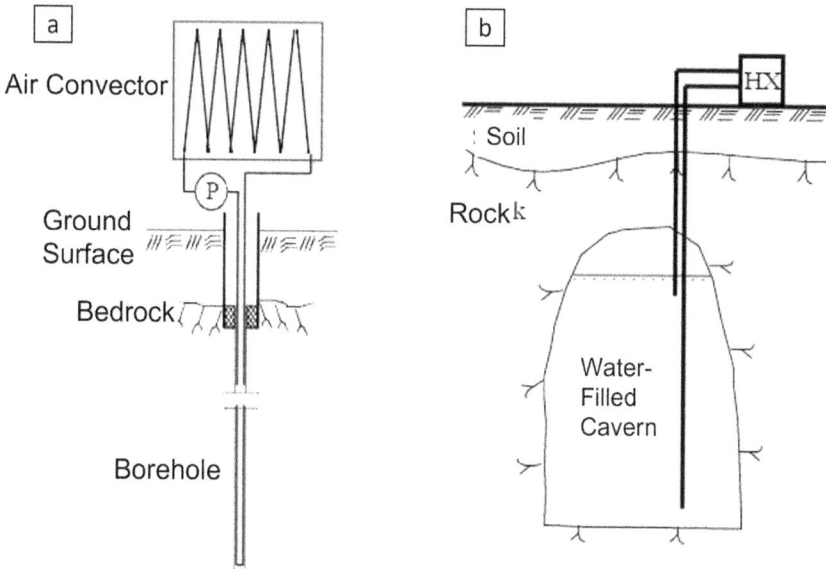

FIGURE 4.5 (a) Cutting through a single borehole system; (b) rock cavern hot water storage (Schmidt *et al.*, 2015).

2015). Water is used in massive, exposed, subterranean caverns called CTES systems to store thermal energy. Natural or human-made caverns can be utilized, such as depleted oil or gas reserves or abandoned mining tunnels and shafts. Although technically conceivable, the actual use of these storage systems is currently constrained since they need particularly specialized site conditions that are frequently absent (Zizzo, 2009).

Figure 4.6 depicts the three most popular thermal storage systems: ATES, BTES, and CTES. The two most promising technologies are borehole heat exchangers and aquifer storage. In several nations, these ideas have already been implemented as operating systems in the energy sector. Commercially, alternative solutions are rarely adopted. Aquifer separation and a reduction in water chemistry issues are benefits of closed systems. The greater capacity of a well for heat transmission than a borehole is a benefit of open systems. Therefore, if the subsurface is geologically and hydrochemical compatible, ATES is often the best alternative (Schmidt, Mangold and Müller-Steinhagen, 2003).

Pit thermal energy storage (PTES) and water tank thermal energy storage (TTES), also called artificial aquifers, are human-made structures that, like underground tanks, are built underground or near the surface to avoid high excavation costs. They must therefore be insulated at the top and the walls, at least to a certain depth. Hydrogeologic conditions at a particular site are not as important as for the other concepts. Table 4.1 summarizes some of the characteristics of main seasonal storage concepts. Figure 4.7 represents the main benefits and drawbacks of UTES technologies.

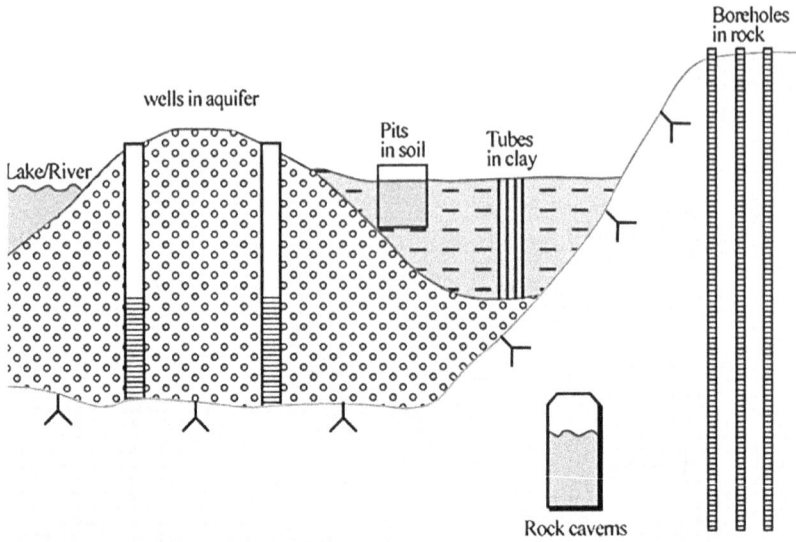

FIGURE 4.6 Schematic representation of the most popular UTES systems (Matos, Carneiro and Silva, 2019).

TABLE 4.1
Comparison of Storage Concepts (Matos, Carneiro and Silva, 2019)

Storage Concept	TTES	PTES	BTES	ATES
Storage medium	Water	Gravel water[1]	Ground material (soil/rock)	Ground material (sand/water or gravel)
Heat capacity (kWh/m³)	60–80	30–50	15–30	30–40
Storage volume for 1 m³ water equivalent	1 m³	1.3–2 m³	3–5 m³	2–3 m³
Geological requirements	Stable ground conditions Preferably no groundwater 5–15 m deep	Stable ground conditions Preferably no groundwater 5–15 m deep	Drillable ground Groundwater favorable High heat capacity High thermal conductivity Low hydraulic conductivity ($K < 1 \times 10^{-10}$ m/s) Natural groundwater flow <1 m/y 30–100 m deep	Natural aquifer layer with high hydraulic conductivity ($K > 1 \times 10^{-5}$ m/s) Confining layers on top and below No or low natural groundwater flow Suitable water chemistry at high temperatures Aquifer thickness 20–50 m deep

Tank TES (water)

+ high thermal capacity (water)
+ good operation characteristics (high (dis-) charging power, usable as buffer store)
+ freedom of design (geometry)
+ thermal stratification
(+) maintenance/repair
(-) limited size (< 100.000 m3)
(-) primary energy demand
- high construction costs

Pit TES (gravel-water or water)

+ reasonable construction costs
+ medium (gravel-water) to high (water) thermal capacity
+ nearly unlimited store dimensions
(+) operation characteristics (medium charging power in case of gravel water)
(-) complex and costly cover (in the case of water)
- limited freedom of design (slope angle)
- maintenance repair difficult/not possible

BTES (soil)

+ low construction costs
+ easily extendable
- low thermal capacity
- operation characteristics (low (dis-)charging paver, buffer required, heat pump recommended)
- limited choice of locations
- no thermal insulation at the side and bottom possible
- maintenance repair difficult/not possible

ATES (saturated sand-water)

+ very low construction costs
(+) medium thermal capacity
(-) operation characteristics (low/medium (dis-) charging power, buffer, and heat pump recommended)
- very limited choice of locations
- no thermal insulation possible, relatively high thermal losses

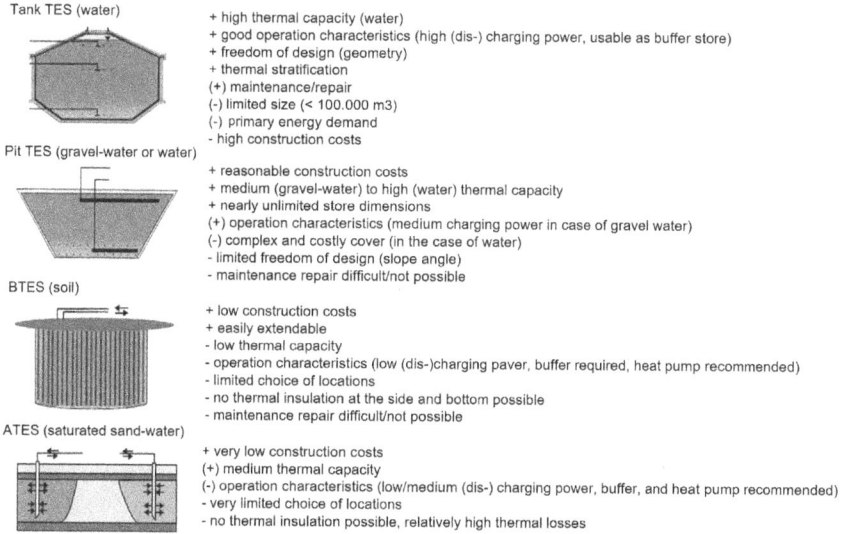

FIGURE 4.7 Construction concepts of large-scale/seasonal UTES systems and their benefits and drawbacks (Dahash *et al.*, 2019).

4.3 APPLICATIONS OF UNDERGROUND THERMAL ENERGY STORAGE SYSTEMS

TES systems may be used as long-term (seasonal) storage to supplement natural and renewable energy sources as well as short-term storage to balance peak loads. In situations where there is an economically advantageous timing mismatch between energy demand and supply, TES systems are employed for heating or cooling. UTES systems have a variety of uses, including in greenhouses, buildings, etc.

4.3.1 Use of UTES in Agricultural Greenhouses

Human-made structures, such as greenhouses, are one possibility for UTES. UTES systems are advantageous because they utilize the extra heat stored in the summer to heat the greenhouses in the winter. UTES systems can meet greenhouses' cooling and heating needs (Gorjian *et al.*, 2020). A thermal energy source that is stored in an ATES can be made from the heat that has collected in a greenhouse. Figure 4.8 depicts this procedure. When the aquifers' conditions are unfavorable, UTES systems are applied. CTES systems protect several subterranean cavities. In the first ATES greenhouse, the experimental setup was constructed in 2002, and the first commercial use followed in 2003 (Paksoy and Beyhan, 2015).

Heat or cold is taken or delivered from the ground via closed-loop fluid circulation. The fluid often contains an antifreeze to allow the system to operate below the freezing point. Similar to the ATES system, the greenhouse distribution system connects the greenhouse with the drilling field via heat exchangers (Paksoy and Beyhan, 2015). The ATES technology was used for the first time in a greenhouse's heating

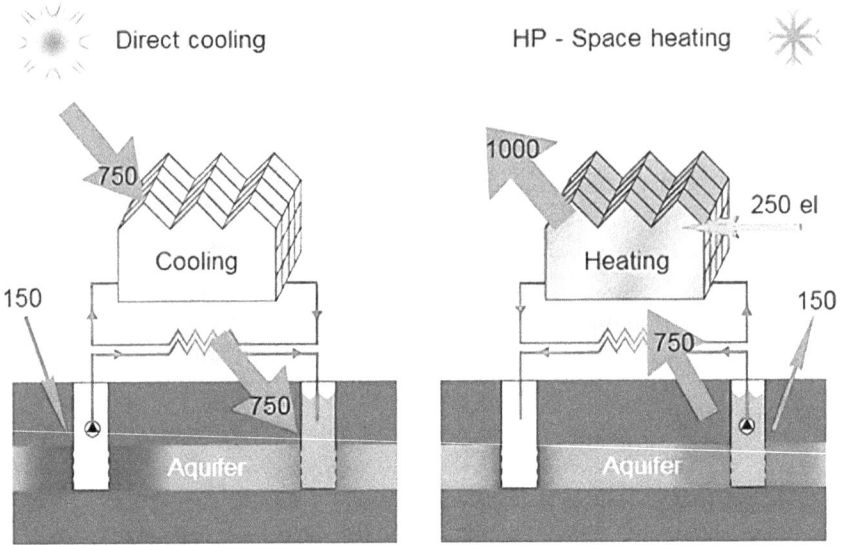

FIGURE 4.8 Diagram of an ATES system for greenhouse application.

Source: Courtesy of IF Tech International BV, Netherlands (Paksoy and Beyhan, 2015).

and cooling system in a Mediterranean environment. This project, which was carried out at Cukurova College, consists of two distinct greenhouses with a combined size of 360 m². While the second greenhouse solely employed the ATES system for the heating system, the first greenhouse incorporated both the cooling and heating systems from ATES. The fundamental design of the ATES system as implemented for usage in a greenhouse is shown in Figure 4.9. The ATES system was employed in this pilot project for each greenhouse in a collection of wells, each of which had a mix of hot and cold wells. Each well was 0.4 m in diameter and 80 m deep (Turgut *et al.*, 2009; Paksoy and Beyhan, 2015).

In order to look at how heating systems affect greenhouses, three various heating system types were examined in Geneva, Switzerland. The heating systems included a standard system with a natural gas boiler, one that used a tank storage system, and another that used an underground borehole storage system. The greenhouses with tanks and the subterranean storage were found to have energy offsets of 13% and 11%, respectively, without having any detrimental effects on product performance. About 25% of the storage capacity is thought to be accounted for by these systems. The energy offset might theoretically be four times higher than it is now. It was discovered that tank storage has a coefficient of performance (COP) of 1.7 while subterranean storage has a COP of 5.8. The latter value was found to be a design flaw and not due to tank storage (Paksoy and Beyhan, 2015).

The Geo-Heat Center in the US conducted research comparing the economic viability of using natural gas heating systems vs. ground-source heat pumps (GSHP) to satisfy greenhouse heating demands. The heating needs of a 4047 m² fiberglass

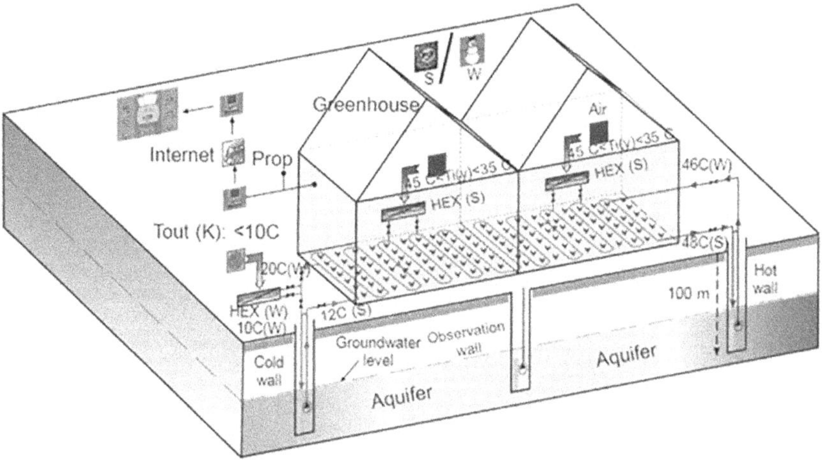

FIGURE 4.9 Schematic view of an ATES greenhouse in Cukurova University, Turkey (Paksoy and Beyhan, 2015).

greenhouse were calculated using climatic data from Boston, Dallas, Denver, and Seattle. BTES systems were shown to be economically viable at natural gas costs of $0.21–0.35/m³, while BTES systems became economically viable at natural gas prices over $0.53/m³ (Mehling and Cabeza, 2015).

The Pacific Agricultural Research Centre in Agassiz, Canada, installed an ATES system to provide thermal energy for its greenhouses. The system included a seasonal energy storage element and was used for both heating and cooling. This design produced supplementary heat using a natural gas boiler rather than cooling towers. It was discovered that the older version's building cost was higher, costing $35 billion less annually. As a consequence, there were about 300 tons fewer greenhouse gas emissions and 6,000 GJ less energy used per year. There was no 'closed' greenhouse company involved in the project (Mehling and Cabeza, 2015). For the GESKAS project, two miniature greenhouses were constructed at the Belgian Horticulture Research Institute. These greenhouses have excellent ventilation since tomatoes were grown there. Two open reference greenhouses outfitted with conventional methods were utilized for comparison, and the three characteristics of harvesting behavior, production rates, and energy were taken into account. Finally, the closed greenhouse produced more (7–11%) and used less (8–34%) primary energy. The closed greenhouse needed a heat pump and a suitable UTES system to operate at its best (Paksoy and Beyhan, 2015). In Shanghai, East China, Lee researched a 2,304 m² solar-heated greenhouse with a seasonal thermal energy storage system. This energy storage system employed 4,970 m³ of subterranean soil to store heat produced by a 500 m² solar collector throughout the winter months using U-tube heat exchangers. It was discovered that 208.9 GJ was later removed for heating the greenhouse, whereas 331.9 GJ was charged during the first year of operation. It was discovered that the system can keep the inside air at a temperature that is 13°C above ambient. Additionally, the

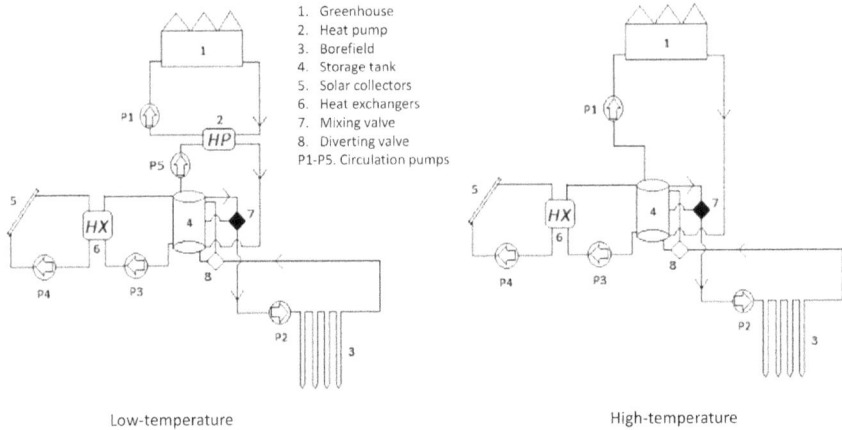

1. Greenhouse
2. Heat pump
3. Borefield
4. Storage tank
5. Solar collectors
6. Heat exchangers
7. Mixing valve
8. Diverting valve
P1-P5. Circulation pumps

Low-temperature   High-temperature

FIGURE 4.10   Schematic of a ground-based TES system with Fresnel lens in a solar greenhouse (Li *et al.*, 2019).

electrical coefficient of performance (ECOP) was higher than the typical heat pump heating system in the first year of operation, at roughly 8.7 (Lee, 2013).

Li *et al.* (2019) investigated a solar-heated greenhouse that used a Fresnel lens concentrator. To provide heat in the absence of sunlight and ensure plant growth, the geothermal storage system was used. The experimental results showed that the duration of heat transfer is five days when the heating pipe is installed at a depth of 1.65 m underground. The increase in the total temperature of the soil is about 4°C. It was shown that the depth of the buried pipe should be more than 2.5 meters to achieve seasonal heat storage in this system. It was also discovered that the usage of TES can guarantee an interior air temperature over 8°C during the coldest season with severe weather conditions, which guarantees the minimum temperature needed for plant development. Figure 4.10 depicts the system under investigation in schematic form.

In another study by Lee *et al.* (1994), the plastic greenhouse's usage of solar energy was made more effective by the subsurface heat exchange system. In a plastic greenhouse with an underground heat exchange system, the properties of heat energy that was stored and released were examined. This system's average thermal energy storage and release rates were 1,484 kJ/m day and 555 kJ/m day, respectively. The heat exchange system's stated average coefficient of performance was 2.86. The air temperature in a greenhouse composed of plastic was also attempted to be predicted. The findings of the forecast and those of the measurement were rather well in accord.

In Shanghai, east China, Xu *et al.* (2014) examined the thermal performance of a solar-heated greenhouse (with a surface area of 2,304 m²) outfitted with a seasonal BTES system. The proposed TES system used U-tube heat exchangers to store thermal energy from a 500 m² solar collector in 4,970 m³ of the subterranean soil (Figure 4.11). During the heating season, the heat exchanger tubes situated on the plant shelves and bare soil discharged the thermal energy that had been stored into the greenhouse. It has been shown that the created system can keep the interior air

FIGURE 4.11 (a) Solar-powered greenhouse; (b) the array of solar collectors; (c) BTES equipment (Xu *et al.*, 2014).

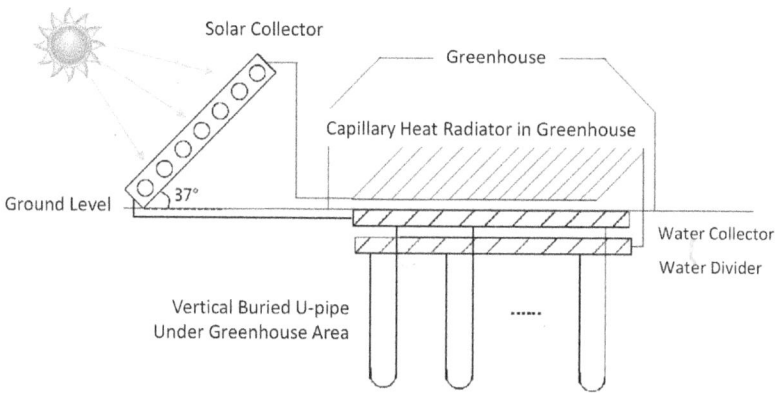

FIGURE 4.12 Schematic representation of the solar greenhouse equipped with an SSSHS (Zhang *et al.*, 2015).

at a temperature that is 13°C above the outside air when the outside temperature is 2°C at night. In the first year, the experiment's electrical COP (ECOP) was almost 8.7, outperforming a typical GSHP unit in terms of performance. A low-cost seasonal solar soil heat storage system (SSSHS) for greenhouse heating was created by Zhang *et al.* (2015). In this design, solar energy was captured and stored in the soil to fulfill the greenhouse's requirement for heating throughout the bitterly cold winter months (Figure 4.12). TRNSYS software was employed in this work to mimic the procedure and outcomes of solar energy and soil heat absorption. Then, using experimental data collected throughout one season, the simulated findings were verified. It was discovered that keeping the greenhouse area's interior air temperature above 12°C led to energy savings of 27.8 kWh/m².

Subsurface heat exchangers were used in a study by Zhou *et al.* (2017), to use hot water produced by solar collectors to heat a greenhouse. As seen in Figure 4.13, the heat storage circuit runs throughout the day to charge the PCM, while the heating circuit discharges the PCM at night (keep valves 1, 2, and 3 open). A project was carried

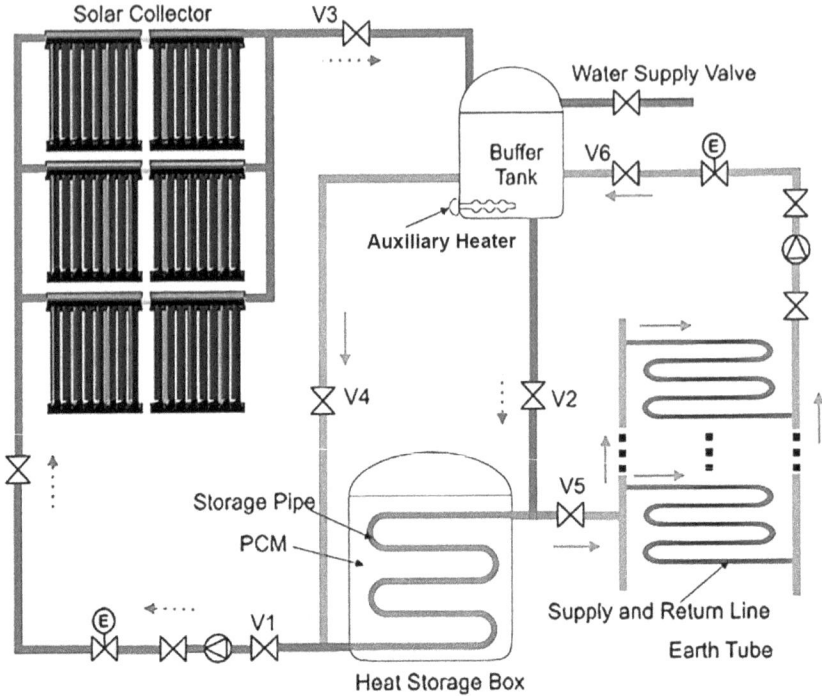

FIGURE 4.13 Schematic view of a solar heating system equipped with LHTES (Zhou *et al.*, 2017). The flow direction in the heating loop is shown by blue arrows, while the flow direction in the heat storage loop is indicated by pink arrows.

out by Turgut *et al.* (2009) at the University of Cukurova in the years 2005–2006 to ascertain the heating and cooling capabilities of ATES systems in greenhouses in Mediterranean climates. Two greenhouses with a combined area of 360 m² were taken into consideration for this project. ATES technology was employed in one of the greenhouses for heating and cooling, and conventional heating and cooling systems were used in the other. The groundwater and exchanger water temperatures, as well as the outside and interior temperatures of the greenhouses, were all monitored during the experiment. To investigate plant development and fruit output, the tomato crop that was cultivated in both greenhouses was used. Also assessed were the greenhouses' energy expenses (oil for normal heating and electricity for ATES). Therefore, these two systems were contrasted from a technological and financial standpoint. The data gathered indicated that the ATES system had high potential for air conditioning greenhouses in the Mediterranean climatic zone, both for heating and cooling. Due to this performance, no heating oil was ever used in the greenhouse heated by the ATES system between October 20 and April 10. The internal temperature of the greenhouse never went beyond the crucial limit of 12°C. As opposed to the conventionally heated greenhouse, the ATES greenhouse saw less temperature variation. When compared to the traditionally (with fuel oil) heated greenhouse, using ATES for heating resulted in energy cost savings of roughly 70%. In the heated greenhouse using the ATES system as opposed to the conventionally heated greenhouse, the tomato experiment's production was around 20% greater.

4.4 USE OF UTES IN BUILDINGS

Since the 1990s, as experience and proof of BTES's cost-effectiveness have expanded, so have the number and variety of buildings utilizing it for combined heating and cooling. Interest in such systems has decreased in favor of the simpler low-temperature BTES after the initial decades of ardent development of high-temperature seasonal storage (Gehlin, 2016). However, there has recently been a resurgence of interest in high-temperature BTES applications with seasonal storage of solar heat or industrial and cogeneration plant waste heat. Although there are not many such uses, the medium-sized Drake Landing Solar Community district heating project in Canada and the industrial waste heat storage system in Emmaboda, Sweden, both of which were successful, clearly demonstrate the potential for such applications. As district heating utilities adjust to a building stock that has been upgraded with low-temperature heating and thermally active building systems, the thermal buffering capabilities of BTES systems has the potential to increase the efficiency and resiliency of urban-scale district heating systems. A tendency toward bigger BTES systems may be seen in the recent construction of some extremely large BTES systems with more than 1,000 wells. According to the total number of meters of boreholes drilled, the top five BTES systems worldwide include (Gehlin, 2016):

- China's Tianjin Cultural Center has 3,789 120-meter-deep boreholes (total 454,680 m)
- Ball State University, Indiana, USA: 1,806 135-meter-deep boreholes (total 243,810 m)
- Lotte World Tower, Seoul, South Korea: 720 200-meter-diameter boreholes (total 144,000 m)
- ELI-NP Marguele in Bucharest, Romania: 1,080 125-meter-deep boreholes (total 135,000 m)
- Zhungguancon International Center in Beijing, China: 1,060 123-meter-deep boreholes (total 130,380 m)

In contrast to the early BTES systems built in the 1980s, which were dug only to a modest depth of less than 100 m, today's boreholes are frequently bored to a depth of 120–200 m or even deeper. New BTES projects sometimes have borehole depths up to 300 m, notably in Scandinavian countries. If BTES can be used to infrastructure applications in the real world, it will be interesting to watch. Future studies should focus further on how solar heat stored in a BTES system may prevent ice from developing on roads, bridge decks, train stations, and parking lots. As long as BTES systems are cost-competitive with less efficient systems and produce fewer greenhouse gases, their usage and development have a promising future (Gehlin, 2016).

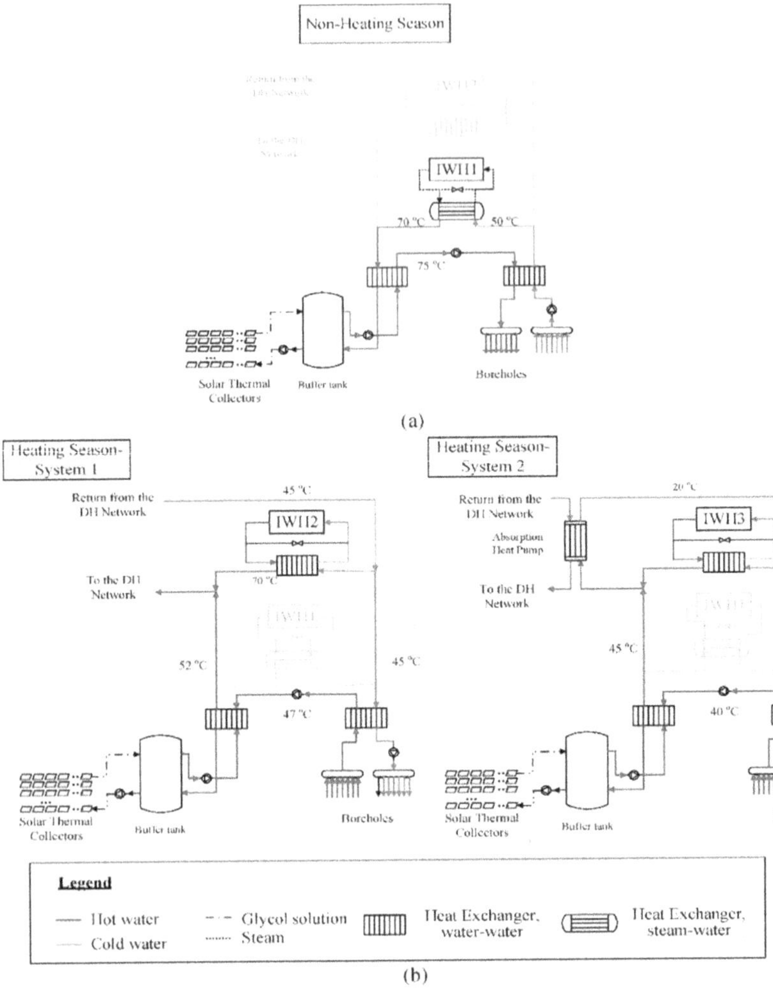

FIGURE 4.14 System diagram under desired operational conditions: (a) non-heating season and (b) heating season (Xu *et al.*, 2018).

Xu *et al.* (2018) evaluated an experimental ADH framework in Chifeng, China, which included a 0.5-m TES system, an on-site solar thermal system, and a lot of heat from a copper plant (Figure 4.14). In this study, two local heating framework circumstances were examined, and the critical implementation of seasonal TES (STES) was examined utilizing a model-based method and Modelica. To assess the effectiveness of the STES system, several performance metrics, including the heat of extraction, the heat of injection, and the storage capacity factor, were chosen. The findings show that improved energy performance is correlated with lower STES discharge temperature. Further research was done on STES's long-term performance. It was discovered that STES stabilization timeframes of one to three days were reduced by discharging at lower temperatures.

A greenhouse with a hybrid adaptive BTES system, a 100-kW heat pump, and a TES system was created for Maere College in Norway. According to the testing findings, this method may reduce fossil fuel usage by 90% in a year. An extra heating system is required to adequately heat the greenhouse at peak heating loads (i.e., extremely cold external temperatures), despite the combined system's overall energy efficiency of 68% and COP of 2.6. The modification of the greenhouse's heat distribution system to a conduction heating system close to the plants was the most economical option to raise the system's efficiency (Paksoy and Beyhan, 2015).

The usage of the ATES system is governed by stringent regulations, one of which is the obligation to maintain the modulation of the total heat exchange in the ground, to ensure environmental and energy sustainability. Typically, this technique combines an air handling unit (AHU) with a heat pump (HP). In either scenario, this reduces the ATES heating/cooling systems' operational efficiency and energy use. This study assesses a useful technique for recuperation called nocturnal ventilation (NV). Further, the performance of NV, which may be utilized for excellent cold recovery with an energy consumption of up to 19.3 MWh per year, is compared to the operational efficiency and energy efficiency of a regularly used HP and AHU, which reaches a 26.4% increase in the COP (Bozkaya and Zeiler, 2019). Ground source heat pumps (GSHP) are utilized in both commercial and residential settings to more effectively and efficiently satisfy the demands for heating and cooling. Additionally, solar thermal components have been added to traditional GSHP systems to boost seasonal heat storage and decrease the estimated size of the ground heat exchanger. Due to the high installation costs, this innovation has so far exclusively been deployed in big commercial or residential structures.

A solar-assisted ground source heat pump (SAGSHP) was developed for use in home heating systems in the study by Naranjo-Mendoza et al. (2019). The method uses a network of irregularly spaced shallow vertical wells (each 1.5 m deep) to occasionally store heat in an underground 'earth energy bank'. After 19 months of use, the findings showed that this system provides outstanding performance for heating the building in the winter. Solar power introduced into the soil functions as a heat storage system in addition to aiding in redressing the thermal imbalance.

The results also indicate that the control method has to be improved, particularly to avoid having a higher feed temperature for the evaporator. In Overberg, Switzerland, a simple solar-heated house was built in 1989. It heats a 130 m^2 house using 84 m^2 of solar panels and 118 m^3 of capacity in three tanks (92, 13, 13 m^3). The residence, constructed in 2006, reportedly receives 95% of its warmth from a 28 m^3 buffer tank and 69 m^2 of solar collectors on the south-facing roof. There is a building that complies with the passive home criteria on the west coast of Ireland (Figure 4.15a). A monitored and integrated subterranean thermal energy storage system is included. A STES system with a 10.6 m^2 evacuated tube solar thermal system, a 300-liter hot water storage tank, a 23 m^3 water STES, and a combined underfloor heating and reheat system is part of the 215 m^2 (treated floor area) standalone passive home.

Scandinavian Houses, a company that creates energy-efficient homes, participated in the FP 7 EINSTEIN project and refurbished a building by installing an STES in the basement of a brand-new structure (Figure 4.15 a, b). The STES is a 50 m^2 solar array that comprises 16 flat-plate collectors (FPCs) with a total aperture area of 32 m^2

FIGURE 4.15 (a) Photo of the passive house with solar panels and STES (under green-house); (b) installation of 600-mm-thick polystyrene insulation on STES (Navarro *et al.*, 2016).

and 10 evacuated tube collectors (ETCs) with an aperture area of 1.8 m^2 (Navarro *et al.*, 2016).

In a study by Paksoy *et al.* (2000), a special design was proposed that was a combination of solar energy and ATES. This system prevented the waste of electricity and oil, which was intended to supply heating and cooling requirements of Cokurova University, Balkali Hospital in Adana. In summer, this system could provide heating and cooling in winter requirement for the hospital by storing solar energy underground. The main source of cooling for this system was the hospital's ventilation air and the surface water of Seyhan Lake. The simulation results showed that a square well field of a maximum of 350×400 m is sufficient for storing about 14,000 MW/year as of cooling and heating at +10°C. In this ATES system, about 7,000 MW/year was stored in the aquifer's cold part with an average temperature of +9°C. This storage was planned to supply heat for about 2,000 hours in winter. As storage losses occur during storage, about 500 MW are lost, increasing the temperature by about 1°C to +10°C. In summer, about 6,500 MW was used to supply cooling for the hospital during the operation of about 3,000 hours (17 h/day). This could save about 3,000 MW of electricity by replacing much of the traditional chiller cooling. It was found that the ATES system consumed about 250 MW of electricity, which was mostly consumed by the pumps in the system. The average COP for conventional cooling is considered 2.0. To preheat the ventilation air through the ATES system, from the warm side of the aquifer, about 7,000 MWh of thermal energy was pumped, for which about 300 MW of electricity was required to drive the involved pumps. However, it was estimated that by using the ATES system for preheating, 1,000 m^3/year of the oil will be saved. The environmental benefit of this project was the reduction in energy consumption in the form of electricity and fuel oil, and the replacement of refrigeration equipment that uses ozone-depleting Freon-12 gas. It was also found that savings of about 1,000 m^3/year in heating oil reduces the CO_2 emissions by approximately 2,100 tonnes/year, SO_x by 7 tonnes/year, and NOx by 8 tonnes/year.

FIGURE 4.16 Falstadsenteret located in Levanger (Midttømme and Banks, 2008).

Additionally, it was proved that replacement of existing chillers with the power of 2 MW that uses Freon-12 results in the saving of about 0.7 tonnes of Freon-12 per year. The largest ATES facility located at Oslo Gardermoen Airport in Norway has heating and cooling capacities of 7 MW and 6 MW, respectively (Sanner, 1998). As an example, a standard BTE was completed at Falstadsenteret, a historical museum with an area of 2850 m² in Levanger (Figure 4.16). In this system, the heating and cooling units consist of a 130-kW heat pump and 13 borehole heat exchangers (BHEs), at a depth of 180 m. The total cost of the GSHP and BTES was calculated as 170,000 euros, and the payback period was estimated as 12 years compared to traditional heating and cooling systems (Midttømme and Banks, 2008).

In winter 2007, a BTES system consisting of 228 wells with 200 m deep was drilled. This system could supply heating and cooling to the new Akershus University Hospital (Ahus), as shown in Figure 4.17. The total area of the building was 137,000 m² with heating and cooling demands of 26 GWh and 8 GWh, annually. This project aimed to supply at least 40% of the energy for cooling and heating from renewable energies. The planned BTES was commissioned in May 2007, but the second phase of drilling was implemented in 2009–2010, expanding the BTES system to a total of 350 wells. The wells were drilled in a layer of dioritic rock with a thickness of 5–40 m. This thick clay layer could increase the drilling costs. In this project, a combined ammonia cooler and heat pump unit was installed (Midttømme et al., 2009). The total cost of this project, including the BTES and GSHP units, was calculated as $19.5 million. The original plan was to drill the boreholes close to the hospital, but preliminary evaluations, including seismic geophysical

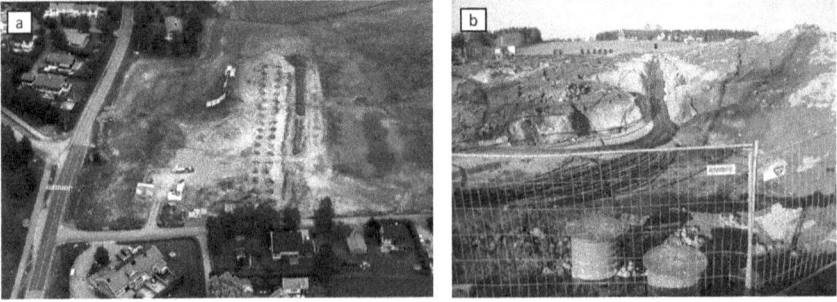

FIGURE 4.17   (a) BTES at Ahus under construction in the summer of 2007; (b) boreholes to manifold pipelines at Ahus (Midttømme *et al.*, 2009).

surveys and test drilling, indicated that there are high-density, clay-filled fracture zones. In this regard, it was found that full-scale drilling is expensive and complex. Therefore, it was decided to relocate the BTES borehole array to a field approximately 300 m from the hospital (Midttømme and Banks, 2008). Today, most of the borehole heads are located completely underground, where the farmers can grow their crops in the field.

4.5 CONCLUSIONS AND PROSPECTS

TES systems have emerged as important tools for increasing energy efficiency since the 1970s, despite traditional fossil-based power generation plants. TES systems are considered alternative heating and cooling supply solutions, which can decrease the consumption of electricity and fossil fuels and also be used instead of mechanical cooling equipment. In this context, TES technology plays a key role to overcome the world's dependence on burning fossil fuels and, consequently, the tremendous release of GHGs. Thermal energy can simply be dissipated into the environment if it is not stored, necessitating more fossil fuels to be burned. As discussed in this chapter, the most common TES systems encountered today are STES systems, TES systems integrated with solar thermal power plants, TES units integrated with distributed TES systems mainly for medium-temperature residential solar thermal or applications of geothermal power, TES systems passively used in buildings, greenhouses, heat and cold storage, and heating, ventilation, and air conditioning (HVAC) systems mostly utilized in buildings, etc.

Among the various TES technologies, UTES systems have proven that they can contribute significantly to reducing greenhouse gas emissions. In this chapter, different types of UTES systems were presented and discussed. However, the most practical UTES among the technologies studied are BTES systems. For ATES systems, there are still no long-term observations for deeper studies, and the simulation analyses are always performed without considering the changing load requirements of the end users. In addition, given the problems with groundwater, more attention should be paid to saline aquifers for TES. BTES systems at low to medium temperature levels can supply heating and cooling efficiently of large buildings, such as commercial and institutional buildings. The BTES technology has a high potential to make a significant contribution to sustainable and energy-efficient systems in several regions around the world. In cooler climates, it offers significant reductions in cooling and heating costs. As demand for energy efficiency increases, more and more buildings are being constructed with modified insulation. The well-insulated buildings, due to increased internal heat gains, lead to high comfort cooling demands in cooler climate conditions. BTES systems have the benefit of the increased cooling demand and are enable to supply both heating and cooling in a very affordable and elegant manner. More scientific research is needed in this area to further explore UTES systems. In addition, future studies could focus on simulating the dynamic coupling of subsystems below and above ground and controlling system operation to enhance system performance.

NOTE

1 Similar in energy density and volume to TTES systems, water may be employed as a storage medium in PTES.

REFERENCES

2030 Targets | European Commission (2022) "2030 climate & energy framework," 2023. Available at: https://climate.ec.europa.eu/eu-action/climate-strategies-targets/2030-climate-energy-framework_en.

Al Shaqsi, A.Z., Sopian, K. and Al-Hinai, A. (2020) 'Review of energy storage services, applications, limitations, and benefits', *Energy Reports*, 6, pp. 288–306. Available at: https://doi.org/10.1016/J.EGYR.2020.07.028.

Alva, G., Lin, Y. and Fang, G. (2018) 'An overview of thermal energy storage systems', *Energy*, 144, pp. 341–378. Available at: https://doi.org/10.1016/j.energy.2017.12.037.

Bozkaya, B. and Zeiler, W. (2019) 'The effectiveness of night ventilation for the thermal balance of an aquifer thermal energy storage', *Applied Thermal Engineering*, 146, pp. 190–202. Available at: https://doi.org/10.1016/j.applthermaleng.2018.09.106.

Cabeza, L.F. *et al.* (2021) 'Introduction to thermal energy storage systems', *Advances in Thermal Energy Storage Systems: Methods and Applications*, pp. 1–33. Available at: https://doi.org/10.1016/B978-0-12-819885-8.00001-2.

Dahash, A. *et al.* (2019) 'Advances in seasonal thermal energy storage for solar district heating applications : A critical review on large-scale hot-water tank and pit thermal energy storage systems', *Applied Energy*, 239, pp. 296–315. Available at: https://doi.org/10.1016/j.apenergy.2019.01.189.

De Rosa, M. *et al.* (2021) 'Prospects and characteristics of thermal and electrochemical energy storage systems', *Journal of Energy Storage*, 44, p. 103443. Available at: https://doi.org/10.1016/J.EST.2021.103443.

Dinçer, İ. and Rosen, M.A. (2010) *Thermal Energy Storage: Systems and Applications*, 2nd ed. Wiley.

Drijver, B., Dinkla, I. and Grotenhuis, T. (2010) 'National research program on the effects of underground thermal energy storage (UTES)', Use of the subsoil in A, Conference: ConSoil 2010 - Salzburg, Austria. pp. 1–9.

Enescu, D. *et al.* (2020) 'Thermal energy storage for grid applications: Current status and emerging trends', *Energies*, 13(2), p. 340. Available at: https://doi.org/10.3390/EN13020340.

Finck, C. *et al.* (2018) 'Quantifying demand flexibility of power-to-heat and thermal energy storage in the control of building heating systems', *Applied Energy*, 209, pp. 409–425. Available at: https://doi.org/10.1016/j.apenergy.2017.11.036.

Gao, L. *et al.* (2017) 'A review on system performance studies of aquifer thermal energy storage', *Energy Procedia*, pp. 3537–3545. Available at: https://doi.org/10.1016/j.egypro.2017.12.242.

Gehlin, S. (2016) 'Borehole thermal energy storage', in *Advances in Ground-Source Heat Pump Systems*. Elsevier, pp. 295–327. Available at: https://doi.org/10.1016/B978-0-08-100311-4.00011-X.

Gorjian, S. *et al.* (2020) 'On-farm applications of solar PV systems', in S. Gorjian and A. Shukla (eds) *Photovoltaic Solar Energy Conversion*. 1st ed. Elsevier, pp. 147–190. Available at: https://doi.org/10.1016/B978-0-12-819610-6.00006-5.

Gorjian, S. *et al.* (2022) 'Solar energy for sustainable food and agriculture: Developments, barriers, and policies', in *Solar Energy Advancements in Agriculture and Food Production Systems*. Elsevier, pp. 1–28. Available at: https://doi.org/10.1016/B978-0-323-89866-9.00004-3.

Gorjian, S., Calise, F., *et al.* (2021) 'A review on opportunities for implementation of solar energy technologies in agricultural greenhouses', *Journal of Cleaner Production*, 285, p. 124807. Available at: https://doi.org/10.1016/j.jclepro.2020.124807.

Gorjian, S., Ebadi, H., *et al.* (2021) 'Recent advances in net-zero energy greenhouses and adapted thermal energy storage systems', *Sustainable Energy Technologies and Assessments*, 43, p. 100940. Available at: https://doi.org/10.1016/j.seta.2020.100940.

Hartmann, N. *et al.* (2012) 'Simulation and analysis of different adiabatic compressed air energy storage plant configurations', *Applied Energy*, 93, pp. 541–548. Available at: https://doi.org/10.1016/j.apenergy.2011.12.007.

Hasnain, S.M. (1998a) 'Review on sustainable thermal energy storage technologies, part I: Heat storage materials and techniques', *Energy Conversion and Management*, 39(11), pp. 1127–1138. Available at: https://doi.org/10.1016/S0196-8904(98)00025-9.

Hasnain, S.M. (1998b) 'Review on sustainable thermal energy storage technologies, part II: Cool thermal storage', *Energy Conversion and Management*, 39(11), pp. 1139–1153. Available at: https://doi.org/10.1016/S0196-8904(98)00024-7.

Heat – Renewables 2019 – Analysis – IEA (2022) Renewable heat – Renewables – Analysis – IEA. Available at: https://www.iea.org/reports/renewables-2022/renewable-heat.

Heating – Analysis –IEA (2022) Heating – IEA. Available at: https://www.iea.org/energy-system/buildings/heating.

Jalili Jamshidian, F., Gorjian, S. and Far, M.S. (2018) 'An overview of solar thermal power generation systems', *Solar Energy Research*, 3, pp. 301–312.

Kalaiselvam, S. and Parameshwaran, R. (2014) 'Thermal energy storage technologies', in *Thermal Energy Storage Technologies for Sustainability*. Elsevier, pp. 57–64. Available at: https://doi.org/10.1016/B978-0-12-417291-3.00003-7.

Lee, C.H. *et al.* (1994) 'Characteristics of the stored and released thermal energy in plastic greenhouse with underground heat exchange system', *Journal of Biosystems Engineering*, 19(3), pp. 222–231.

Lee, K.S. (2013) 'Underground thermal energy storage', *Green Energy and Technology*, 75. Available at: https://doi.org/10.1007/978-1-4471-4273-7.

Li, Y. *et al.* (2011) 'An integrated system for thermal power generation, electrical energy storage and CO_2 capture', *International Journal of Energy Research*, 35(13), pp. 1158–1167. Available at: https://doi.org/10.1002/er.1753.

Li, Z. *et al.* (2019) 'Study on the performance of a curved fresnel solar concentrated system with seasonal underground heat storage for the greenhouse application', *Journal of Solar Energy Engineering*, 141(1). Available at: https://doi.org/10.1115/1.4040839.

Liu, L. *et al.* (2020) 'State-of-the-art on thermal energy storage technologies in data center', *Energy and Buildings*, 226, p. 110345. Available at: https://doi.org/10.1016/j.enbuild.2020.110345.

Matos, C.R., Carneiro, J.F. and Silva, P.P. (2019) 'Overview of large-scale underground energy storage technologies for integration of renewable energies and criteria for reservoir identification', *Journal of Energy Storage*, 21, pp. 241–258. Available at: https://doi.org/10.1016/j.est.2018.11.023.

Mehling, H. and Cabeza, L.F. (2015) 'Integration of active storages into systems', in *Heat and cold storage with PCM. Heat and Mass Transfer.* Springer, Berlin, Heidelberg. https://doi.org/10.1007/978-3-540-68557-9_6.

Midttømme, K. and Banks, D. (2008) 'Ground-source heat pumps and underground thermal energy storage: Energy for the future', *NGU Special*, pp. 93–98.

Morofsky, E. (2007) 'History of thermal energy storage', in *Thermal Energy Storage for Sustainable Energy Consumption*. Dordrecht: Springer, pp. 3–22. Available at: https://doi.org/10.1007/978-1-4020-5290-3_1.

Naranjo-Mendoza, C. *et al.* (2019) 'Experimental study of a domestic solar-assisted ground source heat pump with seasonal underground thermal energy storage through shallow boreholes', *Applied Thermal Engineering*, 162, p. 114218. Available at: https://doi.org/10.1016/j.applthermaleng.2019.114218.

Navarro, L. *et al.* (2016) 'Thermal energy storage in building integrated thermal systems: A review. Part 1. Active storage systems', *Renewable Energy*, 88, pp. 526–547. Available at: https://doi.org/10.1016/j.renene.2015.11.040.

Novo, A.V. *et al.* (2010) 'Review of seasonal heat storage in large basins: Water tanks and gravel–water pits', *Applied Energy*, 87(2), pp. 390–397.

Olabi, A.G. and Abdelkareem, M.A. (2021) 'Energy storage systems towards 2050', *Energy*, 219, p. 119634. Available at: https://doi.org/10.1016/J.ENERGY.2020.119634.

Paksoy, H.Ö. and Beyhan, B. (2015) 'Thermal energy storage (TES) systems for greenhouse technology', in *Advances in Thermal Energy Storage Systems*. Elsevier, pp. 533–548. Available at: https://doi.org/10.1533/9781782420965.4.533.

Paksoy, H.O. *et al.* (2000) 'Heating and cooling of a hospital using solar energy coupled with seasonal thermal energy storage in an aquifer', *Renewable Energy*, 19(1–2), pp. 117–122. Available at: https://doi.org/10.1016/S0960-1481(99)00060-9.

Sadeghi, G. (2022) 'Energy storage on demand: Thermal energy storage development, materials, design, and integration challenges', *Energy Storage Materials*, 46, pp. 192–222. Available at: https://doi.org/10.1016/j.ensm.2022.01.017.

Sanner, B. (1998) 'Underground thermal energy storage with heat pumps', *IEA Heat Pump Centre Newsletter*, 16(2), pp. 10–14.

Sarbu, I. and Sebarchievici, C. (2016) *Solar Heating and Cooling Systems: Fundamentals, Experiments and Applications*. Elsevier.

Schmidt, T., Mangold, D. and Müller-Steinhagen, H. (2003) *Seasonal Thermal Energy Storage in Germany*. ISES Solar World Congress.

Schmidt, T. *et al.* (2015) 'Large-scale thermal energy storage large-scale thermal energy storage', *Renewable and Sustainable Energy Reviews*, 14, pp. 56–72.

Sharma, A. *et al.* (2009) 'Review on thermal energy storage with phase change materials and applications', *Renewable and Sustainable Energy Reviews*, 13(2), pp. 318–345. Available at: https://doi.org/10.1016/j.rser.2007.10.005.

Taylor, P.G. *et al.* (2013) 'Developing pathways for energy storage in the UK using a coevolutionary framework', *Energy Policy*, 63, pp. 230–243. Available at: https://doi.org/10.1016/j.enpol.2013.08.070.

Turgut, B. *et al.* (2009) 'Aquifer thermal energy storage application in greenhouse climatization', *Acta Horticulturae*, 807, pp. 143–148. Available at: https://doi.org/10.17660/ActaHortic.2009.807.17.

Xu, J., Wang, R.Z. and Li, Y. (2014) 'A review of available technologies for seasonal thermal energy storage', *Solar Energy*, 103, pp. 610–638. Available at: https://doi.org/10.1016/j.solener.2013.06.006.

Xu, J. *et al.* (2014) 'Performance investigation of a solar heating system with underground seasonal energy storage for greenhouse application', *Energy*, 67, pp. 63–73. Available at: https://doi.org/10.1016/j.energy.2014.01.049.

Xu, L. *et al.* (2018) 'Application of large underground seasonal thermal energy storage in district heating system: A model-based energy performance assessment of a pilot system in Chifeng, China', *Applied Thermal Engineering*, 137, pp. 319–328. Available at: https://doi.org/10.1016/j.applthermaleng.2018.03.047.

Zhang, L. *et al.* (2015) 'A low cost seasonal solar soil heat storage system for greenhouse heating: Design and pilot study', *Applied Energy*, 156, pp. 213–222. Available at: https://doi.org/10.1016/j.apenergy.2015.07.036.

Zhou, N. *et al.* (2017) 'A study on thermal calculation method for a plastic greenhouse with solar energy storage and heating', *Solar Energy*, 142, pp. 39–48. Available at: https://doi.org/10.1016/j.solener.2016.12.016.

Zizzo, R. (2009) *Designing an Optimal Urban Community Mix for an Underground Thermal Energy Storage System*. Aceee.Org.

5 Thermal Energy Storage

Mohamad Aramesh and Bahman Shabani

5.1 INTRODUCTION

Depending on the end user, whether it is any of the residential, commercial, industrial, or other energy-consuming sectors, the energy demand can be variable on an hourly to a seasonal basis [1, 2]. The supply of energy also can have a variable rate depending on how the energy is being sourced. Knowing this intermittent relation, it is a well-known problem to have mismatches between the supply and demand of energy [3]. This issue is more strongly pronounced when it comes to inherently intermittent renewables that are becoming more and more popular globally. An effective way to overcome this issue is utilising energy storage technologies [4]. Through storage of energy, it is possible to save the excess amount of generated energy and use it when there is a shortage of energy supply [5, 6]. However, this is not the only case where energy storage can be useful. Various energy-consuming technologies are not connected to their energy source permanently. For such technologies to work, they need to store their required amount of energy when connected to the source, and then consume the stored energy when they are operating. A common sample of such cases is the batteries that can be found in many devices that people use on a daily basis. On a larger scale, energy storage methods are employed for peak shaving purposes to maintain the stability of the grid [7].

There are several types of energy and energy storage methods for storing electrical, thermal, and mechanical forms of energy; however, the focus in this chapter is on the storage of thermal energy. The common sources of thermal energy include fossil fuels, nuclear fission reaction, solar energy, geothermal energy, and biomass, the first three of which are the main players in the generation of electricity throughout the world [8]. Various applications can be mentioned for thermal energy, which include but are not limited to: power generation and process heating [9], space heating and cooling [10], water heating and desalination [11], cooking [12], and drying of food and agricultural products [13].

Loss of thermal energy is inevitable, and normally it is tried to minimise it by either insulation or capturing the heat that is otherwise wasted for use in other thermal applications. There are also many cases where thermal energy is not the source of energy, but it is produced during a process, usually being lost in the form of heat. The human body, electronic devices, moving parts of a mechanical system, and exothermic chemical reactions are examples that produce heat [14]. Various approaches have been developed to date in order to take advantage of the waste heat, instead of letting it go [15].

Storage of thermal energy can help with not only balancing the supply and demand of energy, but also utilisation and management of the waste heat. As a result of that,

DOI: 10.1201/9781003345558-6

storing thermal energy can help with increasing the efficiency of various processes that lead to reducing the consumption of the energy and power systems. The fact that less energy would be required when storage of thermal energy is taken advantage of means less harm to the environment, especially if the energy source is a fossil fuel. Therefore, prevention of the heat losses can make storage of thermal energy a more environmentally friendly choice and a technoeconomically favourable solution. For the European Union it is estimated that introducing thermal energy storage (TES) to the industrial and building sectors leads to an annual energy savings of 7.8% as well as 5.5% reduction in CO_2 emissions [16].

The origin of TES technology development goes back to 1955, when it was tried to take advantage of the heat wasted by commercial vehicles' exhaust gas [17]. Garg et al. [18] published the first edition of their book on solar thermal energy storage in 1985. Recently, common applications of TES include storage of solar thermal energy [19], thermal energy management of buildings [20], batteries [21], fuel cells [22] and hydrogen storage systems [23–25]. More on the applications of TES will be presented in this chapter.

TES methods, based on the process of storing thermal energy, can be divided into three main types of sensible heat storage (SHS), latent heat storage (LHS), and thermochemical heat storage (TCHS). Furthermore, certain categories of materials are associated with each of these TES methods. Figure 5.1 summarises these methods along with the specific material types used in each of them.

Apart from the three main TES methods introduced in Figure 5.1, there is also the cold thermal energy storage (CTES) approach, which has the same process as other methods, with the only difference that it is operated at low temperatures. Therefore, the CTES approach cannot be considered as a unique method of TES and is not presented in Figure 5.1. However, having the operating temperatures below 0°C, CTES is briefly discussed later in this chapter.

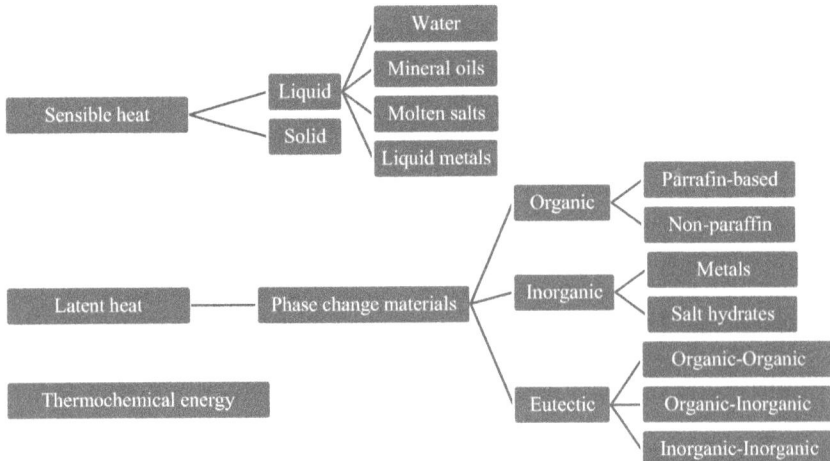

FIGURE 5.1 Thermal energy storage methods based on their process and the materials involved.

5.2 SENSIBLE HEAT STORAGE

5.2.1 THE CONCEPT

SHS is a common TES method that relies on increasing the temperature of the storage medium for storing heat. Then the material in which the energy is stored is either used directly, or its heat is transferred to where it is intended to be used. SHS approaches are usually cheaper and less complicated compared to other TES methods [26]. A simple and very popular example of this method is hot water storage tanks.

SHS systems are mostly known to go through three main phases of charge, storage, and discharge of thermal energy. In the charging phase, the storage medium receives thermal energy. Then depending on the system design, the thermal energy is kept in the storage medium for the duration of the storage phase. Finally, when the stored energy is demanded, the discharge phase begins, and the stored heat is extracted from the storage medium. Based on the specifications of the TES system, these three phases can follow each other in a subsequent manner or can occur simultaneously. These phases are introduced here because they will be used throughout the text to better explain different processes.

5.2.2 MATERIALS

Table 5.1 introduces some of the common SHS materials along with some of their thermophysical properties. It is noteworthy that only a selected number of the SHS materials are given in Table 5.1. More materials and details can be found in the literature [8, 27–30].

Water is one of the main choices for SHS applications because it is cheap, widely available, compatible with well-established technologies, and its thermophysical

---

TABLE 5.1

Thermophysical Properties of Common Sensible Heat Storage Materials (data gathered from [8, 27–30])

Material	Type	Operating Temperature Range (°C)	Density (kg/m³)	Specific Heat (J/kg.K)	Thermal Conductivity (W/m.K)
Water	Liquid	25–90	1000	4184	0.58
Mineral oil	Oil	200–300	770	1966	0.12
Therminol VP-1	Oil	250–400	1068	1546	0.137
Nitrate salt	Molten salt	250–450	1908	1655	0.514
Solar salts	Molten salt	120–565	~1900	~1500	~0.6
Rock	Solid	200–300	2560	960	0.48
Concrete	Solid	200–400	2200	850	1.5
Sand	Solid	200–300	1602	830	1
Brick	Solid	200–1200	3000	1150	5
Liquid sodium	Liquid	270–530	850	1300	204

properties are more suitable for this purpose than those of other common SHS materials [16]. For example, the boiling point of water at atmospheric pressure makes it a preferred choice for SHS in a temperature range of 25–90°C. That is, while limited alternatives can be utilised at this temperature range. As another example, water has one of the highest specific heats amongst common SHS materials. The specific heat determines the capacity of a material in receiving thermal energy at the cost of increasing its temperature. This means with the same mass, the material with higher heat capacity would store more thermal energy. Or in other words, to store the same amount of thermal energy, less mass is required of the material with higher heat capacity. Therefore, water, with its high specific heat, offers superior heat storage performance in terms of the energy storage capacity per unit mass. Several other materials are employed for SHS, such as rocks, oils, and molten salts.

Other common SHS materials are mostly employed for high-temperature applications. Mineral oils and molten salts are commonly used in concentrated solar power plants. As these materials are in liquid form, they can flow through a process, which makes the charge and discharge phases easier. Liquid metals also provide the same benefit, but they must be employed in oxygen and oxide-free environments [27].

When solid SHS materials are employed, there must be a heat transfer fluid involved, mostly air, to charge and discharge the TES medium. Rocks, concrete, and sand are easily available at relatively low costs. This makes them popular choices for use in many solid-based SHS applications, such as process heating [31]. Bricks, which are usually available at higher prices, offer another category of choice and are usually used for heat storage applications in brick structures [29].

Liquid SHS materials must be kept in leak-free pressure vessels as they can reach their saturation points by receiving heat. There is no such concern with solid SHS materials, and they do not need any special equipment for their storage. Solid storage materials are not flowable and cannot be circulated around a process and must be stationary. For this same reason, it is not possible to simultaneously charge and discharge them. That is, whilst liquid SHS materials, because of their flowability and the buoyancy forces, make it possible to establish a thermal stratification so that they can be charged and discharged simultaneously. This drawback of solid SHS materials causes them to experience temperature gradually while being thermally discharged. Another drawback of solid SHS materials is that they normally require a large flow rate of the heat transfer fluid for charging and discharging them while they also cause a significant pressure drop in the flow of the heat transfer fluid [27]. In the next section, water as the most commonly used SHS is reviewed in detail.

5.2.3 WATER-BASED METHODS OF SENSIBLE HEAT STORAGE

5.2.3.1 Water Tank

As discussed, and reported in Table 5.1, while having a relatively low density compared to other options, water offers the highest heat capacity, or in other words, highest storage capacity per unit mass, among the common SHS materials. In addition to that, depending on the application, hot water can be consumed directly whilst the other materials normally must transfer their stored heat to the end user via a heat transfer medium that causes additional heat losses and delay in the process. Because of that, there have been various designs of SHS methods based on water, both on small and large scales.

On the small scale, mostly water tanks are employed as they are easily available and are to some extent portable. Other common water-based SHS methods can be introduced as underground types and include aquifers, boreholes, pits, and rock caverns. Integration of the underground SHS methods with renewable sources and specifically solar energy [32], and heat pumps as an energy-efficient heating technology [33], are gaining more popularity these days. SHS systems that employ water tanks are arguably the most accessible and most common SHS method. Hot water tanks are commercially available in different shapes and sizes at reasonable prices. Moreover, the standards are fairly flexible with installation location of hot water tanks, and even small-scale versions of hot water tanks can be used as portable thermal storage units. The process of heat storage is also straightforward. In the simplest form, hot water is put into and stored inside the tank, and then it is completely removed during the discharge period. In many applications, including the process heating or residential hot water, the charge and discharge phases happen simultaneously. This means there is a continuous hot water flow through the storage tank. The hot water that is removed from the tank is replaced with cold water that is warmed by receiving heat from an external source of heat.

The amount of sensible thermal energy stored inside a water tank can simply be calculated using the following equation, Eq. (1) [34]:

$$Q_{st.} = \int_{T_i}^{T_f} mc_p dT \qquad (1)$$

where $Q_{st.}$ is the stored amount of thermal energy (J), m is the mass (kg), c_p is the heat capacity (J/kg°C), and T is the temperature (°C) of the SHS material. The subscripts of i and f denote the initial and final temperatures of the SHS material during the storage of energy. The mass of the material in which the heat is stored is obviously an independent parameter here, and the usual SHS materials have negligible heat capacity changes in their operating temperature range. Because of that, mostly a simplified form of this equation, Eq. (2), is used to calculate the amount of stored thermal energy [34]:

$$Q_{st.} = mc_p \Delta T = \rho V c_p \Delta T \qquad (2)$$

where V and ρ are the volume (m³) and density (kg/m³) of the SHS material, respectively. ΔT is the temperature change (°C) during the period that energy is stored. Eqs. (1) and (2) can be used for any form of SHS. These equations also support the role of heat capacity of the SHS materials in defining their potential to store thermal energy.

Based on the geometry and specifications of the tank, there can be several methods to thermally model SHS processes. However, the storage efficiency or thermal efficiency ($\eta_{th.}$) is one of the main performance indicators that can be calculated regardless of system specifications, as in Eq. (3) [35]:

$$\eta_{th} = \frac{extracted\,energy}{injected\,energy} \qquad (3)$$

For water tanks with simultaneous charge and discharge phases, by assuming a linear change of temperature and a constant heat capacity, Eq. (3) can be expressed in the following form as Eq. (4):

$$\eta_{th} = \frac{mc_p\left(T_f - T_0\right)}{mc_p\left(T_i - T_0\right)} = \frac{T_f - T_0}{T_i - T_0} \qquad (4)$$

where T_i and T_f are the water temperature (°C) at the inlet and outlet of the storage tank, respectively. T_0 is the initial water temperature inside the tank before getting charged up.

5.2.3.2 Pit

Before introducing pit TES, a general description of underground SHS methods is presented here. Deep in the ground, groundwater temperature is almost constant throughout the year. This is a unique feature of underground water TES. The temperature fluctuations above ground can affect up to 10 m deep shallow waters [36]; however, at depths between 10 m and 20 m below ground, the groundwater temperature would be at an almost constant temperature of about 1–2°C warmer than the local annual average temperature at the ground surface [36]. By going below 20 m though, the groundwater temperature would increase by about 1°C for every 35 m [36]. As for pit TES, its schematics are shown in Figure 5.2.

The basics of pit TES, as shown in Figure 5.2, are very similar to that of water tank TES, with the main difference being that for the pit type, water is stored inside

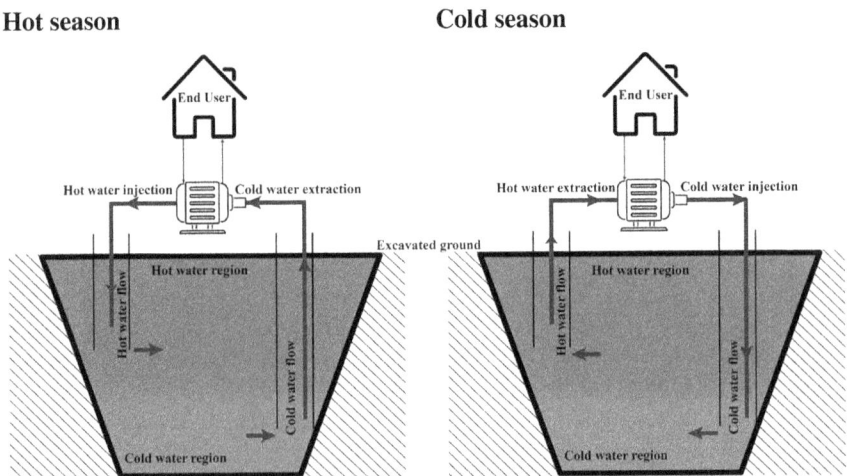

FIGURE 5.2 Schematics of a pit as a sensible heat TES.

an excavated ground [37]. In some cases, water is stored inside an encloser made of concrete or stainless steel that is constructed and placed just below the ground surface, 5–15 m deep into the ground [38]. To obtain better storage efficiency, the pit is normally insulated specifically at the top side [38]. A list of real-world projects mostly located in Europe, along with their specifications, is given in reference [35]. Furthermore, a thermal modelling method of pit TES can be found in the literature [37]; however, to keep this chapter brief, it is not discussed here.

5.2.3.3 Aquifer

Aquifer SHS is one of the most popular methods of underground TES. This method takes advantage of saturated and permeable underground water, which is not rare and can be found around built-up areas [36]. The appropriate aquifers for TES would be somewhere in the range of 10–150 m deep in the ground and surrounded by deposits of sand and gravel and highly fractured rocks [39]. The water inside the aquifer is required to have limited mineral content, relatively high hydraulic conductivity, and consistent concentrations of dissolved oxygen [39]. Figure 5.3 shows schematics of an aquifer TES.

The concept of the aquifer SHS type, as shown in Figure 5.3, is to simultaneously store and use cold and hot water, depending on the needs of the end user during the hot and cold seasons. There are hot and cold wells inside the aquifer. During the cold season, water is pumped from the hot well, where it is warmed up by geothermal energy. The hot water flow then goes through heat exchanging processes to provide heat to the end user. This flow of water is cooled down and sent to the cold well, and cold water is stored for usage during the hot season. During the hot season, an opposite process takes place, where cold water from the cold well is used to meet the end users' cooling needs by using a heat exchanger. Then the warm water is sent to the hot well [40].

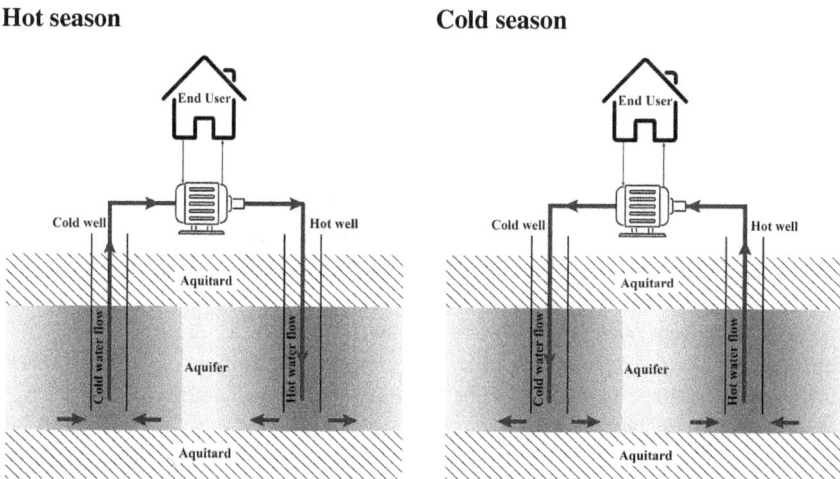

FIGURE 5.3 Schematics of an aquifer TES, recreated from [40] [with permission of reuse from Elsevier].

The main performance indicator of an aquifer SHS is its thermal recovery ratio or thermal efficiency, which is defined as follows in Eq. (5) [40] by using the same concept given in Eq. (3):

$$\eta_{th} = \frac{extracted\ energy}{injected\ energy} = \frac{\int \dot{m}c_p\left(T_p - T_a\right)dt}{\int \dot{m}c_p\left(T_i - T_a\right)dt} \tag{5}$$

where η_{th} is thermal efficiency, $\dot{m}$ is water flowrate (kg/s), c_p is water specific heat (J/kg.K), T is water temperature (°C), and t is time (s). The subscript of p denotes the temperature of water delivered to the end user, and i indicates the initial water temperature at the source well. The subscript a also denotes the ambient temperature. The water flowrate is the same throughout the aquifer TES cycle, and at the operational temperature range of an aquifer SHS, changes of water specific heat are negligible. By considering linear changes in temperature by time, Eq. (5) can be simplified as follows in Eq. (6) [41]:

$$\eta_{th} = \frac{T_p - T_a}{T_i - T_a} \tag{6}$$

A vital consideration when designing an aquifer SHS is to have cold and hot wells inside an aquifer separated enough to prevent mixing of the hot and cold regions. In other words, a thermal breakthrough must be avoided where the hot and cold regions affect each other's temperature. A parameter known as 'thermal breakthrough time' is normally used to quantify the possibility of a thermal breakthrough inside an aquifer SHS. Several relations have been presented to calculate the thermal breakthrough time; some are simple [40] and some are very detailed and complicated [42]. However, the main simple concept is presented as follows in Eq. (7) [41]:

$$t_{breakthough} = \frac{well\ spacing}{water\ front\ velocity} \tag{7}$$

where $t_{breakthough}$ is the thermal breakthrough time, and the waterfront velocity is the relative velocity of the hot and cold water fronts movement towards each other. This parameter determines how long it takes for thermal breakthrough to occur where the hot and cold regions inside the aquifer begin to get mixed. Such a mixing initially reduces the efficiency of the whole TES system, and as it progresses further the temperature difference between the two hot and cold regions can drop significantly such that the TES system becomes unable to operate effectively. To provide an idea of an acceptable thermal breakthrough time for a large-scale aquifer (as a benchmark), this parameter has been estimated to be about 60 years for a large-scale aquifer SHS system practiced in Denmark [41]. In this case, the system supplies 100 m³ of hot/cold water per day, and its hot and cold wells are separated by 1 km.

5.2.3.4 Borehole

A borehole system is another popular method of underground SHS as the technology around it is not complicated. Boreholes are basically underground heat exchangers

Hot season   **Cold season**

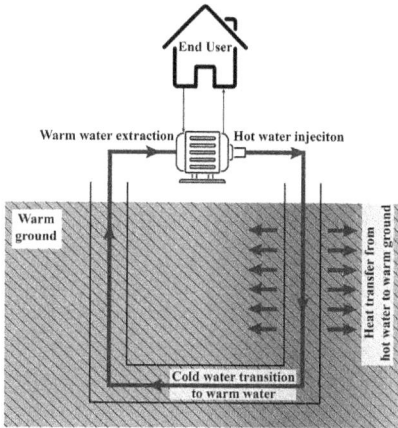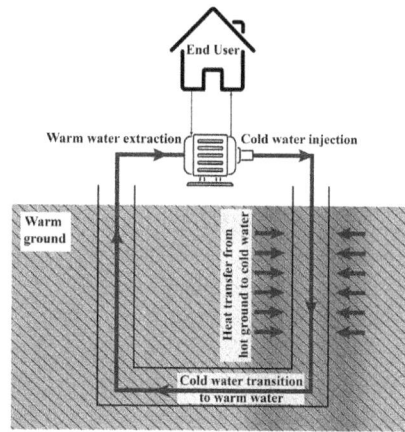

FIGURE 5.4   Conceptual schematics of a borehole TES.

that go 20–300 m deep into the ground [36] and can be constructed almost every-
where in the ground whether it is made of unconsolidated material or rock [43]. General simplicity and good availability have made this method a popular underground
TES option, which is increasingly used together with ground-sourced heat pumps
[44]. A borehole TES is schematically shown in Figure 5.4.

In borehole TES systems, the ground itself plays the role of the storage material,
and water acts as the heat transfer fluid. However, because the input and output of
the TES system is water, it is considered a water-based underground TES method.
The underground materials normally have an overall effective thermal conductivity
in the range of 0.3–6 W/m.K that is considered to be relatively low. A large list of
underground materials and their thermophysical properties can be found in reference
[43]. Given that the borehole is built deep into the ground, there is no need for further
insulating them [43]. In small-scale applications, such as space cooling and heating
of small buildings, one borehole would suffice, but for larger-scale applications, as
shown in Figure 5.4, usually several boreholes are used. A list of some borehole TES
projects operating around the world, along with their number of constructed boreholes, can be found in reference [45].

Borehole TES systems operate in a seasonal manner. During hot months, the
end user is cooled down and its excess generated thermal energy is transferred to
the ground and is stored there through the boreholes. Then during cold months,
the ground discharges the stored thermal energy to the end user. In this period, the
ground gradually cools down and gets ready for the next cycle. More on seasonal
heat storage is presented in section 5.7.

Several modelling methods have been introduced in the literature around borehole
TES systems [36] that are not within the scope of this chapter, specifically because

the boreholes come in different configurations and shapes. However, a general approach to describe the thermal efficiency of this method is given in Eq. (8) [46]:

$$\eta_{th} = \frac{extracted\ energy}{injected\ energy} = \frac{\sum_{j=1}^{N_{ex.}} \left[\rho \dot{V} c_p \left(T_o - T_i \right) t N_B \right]_j}{\sum_{j=1}^{N_{in.}} \left[\rho \dot{V} c_p \left(T_i - T_o \right) t N_B \right]_j} \tag{8}$$

where ρ, $\dot{V}$, c_p, and T are the density (kg/m³), volumetric flowrate (m³/s), specific heat (J/kg.K), and the temperature (°C) of the heat transfer fluid (i.e., water in most cases), respectively. t is the duration of the period that either injection or extraction of heat is occurring (s). $N_{ex.}$ and $N_{in.}$ are the number of injection and extraction processes happening during a year, and N_B is the number of boreholes involved in each process. The subscripts i and o denote the input and output temperatures of the heat transfer fluid, and the subscript j represents the number of instances that each process occurs during a year.

5.2.3.5 Rock Cavern

The unique feature of rock caverns is that the rock itself, because of its relatively low thermal conductivity of around 2–6 W/m.K [47], acts as a thermal insulator. This thermal insulation layer is effective enough that SHS conditions inside rock caverns is not affected by the weather conditions and the seasonal changes on the ground surface [48]. Rock caverns have a specific advantage over other underground TES methods as well, and that is they either already exist naturally or were previously built for other purposes. Depleted oil or natural gas fields, or abandoned mine tunnels and shafts, all can be used as rock cavern SHS arrangements. A rock cavern TES is schematically shown in Figure 5.5.

Hot season   **Cold season**

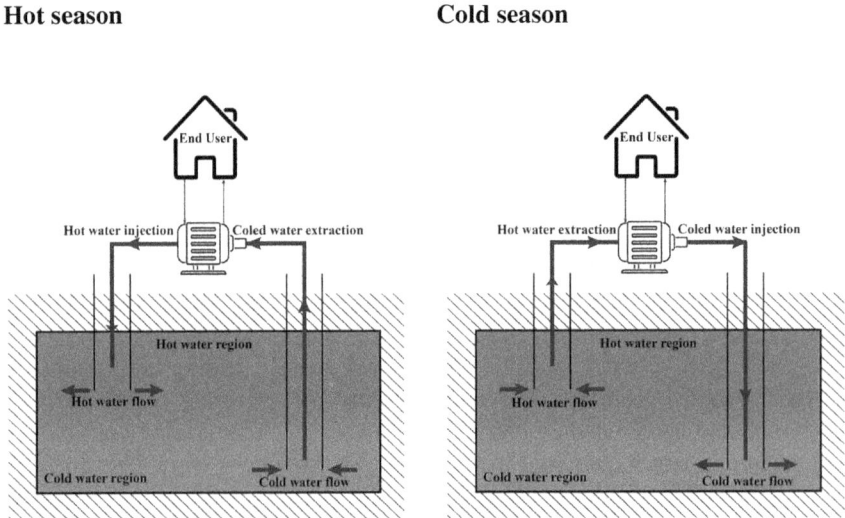

FIGURE 5.5   Schematics of a rock cavern TES.

Normally, to enhance the efficiency of a rock cavern SHS, a thermal stratification is established inside the storage medium that takes advantage of buoyancy forces. Naturally, the warmer water remains at the higher levels inside the cavern. The charging and discharging of the thermal energy can also be designed in a way that amplifies this phenomenon. As shown in Figure 5.5, charging, storage, and discharging phases of SHS inside the rock cavern occur simultaneously. In other words, charging and discharging of hot water are both from the top water levels inside the cavern. The discharge flow releases its stored heat to the end user and returns to the cavern as cold water. This cold water enters the rock cavern at lower levels, and then it is taken out from there to get warmed up by the energy source and get back to the cavern as the charging hot water.

As mentioned earlier, the cavern can be pre-existing, and that would decrease the investment costs of rock caverns significantly. However, pre-existing or pre-built caverns can rarely be found in built-up areas, where the end users are usually located. Furthermore, the construction of caverns is too costly, which makes this method of TES unpopular. Consequently, there are few sources available on energy analysis of rock cavern TES. A detailed method for modelling cavern TES systems is presented by Lee [36].

5.2.3.6 Salinity-Gradient Solar Ponds

A salinity-gradient solar pond, or in short solar pond, is another water-based SHS technology, which is rather more recent than the other methods mentioned in this section. It is under research and development, and although it has already been implemented throughout the world, its applications have been very limited. It is noteworthy that solar ponds are different in nature compared to the other methods introduced earlier. The difference is that a solar pond is a combined solar energy absorber and heat storage method. Therefore, normally a solar pond is not categorised in the same group as other hot water storage methods, and it is mostly recognised as a solar thermal collector rather than a hot water storage method. However, because its operation is based on SHS, here an introduction to solar ponds is also presented.

The idea of solar ponds originally came from brine lakes that would naturally trap, or in other words store, hot water at temperatures around 70°C at the bottom of the lake [49]. The discovery of this phenomenon resulted in establishment of artificial solar ponds throughout the world, with the USA, Israel, and Australia implementing them on large scales [49]. The El Paso solar pond located in Texas, USA, and the Pyramid Hill solar pond located in Victoria, Australia, each with an area of 3,000 m², are among the world largest artificial solar ponds [50].

The mechanism of hot water storage by solar ponds is to some extent opposite to the thermal stratification employed to enhance the performance of pits and rock caverns. For the two latter cases, hot water enters the storage medium at the top and cold water enters it at the bottom. Because of its lower density, hot water remains at the top and cold water, having a higher density, stays at the bottom. Solar ponds, as the name suggests, use solar energy to produce hot water. To do so, an absorber is normally used, which is placed at the bottom of the pond [50]. The absorber is heated up by the sunlight that reaches the bottom and consequently warms up the water above it. In a normal situation, natural convection, which is a result of density difference between hot and cold water, would cause the heat to reach the top of the pond and then lose to the ambient. However, in solar ponds, there is a density gradient,

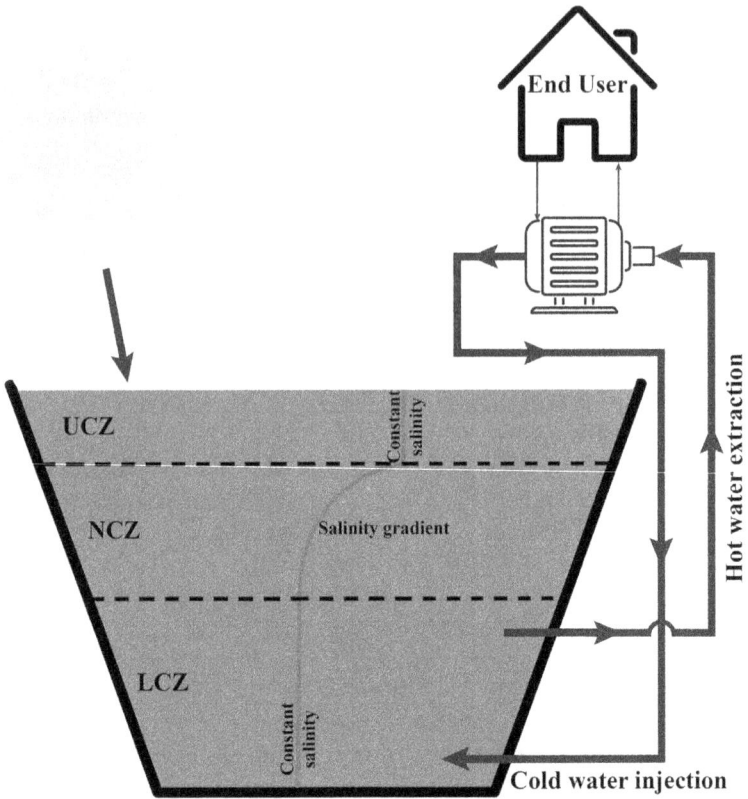

FIGURE 5.6   Schematics of SHS mechanism by solar ponds.

created by adding salt or salty water at the bottom of the pond, which prevents that from happening [51]. Figure 5.6 provides more details on the mechanism of SHS by solar ponds.

As shown in Figure 5.6, based on the salt gradient, there are three main regions inside a solar pond, the upper convective zone (UCZ), the non-convective zone (NCZ), and the lower convective zone (LCZ) [52]. The density at different regions is controlled by salinity. The UCZ is mainly freshwater with an almost constant density, and its presence helps with maintaining the salinity gradient inside the pond. Without the UCZ, the water vaporisation would increase salinity at the top and subsequently disturb the salinity gradient. UCZ itself vaporises over time, and it must be maintained by regularly adding freshwater to the zone. LCZ also has an almost constant density as there is no salinity gradient in this region. This region is where the thermal energy is stored, and for this method to be practical, a rather considerable portion of the pond must be allocated to this region [53, 54]. On top of that, natural convection must be operational in this region so that the whole portion of the pond can play a part in storing the heat. NCZ, on the other hand, through its salinity gradient, which increases by depth, prevents heat transfer by natural convection from LCZ to UCZ (i.e., the thermal conductivity of water is very low).

Solar ponds, because they store heat inside brine with a high salinity, do not directly produce hot water. A heat exchanger is usually needed to transfer the heat stored inside solar ponds to whatever application it is being employed for. A much more detailed introduction and discussion around solar pond technology can be found in reference [55].

5.2.3.7 Technoeconomic Comparison of Common Water-Based TES Methods

Table 5.2 provides a general thermal performance comparison of the discussed water-based SHS methods. As mentioned in section 5.2.3.6, solar ponds are more of a solar thermal collector rather than a water storage method and are different in nature from other water-based SHS methods. Because of that inherent difference, solar ponds cannot be directly compared with the other SHS methods introduced in this chapter. Therefore, in this section, solar ponds are excluded from the comparisons.

Among the methods discussed here, water tanks and an insulated pit, as suggested by Table 5.2, offer the highest storage efficiency and volumetric capacity of energy output. However, these two methods are normally employed for water storage volume of smaller than 100,000 m³, which can be considered for many small- to medium-scale applications. The other three methods, which are obviously associated with high costs and lower efficiencies, are specifically suitable for larger-scale applications and storage volumes in the range of 300,000 m³ to 500,000 m³.

Economic evaluation of these methods is presented in the literature [56]. Cost figures per water storage volume ($/m³) or per amount of thermal energy storage capacity ($/kWh) have been the basis for economic comparison of these methods as discussed here. From the economic point of view, the rock cavern type stays somewhere in the middle, both in terms of investment and operational costs per unit volume [56]. That is because for this type of SHS, as opposed to the water tank and the pit, the cavern may already exist, and there is no need to construct or manufacture the storage medium [36]. The same goes for aquifers as well. However, as natural or previously built artificial caverns are not normally close to built-up areas, they have had limited real-world applications [36]. That is, whilst aquifers are much more

TABLE 5.2
Thermoeconomic Characteristics of Common Water-Based SHS Methods [56]

Specification	Water Tank	Insulated Pit	Aquifer	Borehole	Rock Cavern
Specific heat capacity (kWh/m³.K)	1.16	1.16	0.75	0.63	1.16
Input/output temperature difference (°C)	55	55	55	55	55
Average storage efficiency (%)	90	85	75	70	80
Energy output capacity (kWh/m³)	57	54	31	24	51
Size (1000 m³)	≤100	≤75	50–500	50–400	50–300

accessible, boreholes can be constructed almost anywhere [36, 43]. The latter two technologies are also the cheapest ones per water output volume, as their establishment involves the least amount of construction and manufacturing activities. Between the underground TES methods, the majority of the real-world applications have chosen the aquifer and borehole types as their preferred method of SHS, with the aquifer type being less applicable around the world [39]. As for boreholes, it is almost always practically possible to employ them. They can even be constructed under the building grounds given that they are built before the building is erected [36]. About aquifers, it is noteworthy that in some cases the water might be used to supply drinking water. In that case, it can be irrational to use the aquifer for TES purposes.

When comparing the prices of these SHS methods, care must be given to the fact that these comparisons are per water output volume. Thus, it is not accurate to say water tanks are the most expensive SHS method out there. As mentioned in Table 5.2, water tanks are normally up to 100,000 m^3. In fact, in many applications, much smaller tanks are employed. Residential applications, for instance, use water tanks of up to 0.3 m^3. On the other hand, the smallest aquifer TES would have a volume of around 50,000 m^3. The cost of creating a TES system for an already existing aquifer would be much less than constructing the whole water tank TES system with a tank volume of 50,000 m^3 [57].

Overall, in small-scale applications, a water tank is the best choice for SHS as it is much more accessible and affordable. The maintenance of small-scale water tanks is much easier and cheaper than that of other methods. Also, the water tank method outperforms all other methods by having the highest energy storage efficiency and energy output capacity. In large-scale applications, boreholes and aquifers win the competition because of their much lower investment and operational costs. Figure 5.7

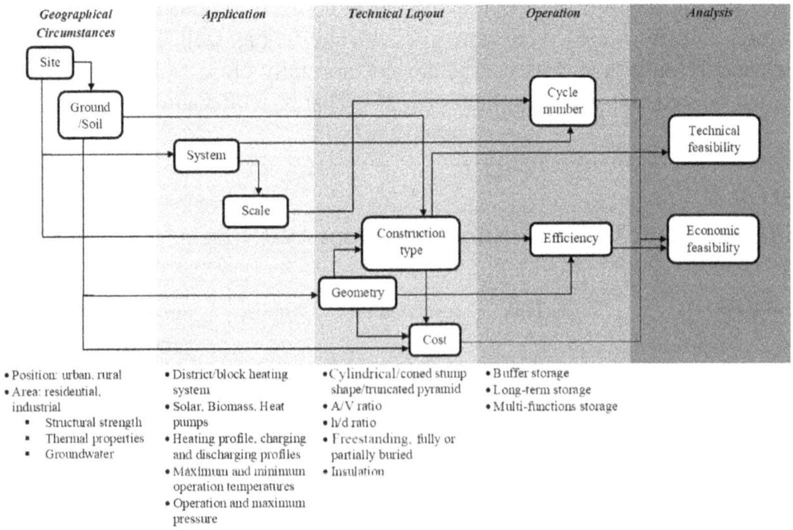

FIGURE 5.7 The main decision-making criteria when choosing a large-scale TES method [35], [with permission of reuse from Elsevier].

provides more information in the form of a decision-making flowchart by using the main criteria for selecting a large-scale underground water-based TES method.

5.3 LATENT HEAT STORAGE

5.3.1 THE CONCEPT

As the name suggests, in this type of TES, thermal energy is stored in the form of latent heat when a material is going through phase change. Because of that, the materials that are proper for LHS are called phase change materials (PCMs). To be proper for LHS purposes, a PCM must receive/release a large amount of latent heat when it goes through a phase change, so that a large amount of thermal energy can be charged/discharged into/out of that in the form of latent heat. Most materials can go through solid-liquid [58] and liquid-gas phase changes [59]. However, there is also a solid-gas phase change, but that can only happen in very specific conditions, especially during discharge phase when the material is in the gas state and needs to become solid. There is a solid-solid phase transition as well, for the materials with crystal structures, when they are changing their crystallisation forms [60].

The solid-gas phase change, besides the difficulties in terms of the phase change from gas to solid, comes with an extreme volume change of the material, which is not usually favourable for the container containing the PCM. The storage and harvesting of stored thermal energy through solid-solid phase transition is also not very efficient and practical. For these reasons, LHS through solid-gas and solid-solid phase transitions is rarely practised [59]. The liquid-gas phase change, also because of the significant volume changes of the materials associated with that, have had limited applications. Known applications of liquid-gas LHS are in heat pipes, thermosyphon, and capillary pump loop [59]. Therefore, LHS methods are mostly based on liquid-solid phase change, and hence this type of LHS is particularly discussed in detail here.

LHS and PCMs, compared to other TES methods, offer several major advantages as follows. LHS materials, i.e., PCMs, are mainly chosen because of their high latent heat so that they can store high amounts of thermal energy in the form of latent heat. That is, when a PCM is at its melting/solidification point, because of its high latent heat of fusion, it can receive/release a high amount of thermal energy. For this very fact, PCMs are capable of transferring heat at an almost constant temperature, also known as an isothermal condition [61]. This characteristic is very helpful in applications in which the temperature must be maintained in a narrow range, such as thermal management applications. It also helps with mitigating any temperature fluctuations in the process, caused by changes in ambient temperature or the heat source conditions [62].

The amount of energy stored in a PCM when it is charged can be calculated using the following equation, Eq. (9):

$$Q_{st.} = Q_s + Q_{latent} + Q_l \tag{9}$$

where $Q_{st.}$ is the total amount of stored thermal energy (J), Q_{latent} is the amount of energy stored in the form of latent heat (J), and Q_s and Q_l are the amount of energy storage in the form of sensible heat (J) when the PCM is in solid and liquid forms, respectively. Eq. (9) can, however, be expanded as follows in Eq. (10) [63]:

$$Q_{st.} = m_{PCM} \times \left[c_{p,PCM,s} \left(T_{PCM,melt.} - T_{PCM,i} \right) + q_{PCM,latent} + c_{p,PCM,l} \left(T_{PCM,f} - T_{PCM,solid.} \right) \right] \quad 10)$$

where the subscript PCM indicates the parameter belongs to PCM; m, c_p, and T are the mass (kg), specific heat (J/kg.K), and temperature (°C), respectively; $q_{PCM,latent}$ is the specific latent heat of PCM (J/kg); the subscripts i and f denote the initial and final status of PCM during the charging period; and $T_{PCM,melt.}$ and $T_{PCM,solid.}$ are the PCM melting and solidification points (°C). It is noteworthy that some materials, specifically paraffins, rather than a specific temperature, have a temperature range at which they melt or solidify. Thus, $T_{PCM,melt.}$ and $T_{PCM,solid.}$ are the starting points of melting and solidification, respectively.

Sensible heat storage inside PCM, in both solid and liquid states, is included for calculating the total amount of energy stored in it. That is because, for example, through a full charging cycle, PCM is initially in solid state at a temperature below its melting point. Then it receives thermal energy and stores it in the form of sensible heat until it reaches its melting point. At that point the thermal energy is stored in the form of latent heat till the PCM is fully melted. After that, the receiving thermal energy is again stored in the form of sensible heat and raises the temperature of the liquid PCM. In cases where the PCM is not fully charged, i.e., it is not completely melted, the SHS term for the liquid PCM is eliminated from Eqs. (9) and (10), and the amount of energy stored is in the form of latent heat only, as described in Eq. (11):

$$Q_{latent} = f m_{PCM} q_{PCM,latent} \quad (11)$$

where f is the liquid fraction of the PCM.

Despite the PCMs' great capacity for storing thermal energy, they are well known for the major drawback of having a low thermal conductivity that results in a slow heat transfer rate [64, 65]. PCMs are normally confined in a storage unit and during melting or solidification, thermal conduction plays an important role in transferring the heat inside the storage medium [66]. Most of the common PCMs have low thermal conductivities, making their charge and discharge processes undesirably slow. A proper solution for this drawback can make LHS-based PCMs more effective in their applications. Several methods have been introduced and tested to address this issue, which are discussed later in section 5.3.3.

5.3.2 Materials

As shown in Figure 5.1, PCMs can be categorised into three main types of organic, inorganic, and eutectic materials. Organic-type PCMs mostly consist of different grades of paraffin, and there are some non-paraffin materials of this type as well. The inorganic materials are mainly specific metals and salt hydrates. Finally, the eutectic type is a combination of organic and/or inorganic materials. Table 5.3 lists some of

TABLE 5.3
Thermophysical Properties of Common Phase Change Materials for Thermal Energy Storage by Latent Heat [58, 63]

Material	Type	Melting Point (°C)	Latent Heat of Fusion (kJ/kg)	Specific Heat (kJ/kg.K)		Thermal Conductivity (W/m.K)	
				Solid	Liquid	Solid	Liquid
Octadecane	Organic – paraffin	28.1	244	0.480	0.560	0.15	0.15
n-Eicosane	Organic – paraffin	37	247	0.602	0.664	0.23	0.14
n-Hexacosane	Organic – paraffin	56	250	0.680	0.870	0.21	0.21
n-Tritricontane	Organic – paraffin	72	256	0.870	1.110	0.21	0.21
Lauric acid	Organic – non-paraffin	43	180	1.950	2.400	0.15	0.15
Stearic acid	Organic – non-paraffin	58	191	2.830	2.380	0.3	0.1
Acetamide	Organic – non-paraffin	81	263	1.940	1.940	0.5	0.5
Erythritol	Organic – non-paraffin	118	339	1.380	2.760	0.73	0.33
Sodium carbonate decahydrate	Inorganic-salt hydrate	32	267	1.920	3.260	0.51	0.22
Sodium sulphate decahydrate	Inorganic-salt hydrate	32.4	241	1.760	3.300	0.7	0.54
Sodium acetate trihydrate	Inorganic-salt hydrate	58.9	173	4.200	3.680	0.01	0.01
Magnesium nitrate hexahydrate	Inorganic-salt hydrate	89	162	1.840	2.510	0.61	0.49
Magnesium chloride hexahydrate	Inorganic-salt hydrate	117	167	2.250	2.610	0.70	0.57
Gallium	Inorganic-metal	29.8	80.16	0.372	0.397	33.68	33.49
Potassium	Inorganic-metal	63.2	59.59	0.780	0.840	108.3	102.4
69%Lauric+31% palmitic acid	Eutectic	35.2	166.3	-	-	-	-
%66AlCl3+% 34NaCl	Eutectic	93	213	-	-	-	-
Solar salt	Eutectic	220	161	1.050	1.500	0.76	0.52

the most common PCMs along with their selected thermophysical properties. There are numerous PCMs introduced in the literature, and more comprehensive lists can be found in references [27, 58, 63].

The data given in Table 5.3 clearly shows the significance of latent heat compared to specific heat. PCMs are normally chosen because of their high latent heat. Paraffins listed here have a latent heat of fusion of around 250 kJ/kg. Water has a specific heat of around 4.18 kJ/kg.K, which is one of the highest values among SHS materials. Considering the paraffins' latent heat of fusion, the energy needed for melting 1 kg of solid paraffin is equivalent to that needed to warm up 1 kg of water by about 60°C.

The PCM melting point determines the operating temperature at which the PCM is employed. Hence, the process for which the PCM is used is expected to have an operating temperature range in a narrow band above and below the PCM melting point. Therefore, in the LHS context presented here, the PCM melting point range and operating temperature can be considered interchangeable.

In most cases of LHS applications, paraffins are chosen as the heat storage material. There are several reasons for paraffins to be popular PCM options: Paraffins are alkanes with a chemical formula of C_nH_{2n+2}, with fusion latent heat in the range of 120–270 kJ/kg.K, averaging around 200 kJ/kg.K [67]. Paraffins, at a wide operating temperature range of 5°C to above 100 °C, are commercially available at much lower prices compared to other PCMs [63]. Moreover, paraffins are non-toxic, non corrosive, durable, physically and chemically stable even at high temperatures of around 100–200°C, and are not known to cause health or safety hazards [63]. All of these specifications have made paraffins the popular choice to be used as PCMs. However, as mentioned in section 5.3.1 and given in Table 5.3, paraffins have a low thermal conductivity of around 0.2 W/m.K or even less than that. This means that although paraffins can store and release large amounts of thermal energy, they are very slow in doing that. This has been a huge barrier facing the paraffin-based LHS solutions; hence, many efforts have been made so far to address that issue. More on the heat transfer enhancement methods of paraffins is presented in section 5.3.3.

Fatty acids are amongst non-paraffin organic PCMs and are mostly available in low to medium melting point range, i.e., 10–70°C [27]. However, they are relatively more expansive than commercial paraffins, mildly corrosive, and combustible, while having an unpleasant odour [27]. Sugar alcohols such as erythritol are another type of non-paraffin organic PCMs, which are mostly available in the medium to high melting temperature range of about 90–200°C. They normally have a high latent heat of fusion of about 340 kJ/kg.K, compared to that for paraffins (i.e., the average of 200 kJ/kg.K). They offer a high energy density, and because of that they are combustible and even explosive in extreme cases [27].

Inorganic PCMs, i.e., salt hydrates, as indicated by Table 5.3, offer up to four times higher thermal conductivities of around 0.5–0.6 W/m.K, which is still quite low [68]. Salty environments are also associated with corrosion. Weak natural convection, supercooling, and phase separation are other challenges associated with hydrated salts [68]. High leakage in the liquid state is also mentioned as another drawback of this option [69].

Metallic PCMs, as stated in Table 5.3, offer the highest thermal conductivity while showing relatively low heat storage capacities, i.e., low latent heat of fusion. However, when compared to other PCMs, because of their high density, in the same volume metals can store more thermal energy [27]. Metallic PCMs are relatively expensive and highly corrosive, and specific equipment is required for their employment.

Eutectic PCMs, as discussed earlier in this section, are a combination of organic and/or inorganic PCMs. By varying these combinations, a wide variety of eutectic PCMs can be manufactured, many of which have been extensively tested and reported in the literature. When combined, the properties of the final material would be a mixture of those of its composing materials. Because of that, eutectic PCMs come in a wide range of specifications, and there is not a single set of specifications associated with them. On top of that, application of eutectic PCMs has been very limited compared to the other types of PCMs introduced in this section. Detailed information on this type of PCMs is available in reference [70]. Moreover, detailed information on different types of PCMs can be found in references [27, 58, 68, 71].

5.3.3 Methods for Enhancing Heat Transfer Inside PCMs

5.3.3.1 Thermal Conductivity, the Basis for Enhancement

It was explained in section 5.3.1 that common PCMs (specially paraffins) suffer from notably low thermal conductivity, meaning that although they are superb in terms of their capacity for storing heat, their TES process is rather slow. For this reason, many attempts have been made to enhance the heat transfer performance of PCM-based TES systems, as they are further discussed in this section.

Materials other than the PCM itself are often added to PCMs to make PCM composites with enhanced heat transfer performance. It is important though to be able to measure or calculate the thermal conductivity of such composites. Several measurement and estimation methods have been developed for this purpose.

The transient plane source (TPS) method has been a popular approach for measuring thermal conductivity of enhanced PCMs, and commercial equipment working based on this method is available [66]. Through the TPS method, a planar probe is placed inside the testing sample, when the PCM is in solid state. The probe acts both as a heat source and temperature sensor and does the measurement based on the aggregation of the thermal energy. This method is based on the general differential heat conduction equation, Eq. (12):

$$\frac{k_{eff}}{\left(\rho c_p\right)_{eff}} \nabla^2 T_c + \frac{q_{TPS}}{\left(\rho c_p\right)_{eff}} = \frac{\partial T_c}{\partial t} \tag{12}$$

where k_{eff} is the effective thermal conductivity (ETC) of the composite (W/m.K); $\left(\rho c_p\right)_{eff}$ is the multiplication of the composite effective density (kg/m³) and effective specific heat (J/kg.K); T_c is the composite temperature (°C) measured by the planar sensor; Q_{TPS} is the volumetric heat generation rate of the probe (W/m^3); and t denotes the time (s).

The effective density and heat capacity of the enhanced composite can be measured as well. For measuring the heat capacity, differential scanning calorimetry (DSC) is the main method widely employed, with its required equipment being available commercially [72]. However, because there are no chemical reactions or mass transfers involved with the typical PCM enhancement methods, the parameter $\left(\rho c_p\right)_{eff}$ is normally calculated by the general mixtures rule, as in Eq. (13) [66]:

$$\left(\rho c_p\right)_{eff} = \frac{m_{PCM}\,\rho_{PCM}\,c_{p,PCM} + m_{enh}\,\rho_{enh}\,c_{p,enh}}{m_{PCM} + m_{enh}} \tag{13}$$

where the subscripts *PCM* and *enh* indicate the parameters belong to the PCM and the material used for its heat transfer enhancement, respectively.

Other measurement methods are available, such as the laser-flash method [73] for ETC and the transient guarded hot plate technique (TGHPT) [72] for heat capacity. More information on different methods for measuring thermophysical properties of PCMs can be found in references [66, 72–74].

When dealing with ETC, the measurement methods can be generally employed regardless of the enhancement type; however, the estimation methods are mostly specific to each type. That is because how the enhancement materials are spread through the PCM body can affect the heat transfer by conduction mechanism. In the following sections, wherever possible, the relations for estimation of the ETC for each method are presented. However, when there is no other choice, with a very low accuracy, the general rule for mixtures, as presented by Eq. (13) for $\left(\rho c_p\right)_{eff}$, can be used to estimate the ETC of the enhanced PCM as well [66].

5.3.3.2 Thermally Conductive Additives

Introducing materials with high thermal conductivities has been one of the main methods for enhancing the thermal conductivity of PCMs, with nanoparticles being the pioneers of this method. For heat transfer purposes, nanoparticles are normally made of materials with high thermal conductivity. For instance, considering the low paraffin thermal conductivity of around 0.2 W/m.K, that of CuO, as a common and commercially available type of nanoparticle, is around 33 W/m.K [75]. The nanoparticles, which are scattered throughout the host materials, can provide a significantly larger surface area compared to the same materials in bulk form. That is how nanoparticles can effectively improve thermal conduction inside the host medium [76].

In the case of PCM thermal performance enhancement, nanoparticles, with a weight fraction of around 0.1–1%, are normally added to a PCM when it is in liquid phase [77]; however, fractions up to 20% are also reported in the literature [78]. Many studies have been presented in the literature on PCMs enhanced by adding nanoparticles, with a wide variety of PCM materials and at different nanoparticle concentrations. It is not possible to provide a definitive evaluation of the best percentage of nanoparticles to be used for achieving the maximum ETC improvement. The literature reports a wide range of ETC improvement by adding nanoparticles from 13% [62] to even 90% [79].

A variety of theoretical methods have been developed to estimate the ETC of a PCM enhanced by adding nanoparticles. A list of such methods and their respective equations can be found in references [80, 81]. Despite being relatively old, the Maxwell Eucken model is reported to be very reliable for estimating the ETC of nano-enhanced PCMs [81, 82], as shown in Eq. (14):

$$k_{eff} = k_{PCM} \frac{2k_{PCM} + k_{np} + 2\phi\left(k_{np} - k_{PCM}\right)}{2k_{PCM} + k_{np} - \phi\left(k_{np} - k_{PCM}\right)} \tag{14}$$

where k_{eff}, k_{PCM}, and k_{np} are the thermal conductivities (W/m.K) of the composite, the PCM, and the nanoparticle material, respectively, and ϕ is the volume fraction of the nanoparticles inside the nano-PCM composite.

Nanoparticles, with their nano-scale sizes, can well be scattered through a liquid body and establish a large surface area of a highly conductive material inside the PCM body. Because of this behaviour, a small amount of nanoparticles can result in significant improvement in the ETC of the composite [63]. However, nano-enhanced PCMs suffer from a serious issue, which is instability of the composite and sedimentation of the nanoparticles when the composite goes through multiple charge and discharge cycles. Figure 5.8 shows a small sample of a nano-enhanced PCM gone through several charge and discharge cycles. As Figure 5.8 shows, after only two charge and discharge cycles, the nanoparticles begin to significantly settle down inside the liquid composite.

(a)   (b)

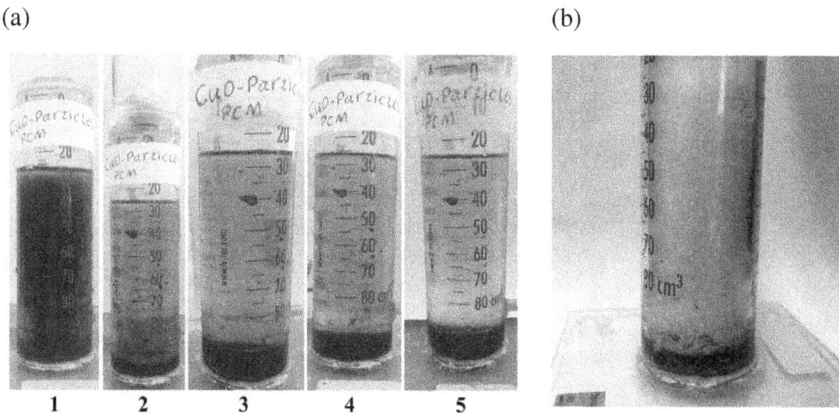

FIGURE 5.8   A sample of a nano-enhanced PCM, (a) in liquid state and after one to five charge and discharge cycles, and (b) in solid state after five charge and discharge cycles [62], [with permission of reuse from Elsevier].

5.3.3.3 Micro/Nano Encapsulation and PCM Slurry

Micro or nano encapsulation of PCMs is another common method for enhancing their heat transfer properties. Micro/nano encapsulated PCMs (M/NPCMs) are composed of a PCM core and a highly conductive shell on the micro/nano scale. However, being on micro/nano scale, compared to a normal PCM, the composite would be capable of a higher heat transfer rate, which is obtained by its much larger heat transfer surface area. Figure 5.9 shows the structure of typical M/NPCMs and their different types.

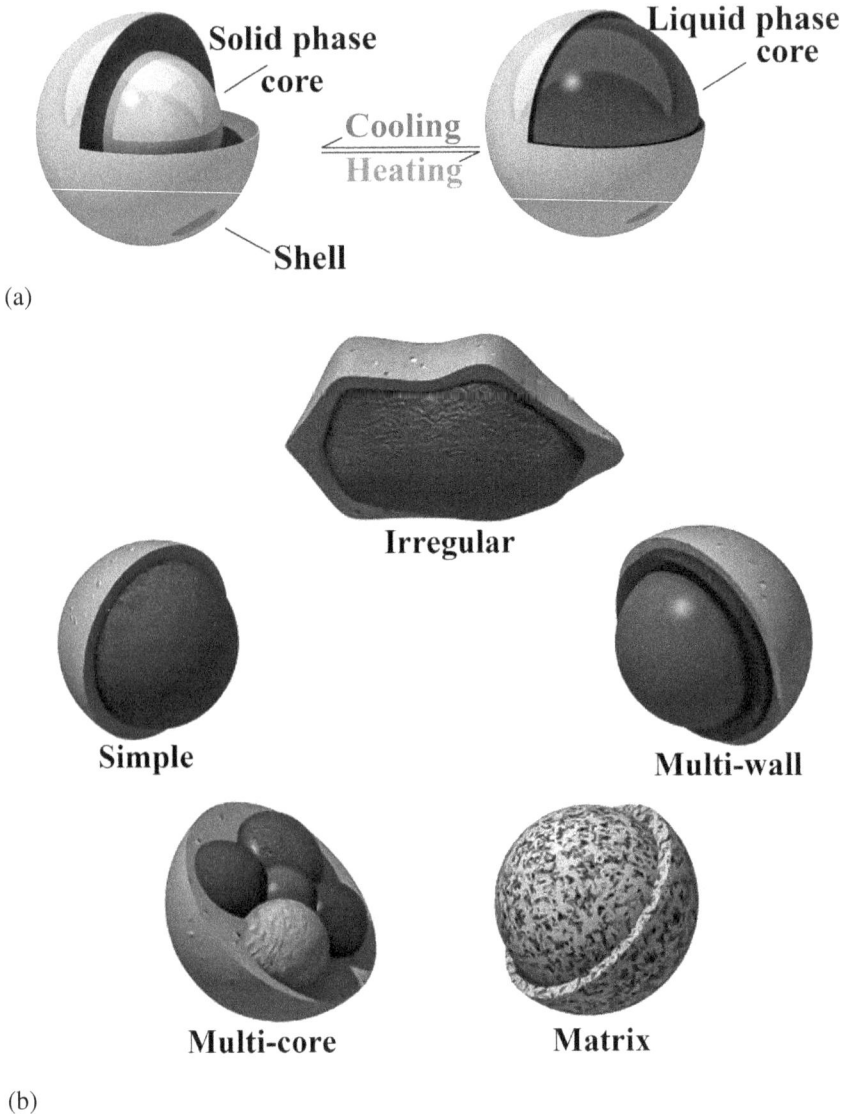

(a)

(b)

FIGURE 5.9   The concept of micro/nano encapsulation: (a) the structure, (b) the types [82], [with permission of reuse from Elsevier].

For the core material, any of the common PCMs introduced in section 5.3 can be used and, as suggested by the literature, paraffin again has been the most common choice for M/NPCMs [58]. As for the shell material, formaldehyde derivatives and polymers are among the common ones. A list of different materials tested for M/NPCMs can be found in the literature [58].

PCMs, and especially paraffins, generally may experience large volume changes when going through phase change. In the case of paraffins, the volume change is around 12% [62]. This significant change in volume can cause serious challenges to maintain TES equipment, specifically considering that PCMs are normally kept in confined spaces. M/NPCMs, in addition to enhancing PCM heat transfer properties, solve the volume change issue associated with PCMs as well. As shown in Figure 5.9a, there is enough space available inside the shell to deal with the PCM volume changes. When a PCM is in its solid state, there will be empty space inside the shell, and when the PCM is in liquid state, this additional space is almost occupied and the entire space inside the shell is utilised.

In most cases, M/NPCMs are of the simple type shown in Figure 5.9a and b because it is easier to fabricate while they offer a good level of enhancement in heat transfer properties of PCMs. More than 15 methods have been employed for fabrication of M/NPCMs with different shapes. Those methods can be categorised into three main types of physical (specific to micro encapsulation), physicochemical, and chemical. The literature suggests that for the mentioned fabrication types, with the same order respectively, the methods called spray drying, coacervation, and interfacial polymerisation are the most common methods employed to date [58, 82]. Here, to avoid unnecessary complexities, only one of the simple methods is briefly explained.

In the coacervation method, a polymer such as gelatine is dissolved in an aqueous or organic media. Then the PCM is dispersed into the solution, and by controlling the solubility of PCM through pH variations, an emulsion is formed. Then by evaporating the solvent, the M/NPCM particles are shaped [82]. The references [58, 74, 82–84] are recommended to find more information on different methods of fabricating M/NPCM.

Apart from the enhancements it brings, encapsulation can also make PCMs flowable through what is called phase change slurry (PCS). Considering that PCMs and PCSs are mainly involved in heat transfer applications, PCS is composed of a heat transfer fluid, normally water, mixed with a micro/nano encapsulated PCM [85]. However, PCSs suffer from their own drawback, which is stratification of the heat transfer fluid mixture, caused by the density difference between the heat transfer fluid and the M/NPCM particles. Through stratification M/NPCM particles can even cause clogging in the pipeline or the pumps [86].

M/NPCMs, when employed in the form of slurry, would have behaviours similar to nanofluids, especially for the nano encapsulation type. For this reason, the Maxwell Eucken model introduced by Eq. (14) can be used to estimate the ETC of the slurry with a good accuracy. But when the M/NPCM is stationary, the ETC of the composite can be estimated by Eq. (15) [82]:

$$\frac{1}{k_{eff}d_{eff}} = \frac{1}{k_{PCM}d_{PCM}} + \frac{d_{eff} - d_{PCM}}{k_{PCM}d_{PCM}d_{eff}} \tag{15}$$

where d_{eff} and d_{PCM} are the M/NPCM and PCM core diameters (m). More on the prior equation and further relations with regards to thermophysical properties of M/NPCMs can be found in reference [82].

5.3.3.4 Introducing Metallic Structures into the PCM Body

Another common and effective method of thermally enhancing PCMs is introducing highly conductive metallic structures, such as fins, metal foams, metallic honeycombs, and heat pipes. Each of these structures has its own specifications that are briefly introduced here. Figure 5.10 shows some samples for each of the aforementioned metallic structures.

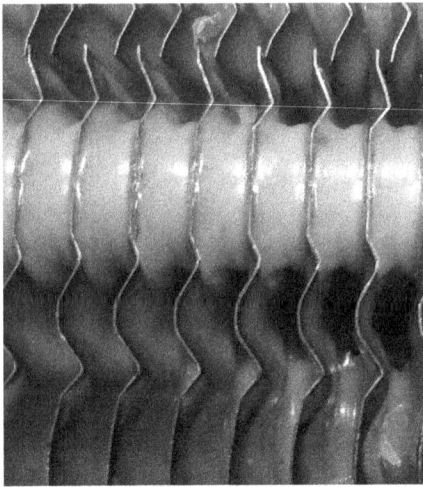

(a)

(b)

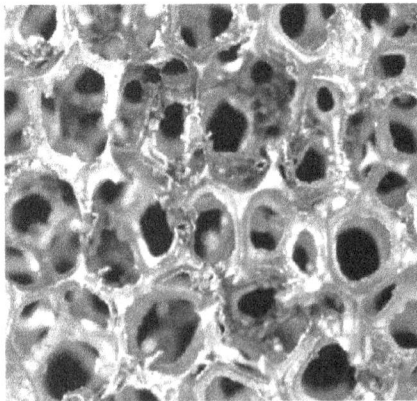

(c)

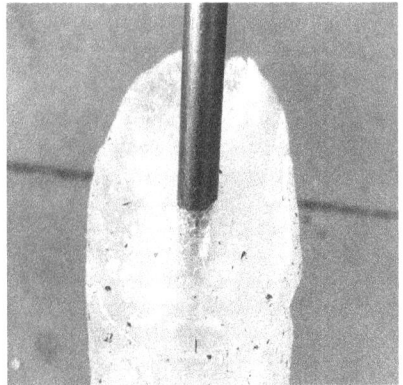

(d)

FIGURE 5.10 Samples of the common metallic structures used for enhancing heat transfer inside a PCM body: (a) fin [87], (b) honeycomb [88], (c) metal foam [89], (d) heat pipe, [with permission of reuse from Elsevier].

Employing fins has been a popular solution to enhance heat transfer, not only for PCMs but also for any other heat transfer medium. Fins are normally thin metallic plates manufactured in different shapes and sizes that can improve the heat transfer in a medium by providing an extended surface area and heat transfer network [90]. Being available at relatively low prices compared to other metals, being easy to work on and to shape, and most importantly, having a high thermal conductivity, aluminium, copper, and steel are the common materials used to manufacture fins [91]. The thermal conductivities of aluminium, copper, and steel are around 220 W/m.k, 400 W/m.k, and 45 W/m.k [92], respectively, which even in the case of steel are much higher than that of common PCMs such as paraffin, that is around 0.2 W/m.k.

Arrangement of the fins, number of fins, the length and width of fins, the fins' thickness, and fin pitch are the design parameters that need to be considered when using fins to thermally enhance a PCM [93]. All of these design parameters can affect the effectiveness of fins in enhancing heat transfer inside PCMs. In the meantime, it must also be considered that the presence of fins can affect the PCM heat storage capacity as well. Except for the case of PCM slurry, whose applications are minimal compared to other cases, PCMs are kept in a confined storage unit. That is, in most cases, there is a specific amount of PCM inside a unit, and to add other materials or structures, there would be no choice other than reducing the PCM mass. Consequently, the reduction in the mass of PCM would result in reducing its heat storage capacity. Therefore, when designing the fins' configuration, a balance must be made between the enhancement in heat transfer and reduction in heat storage capacity.

Another metallic structure used for enhancement of PCMs is called honeycomb, which is in fact a specific fin configuration [94, 95]. The honeycomb structure, because of its strong structural characteristics, is mostly employed for passive thermal management applications [96], which are introduced in section 5.7. Because of their strong mechanical features, they can tolerate the stresses caused by the changes in the volume of PCM when going through a phase change [97]. The honeycomb structures are mostly made of aluminium [98], and the shape of their cells is commonly triangular, trapezoidal, rectangular, hexagonal, and circular [96].

Fins come in different shapes, namely rectangular, longitudinal, radial, annular, tree, planar, spiral, triangular, trapezoidal, helical, etc., and their surface can have different profiles such as a normal flat surface or corrugated or perforated surfaces [90]. All of these different shapes and surface profiles, combined with other design parameters mentioned earlier, create a wide variety of fin and honeycomb designs. Therefore, it is not possible to provide a definitive method to estimate ETC of PCMs enhanced with fins or honeycombs, and it is recommended to use measurement methods in such cases.

Open-cell metal foam is another popular metallic structure that has been employed to enhance heat transfer inside a PCM body. Metal foams are porous metal structures with normally a high volume fraction of empty space (voids) inside their structure. Metal foams are commonly made of copper, aluminium, and nickel, with the thermal conductivity of the latter being around 90 W/m.K [92].

Apart from the material from which the metal foam is made of, there are two main specifications defined to identify the general structure of a metal foam, namely porosity, and pore density or pore size. Porosity is the ratio of the empty volume to metal foam total volume and is normally given in percentage. Pore density, on the other hand, identifies the number of pores inside the structure for a specific length. Pore density is usually given in pores per inch (PPI), which indicates the number of pores in one inch of the metal foam. Pore density has an opposite relationship with pore size. A higher pore density equates to smaller pore sizes and vice versa.

The porosity of open-cell metal foams employed for PCM enhancement is usually in the range of 90–98% and does not go below 70% [66]. This is discussed further in section 5.3.3.5, but to briefly explain it, although lower porosities would result in faster heat transfer, they also can lead to less heat storage capacity. Porosities lower than 70%, and even below 90%, can significantly decrease heat storage capacity. As for the pore densities, because of manufacturing requirements, usual values are in the range of 5–40 PPI, and it has minimal effects on the conductive heat transfer inside the PCM–metal foam composite [66]. Having only the material, porosity, and pore density as the main variables, it is much easier to identify overall thermal behaviour of metal foams.

A PCM–metal foam composite, which like other metallic structures enhances heat transfer inside PCMs through conduction, can be highly effective in their enhancement. With the porosities in the range of 90–98%, up to 27 times higher thermal conductivity compared to that of pure PCM is reported in the literature [66]. Even more enhancement is achievable at lower porosities, such as in one case with the porosity of 71%, a thermal conductivity of 218 higher is reported for the PCM–metal foam composite, compared to its pure PCM [99].

Again, because there are few parameters involved in determination of metal foam specifications, several relations have been introduced in the literature to estimate the ETC of PCM–metal foam composites. One of those relations with a good accuracy is as follows in Eq. (16) [100]:

$$k_{eff} = \frac{\left[k_{MF} + \pi \left(\sqrt{\frac{1-\varepsilon}{3\pi}} - \frac{1-\varepsilon}{3\pi} \right) \left(k_{PCM} - k_{MF} \right) \right] \left[k_{MF} + \frac{1-\varepsilon}{3} \left(k_{PCM} - k_{MF} \right) \right]}{k_{MF} + \left[\frac{4}{3} \sqrt{\frac{1-\varepsilon}{3\pi}} \left(1-\varepsilon \right) + \pi \sqrt{\frac{1-\varepsilon}{3\pi}} - \left(1-\varepsilon \right) \right] \left(k_{PCM} - k_{MF} \right)} \tag{16}$$

where k_{MF} is the metal foam material thermal conductivity (W/m.K) and ε is the metal foam porosity. A detailed discussion on the methods for estimating the PCM–metal foam ETC, along with a list of relations, can be found in references [66, 101].

Heat pipe is another type of metallic structure used for PCM heat transfer enhancement. There is a difference between heat pipe and other types of metallic structures used for this purpose though. The other three methods mostly aim to enhance the heat transfer and provide a good distribution of thermal energy, all inside the PCM body, but the heat pipe aims to transfer the heat between the PCM body and the heat source/sink.

A heat pipe is a sealed and evacuated metallic container with a wick structure on its internal wall that also contains a working fluid inside [102]. The working fluid exists both in liquid and vapour phases and through its phase change and vapour diffusion, and also with the help of convection, it transfers the heat through the heat pipe body [103]. The common materials for the metallic casing are aluminium, copper, and steel, and the commonly used working fluids are water, ammonia, acetone, methanol, and ethanol. A list of casing and working fluid materials along with details on operation principles of heat pipes can be found in references [102, 103].

Heat transfer in heat pipes is passive by nature, meaning that without the need for any external force, the input heat is transferred throughout the heat pipe body. Heat pipes are considered as one of the most efficient technologies of passive heat transfer, capable of maintaining almost a uniform temperature throughout the whole body, even when being heated and/or cooled at several points [103]. Heat pipes have a wide operating temperature range between −70°C and 270°C [103], which covers the operating temperature of most PCMs.

As it was briefly explained previously, inside the heat pipe both phase change and convection heat transfers are going on. On the metallic casing though, thermal conduction is in charge of transferring the heat. Modelling and estimation of all these different forms of heat transfer altogether can be very complicated. Because of that, when accurate modelling of the temperature profile inside the heat pipe is not concerned, normally the whole heat pipe is considered as a solid body with an effective thermal conductivity (not to be confused with ETC of enhanced PCMs). For the heat pipe alone, effective thermal conductivities in the order of 10^2-10^5 W/m.K are reported in the literature [104].

Such incredibly high thermal conductivities can greatly help with rapid transfer of heat inside a PCM body. Similar to fins, heat pipes come in a wide variety of shapes and configurations, and thus it is not possible to provide general relationships for estimating the ETC of PCMs that are thermally enhanced using heat pipes.

5.3.3.5 Comparison of PCM Enhancement Methods

Limited experimental studies on comparison of the aforementioned PCM enhancement methods are reported in the literature. However, using the observations of those studies and the specifications of each method, it is possible to provide a general picture of how these methods are compared to each other.

The addition of nanoparticles into PCMs or micro/nano encapsulation of PCMs are both effective ways to enhance heat transfer inside the PCMs. However, at the same time, both methods suffer from sedimentation and instability of the composite. The melting/solidification cycles that a composite goes through exacerbate this issue. Moreover, compared to utilising nano-PCMs and M/NPCMs, introducing metallic structures into the PCM body has been identified to be more effective for enhancing heat transfer properties of the PCM [92].

About the metallic structures, as mentioned in section 5.3.3.4, their presence can reduce the PCM heat storage capacity by occupying extra space inside the storage unit. Of course, this is not a concern when it is possible to compensate for that by using a larger unit. However, this is a factor to consider when there is no flexibility in terms of enlarging the storage size. In the case of using fins, given that they are usually thin metallic plates, if designed properly, their impact on the capacity of the storage unit would be minimal. As for metal foams, the situation can be worse. At the porosity of 95%, which is normally the highest porosity available commercially, the presence of metal foam would reduce the amount of PCM and subsequently the storage capacity of the PCM unit by 5%. This reduction is not negligible, but given that metal foams provide the best distribution of thermal energy inside the PCM body, depending on the application, it can be an acceptable loss, especially considering the fact that the rest of the PCM is utilised more effectively. Honeycomb structures and heat pipes can also take a considerable space inside the unit, and again there would be a trade-off between the heat storage capacity and the enhancement in heat transfer. However, normally with a small loss of the storage capacity, a significant improvement in the heat transfer performance is achievable.

Apart from the heat storage capacity, introducing metallic materials into PCMs can have another negative effect. While enhancement of thermal conductivity can improve the heat transfer performance of PCMs, conduction is not the only mechanism through which the heat is transferred inside a PCM. PCMs go through a solid-liquid phase change, and when in the liquid state, natural convection also plays an important role in transferring the heat through a PCM body. In fact, literature suggests that for a non-enhanced PCM, natural convection is the dominant heat transfer mechanism, when a PCM is in the liquid state [66]. The presence of a metallic structure inside the PCM can cause movement resistance against liquid PCM, which consequently suppresses the heat transfer by natural convection. In the case of the heat pipe type, given that there are not usually too many heat pipes involved, the PCM surrounding the heat pipes would have enough space almost in any direction to move freely. Therefore, the heat pipe would have minimal effects on the convective heat transfer. As for the fins and honeycomb structures, their arrangement must be designed to minimise their effect on the movement of liquid PCM. As in convective heat transfer, the main direction of liquid movement is upwards and downwards, so it is better to take advantage of vertical arrangements rather than horizontal ones. The metal foam though inevitably puts movement resistance all around the PCM body. Minimising that resistance is only achievable by using a highly porous metal foam with a low pore density, i.e., large pore size. However, the suppression of convective heat transfer caused by the metallic structures is not usually significant enough to offset the positive effect of enhancement in conductive heat transfer [66].

For fin, metal foam, and honeycomb structures. there are a large variety of designs, especially for the fin type, making it a bit hard to compare the effectiveness of these three types of thermal performance enhancement methods. This is while even similar heat transfer performance have been reported in some published resources for the PCMs enhanced by these metallic structures [92, 98]. In such a condition, other parameters, such as economic performance, availability of these options, mass, volume, mechanical strength, and installation aspects, must be considered to select the best structure for a specific application.

5.4 THERMOCHEMICAL HEAT STORAGE

The TCHS method takes advantage of reversible endothermic/exothermic reactions. The concept of TCHS is schematically shown in Figure 5.11.

Using the concept schematically presented in Figure 5.11, in the charging phase, a molecular compound of AB is heated, and through an endothermic reaction, is split into its composing reactants of A and B. Then during the storage phase, these two reactants are kept separated from each other. The storage phase duration would be as

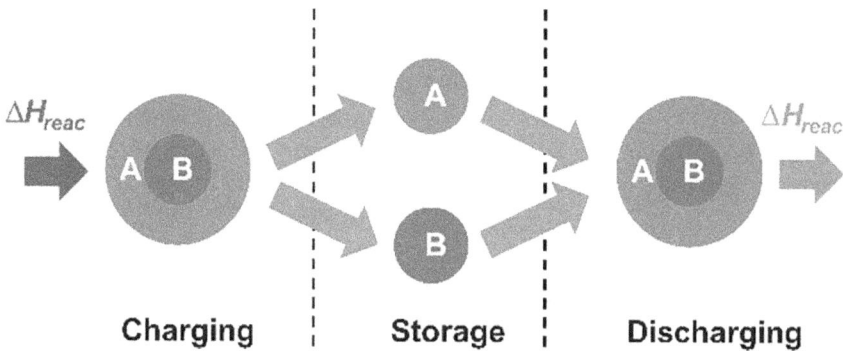

FIGURE 5.11   The concept of thermochemical heat storage [105], [with permission of reuse from Elsevier].

long as these two reactants are kept separated. Whenever needed, the discharge phase can get started when the two reactants are put together. The reactants recombine through an exothermic reaction and release heat.

One type of reaction that has the potential for this purpose is thermal decomposition of metal oxides [106], a sample of which is decomposition of potassium oxide, Eq. (17) [29]:

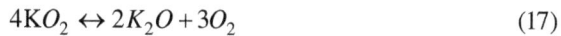

$$4KO_2 \leftrightarrow 2K_2O + 3O_2 \tag{17}$$

This reaction takes place over a temperature range of 300–800°C and has a decomposition heat of 2.1 MJ/kg of potassium oxide [29]. This capacity of heat storage/release for 1 kg of potassium oxide is equivalent to latent heat storage of 10.5 kg of a paraffin with a fusion latent heat of 200 kJ/kg; or in the form of SHS, it is equivalent to warming up 10 kg of water by 50°C. This is very impressive; however, not all such reactions come with a high heat of decomposition such as this. For instance, decomposition of lead oxide requires 0.26 MJ/kg of lead oxide [29].

A unique specification of metal oxides is that their decomposition also releases oxygen, which can be utilized for other purposes, and for the reverse reaction it can use the oxygen available in the ambient. There are other materials with decomposition reactions appropriate for TCHS, such as metal hydroxides, metal chlorides, salt hydrates, and sulphates.

As presented in Table 5.4, high energy densities, much higher than that of the LHS and SHS methods, are common in the TCHS method. The reactions given in Table 5.4 also suggest that apart from oxygen, hydrogen and water can also be side products of TCHS methods. an expanded list of TCHS materials, along with their decomposition reactions and their heat of decomposition, can be found in references [8, 30, 106–108].

In the charging phase of TCHS, for the decomposition reaction to begin, it is first needed to heat a material up to certain temperatures. As suggested by Table 5.4, this temperature is normally above 200°C. When this required temperature is reached, the

TABLE 5.4

Specifications of Some TCHS Materials [29]

Reaction		Temperature (°C)	Energy Density (MJ/kg)
Methane steam reforming	$CH_4 + H_2O \leftrightarrow CO + 3H_2$	480–1195	6.053
Ammonia dissociation	$2NH_3 \leftrightarrow N_2 + 3H_2$	400–500	3.940
Magnesium hydride dehydrogenation	$MgH_2 \leftrightarrow Mg + H_2$	200–500	3.079 + 9 in form of H_2
Calcium hydroxide dehydration	$Ca(OH)_2 \leftrightarrow CaO + H_2O$	402–572	1.415
Catalytic dissociation	$2SO_3 \leftrightarrow SN_2 + O_2$	520–960	1.235

remaining input of thermal energy would be consumed for the decomposition reaction itself. The following equation, Eq. (18), explains this process [74]:

$$Q_{charge} = Q_{sens,ch} + Q_{bind} \tag{18}$$

where Q_{charge} is the total energy (J) consumed to complete the reaction, i.e., the charging phase; $Q_{sens,ch}$ is the amount of energy (J) spent in the form of sensible heat, to heat the material up to the temperature required for the reaction to begin; and Q_{bind} is the binding energy (J) between the molecules of composing the materials. As shown in Figure 5.11, Q_{bind} is introduced as ΔH_{reac} as well.

Because of the need for reaching high temperatures for decomposition of the TCHS materials, the charging phase is mostly done at high temperatures. The reverse reaction occurring during the discharge phase, however, does not need such high temperatures. Therefore, the discharge phase of TCHS is usually at a lower temperature than that of the charging phase [74]. However, there is still a temperature requirement, depending on the materials involved, for the reverse reaction to begin. The relation for calculating the energy released during the discharge period would be similar to Eq. (18), as seen in Eq. (19), with the difference that the sensible heat consumed to heat up the materials would not be a part of total discharged energy:

$$Q_{discharge} = Q_{bind} - Q_{sens,dis} \tag{19}$$

where $Q_{discharge}$ is the total amount of energy (J) released during the discharge period and $Q_{sens,dis}$ is the amount of energy spent to heat up the reactants to begin the reverse reaction.

The discussion following Eq. (17) and the data given in Table 5.4 support the point that compared to SHS and LHS, generally, TCHS offers a higher storage density, i.e., energy storage capacity per mass unit. TCHS also has minimal heat losses as the separated materials are kept in the ambient condition, and thus it does not need heavy thermal insulation as the other two methods do [109]. The release of thermal energy in the discharge phase is a product of chemically combining two or more materials, rather than a release of a directly stored thermal energy. This method of keeping materials separated, to some extent makes TCHS portable as the materials can be transported and recombined in another location. However, currently TCHS is still under research as it comes with its own technical complexities. Currently TCHS is rather costly and has its own technical complexities such as toxicity, corrosiveness, and safety issues. When the reactions are in process, normally the temperature must be controlled. The reactions can also occur at variable rates or not at a desirable rate that makes it difficult to match TCHS with the heat transfer requirements of the intended application [110]. More details on TCHS methods can be found in references [108, 109, 111].

A reversible endothermic/exothermic reaction is the main mechanism for TCHS. Some recent technologies can also be taken advantage of for this kind of TES. Metal hydrides, for instance, which are mainly employed for hydrogen storage applications, have this potential. However, further investigation is required to evaluate this potential. More information on metal hydrides can be found in the literature [112–114].

5.5 COLD THERMAL ENERGY STORAGE

Process-wise, cold thermal energy storage (CTES) is not an individual method of TES different from the ones introduced in earlier sections. In practice, CTES approaches are mostly based on SHS and LHS methods [115]; however, thermochemical CTES is also possible [116]. What distinguishes CTES from other methods is its operational temperature, which can be anywhere below 0°C [116]. This temperature range makes CTES unique and worthy to be briefly discussed in this chapter. A proper example to support this idea is water, which is used in SHS methods that can be employed as a PCM in CTES applications. Water has a latent heat of fusion of 333.6 kJ/kg [117]. As introduced in section 5.3.2, most conventional PCMs have a latent heat in the range of 200–250 kJ/kg, which is considerably less than that of water. A list of some of the materials appropriate for CTES applications is given in Table 5.5.

TABLE 5.5
Materials Appropriate for Employment in Cold Thermal Energy Storage [116]

Material	Heat Storage Approach	Heat Storage Capacity (kJ/kg.K) at Below 0 °C
Water	SHS	Specific heat: 1.390–2.050
Rocks	SHS	Specific heat: 0.740–2.150
Concrete	SHS	Specific heat: 0.830–1.160
Sodium chloride	SHS	Specific heat: 0.600–0.850
Aluminum oxide	SHS	Specific heat: 0.100–0.720
Water	LHS	Latent heat: 333.600
Methane	LHS	Latent heat: 58.430
Propane	LHS	Latent heat: 79.600
Isopentane	LHS	Latent heat: 71.100
Ethanol	LHS	Latent heat: 108.530

A large list of possible CTES materials based on SHS and LHS approaches, including enhanced PCMs, along with their thermophysical properties, is presented in reference [116]. The need for CTES for materials with melting points below 0°C opens the door for employing materials different from those conventionally employed in SHS and LHS approaches. This can create unique opportunities for CTES applications: a great example of this is the water/ice slurry. A mixture of water and ice means that it is being used in the form of LHS. In normal LHS applications, as discussed in section 5.3.3.3, PCM slurries suffer from sedimentation and stratification, which can clog the process pipeline. Both sedimentation and stratification occur due to the slurry being composed of different materials with significant density differences. This is whilst in the case of water/ice slurry the sedimentation and stratification are not a major concern.

As CTES systems operate very similarly to SHS and LHS methods, here they will not be discussed more. The applications of CTES, which are introduced in section 5.7.4, make it unique.

5.6 OVERVIEW OF TECHNOLOGIES, CHALLENGES, AND THE RESEARCH TREND

Many studies have been published on TES methods as introduced so far in this chapter. It is not possible to compare all those variations of the TES methods, but some rough estimations are given in Table 5.6 that were originally presented in a workshop by the International Energy Agency (IEA) in 2011 [118] and are still being supported by the literature [26, 29, 119, 120] to be used for comparing various TES methods.

Before discussing the information given in Table 5.6, it must be noted that in this table the storage capacities are reported as kWh per ton of the storage material just to provide a common base for this comparison. The 'application power capacity' denotes the output power range of the application that the TES methods are normally employed in. Furthermore, water, being the most common material employed in SHS, represents this type of TES.

Based on Table 5.6, SHS offers the lowest heat storage capacity. Considering the maximum values given in Table 5.6, the storage capacities of TCHS and LHS are up to 4.2 times and 2.5 times more than that of the SHS method, respectively. In addition to that, SHS also has the lowest range for efficiency. A medium range and

TABLE 5.6
Comparison of the Three Main TES Methods [118]

TES Method	Storage Capacity (kWh/t)	Application Power Capacity (MW)	Efficiency (%)	Maximum Storage Capability (Duration)
SHS (water)	10–60	0.001–10	50–90	Days to months
LHS (PCM)	50–150	0.001–1	75–90	Hours to months
TCHS	120–250	0.01–1	75–100	Hours to days

a high range can be considered for the efficiency of the LHS and TCHS methods. Therefore, it can be concluded that TCHS offers the best TES performance and LHS somewhat closely follows it. However, as the information given in Table 5.6 suggests, SHS is employed in the widest range of applications. That is because water is easily accessible, cheap, and safe to operate. The equipment for hot water storage are also not complicated and widely available in different shapes and sizes. That is whilst the other two TES methods require their specific equipment and have more limitations than water to work with. Moreover, the price of the materials used in TES methods is an important factor to be considered for selecting them for TES applications. Based on the same report presented through an IEA workshop, for the same energy storage capacity, LHS and TCHS materials can be five to ten times more expensive than water [26, 29, 119]. Another drawback for those two methods, and specifically for the TCHS method, is that they are capable of keeping the stored thermal energy for shorter periods compared to a water-based SHS method. This factor can become important based on the application requirements. In addition to that, stability and cyclability of the materials are currently an ongoing issue for LHS and TCHS methods [121]. Overall, the SHS approach, especially when water is employed, is a mature method. That is, whilst the LHS is under significant research, which has offered rooms for their development; and finally TCHS is considered to be at its early stages of research and development [121].

There is a challenge specific to PCMs used in the LHS method. As presented in Table 5.3, except for metal-based ones, which have very limited applications, PCMs suffer from low thermal conductivities. PCMs are mainly stored in a confined space, and during the TES process, they are either in solid or liquid phases or have a combination of both while one is transitioning to another. To store heat or release the stored energy effectively, the entire material must be involved; otherwise, some parts may not functionally participate in the aforementioned processes. For thermal energy to be properly transferred through a PCM and hence engage its entire mass involved, they need to have a high thermal conductivity, which is not the case for many PCMs [61, 62]. There have been many efforts to overcome this issue, amongst which introducing metal foams, nanoparticles, and fins into the PCMs or encapsulating PCMs with highly conductive materials [66, 92] have gained popularity. However, these solutions, as already discussed in sections 5.3.3.2 to 5.3.3.4, have their own drawbacks and may need further research and development for them to unleash their full potential.

As for TCHS, this method is also in the research phase, with many obstacles to be removed before it can be widely practised. Although TCHS normally offers much higher energy storage density compared to the other two methods, it is very costly and comes with its own technical complexities such as toxicity, corrosiveness, and safety issues. In addition to that, controlling the reactions to obtain a specific heat transfer condition required by the application process is another major technical challenge associated with TCHS.

Implementation of TES methods may not be cost effective in many cases, mostly due to their high initial costs; this makes their payback period undesirably long, which barely justifies the energy savings offered by TES solutions [56]. Hence, TES technologies generally require further developments, with particular attention to making them more cost effective. TES systems are mostly used for hot or cold water applications. In these cases, specifically when LHS and TCHS methods or the SHS methods without a conventional water tank are employed, there are still concerns about the quality of the output hot or cold water.

Apart from the technological advancements required, the economic viability of TES solutions, specifically the LHS and TCHS approaches, can be boosted through the allocation of government incentives. Otherwise, in the absence of a strong economic completeness, the environmental benefits of TES solutions remain the key driver behind their adoption. This becomes a stronger proposition though in the presence of supportive environmental legislative policies.

5.7 APPLICATIONS

5.7.1 RENEWABLE ENERGY TECHNOLOGIES

The application of TES systems can be categorised into two main types of active and passive systems. In active systems there is a component, usually a heat transfer fluid, that connects the TES unit to other parts of the process and helps with charging and discharging the unit, whereas in passive systems, the TES unit has a direct or indirect thermal contact with wherever the heat has aimed to be transferred. In such systems there is no need for an external component to transfer the heat where it is needed [122]. Passive methods are mostly employed in thermal management applications, which are discussed in section 5.7.3.

In many TESs that use SHS approaches, renewables such as solar energy and geothermal energy are perfect candidates for being the heat source. Among the water-based SHS methods introduced in section 5.2.3, water tanks, pits, aquifers, and solar ponds can use solar energy to get warmed up. The borehole and rock cavern methods inherently use geothermal energy as their heat source. LHS methods are a perfect match with solar energy technologies, because solar energy is intermittent by nature. It is not available all day long and, when it is available, ambient conditions such as the weather can significantly affect its availability. This intermittency, along with the variations in demand for energy, results in large mismatches between the supply and demand of the energy. LHS systems not only can help with increasing the efficiency of solar energy harvesting technologies, but also can mitigate those mismatches. Hence, application of PCMs with solar thermal and even solar electrical technologies has been very popular. Integration of PCMs with solar thermal collectors such as flat plate [123], evacuated tube [63], trough [124], heliostat [125], and solar stills (solar desalination) [126] have been widely studied in the literature.

PCMs have also been used for thermal management of photovoltaic (PV) systems for simultaneous harvesting of electricity and heat, also known as PV/T [127]. The inefficiencies of PV panels in generating electricity appear in the form of heat, which increases their temperature [128]. PVs show lower efficiencies at higher temperatures. Being able to transfer heat almost isothermally, PCMs can help with the operation of photovoltaic panels at an almost constant (low) temperature. The heat generated by PVs can also be captured using SHS units.

The potential offered by TES arrangements to enhance the viability of renewable energy technologies has made them popular solutions. More information and a list of some major real-world applications of solar energy technologies integrated with TES can be found in reference [34].

5.7.2 District Heating and Cooling

District heating and cooling (DHC) systems integrated with thermal storage have been highly popular globally. Different scales of DHC integrated with TES are currently operating in Germany, Denmark, France, Sweden, Spain, Britain, Belgium, Canada, and the USA [8, 121]. TES in DHC is used in both short-term and long-term (i.e., seasonal) storage applications. Short-term applications are mainly about compensating real-time mismatches between supply and demand in either heating or cooling modes. The whole idea of the seasonal storage is to store the excess heat during the hot season and/or the excess cold energy during the cold season for later use in opposite seasons. This aspect of their application is receiving interest these days as renewable energy technologies such as solar, geothermal, and biomass are gaining popularity [129]. Figure 5.12 schematically shows the concept of a district heating system powered by solar energy integrated with a TES tank.

The TES method being employed in DHC applications is mostly the SHS one with water used as the main material for that. However, in limited instances, employment of LHS and TCHS has also been reported, where a heat exchanging mechanism is employed to transfer the stored energy to a flow of water [16]. However, considering that DHC is a large-scale application in nature, availability and ease of operation make water the main choice for use as a TES material. Water is commonly stored in water storage tanks, a pit, boreholes, or an aquifer with water tank being mostly used

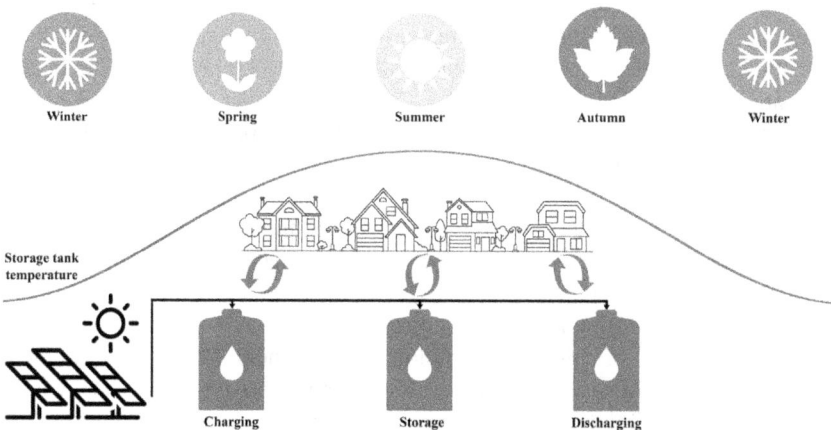

FIGURE 5.12   General overview of a solar-powered district heating system integrated with TES, recreated from [130], [with permission of reuse from Elsevier].

for short-term applications and all of them being appropriate for seasonal storage. The space required for a water storage tank in short-term applications can be any size around 10–100 m^2, depending on the designed capacity of the DHC system. As for the seasonal storage, a tank or a pit-based system requires 1,000–10,000 m^2, and a borehole or an aquifer needs about 10 m^2 [16].

Integration of DHC with TES has been a popular subject for the researchers, with several valuable studies reported in the literature on this topic. The following references in particular are recommended by the authors to learn more about DHC and their integration with TES units and renewable energy sources: [16, 121, 129, 131]. Yang et al. [121] have also done a great job in providing a comprehensive technoeconomic review of seasonal thermal energy storage systems, including those based on SHS, LHS, and TCHS methods.

5.7.3 Thermal Energy Management

The applications discussed earlier in this section can be categorised as active TES systems, in which an external heat source charges a storage medium, and then the stored energy is delivered to an end user. Thermal energy management applications are specific to LHS methods in which a TES component operates in a passive manner. In thermal energy management applications, the goal is to keep the end user at an almost constant temperature. As discussed in section 5.3, PCMs with their high latent heat make it possible to receive and release thermal energy at an almost constant temperature. Thermal management of electrical batteries and buildings are the most common applications of this type.

Electrical batteries generate heat when operating, and in the meantime their performance and even their lifespan are depleted if operated at elevated temperatures. Lithium-ion batteries specifically produce large amounts of heat whilst operating that can accelerate their aging and affect their charging/discharging capacities [21]. That is why electrical batteries must usually be operated within a narrow temperature range to maintain their health and performance. PCMs facilitate the operation of batteries at almost constant temperature points, making them prime candidates for thermal management of electrical batteries. Reference [132] provides more detailed information on this application.

Thermal management of buildings is another application in which PCMs can be very effective. Thermal comfort or a comfortable temperature range is defined for most buildings that are used for different purposes. For instance, in the case of buildings with human occupants, based on the function of the building, the comfort temperature range is around 20–32°C [133]. PCMs, again because of their unique ability to transfer heat almost isothermally, are great candidates to help with keeping building temperature at a desirable range. Wallboards, being easy to work with, have been a popular component of buildings to employ PCMs in them. Many studies in the literature and real-world applications have used PCM wallboards for thermal management of buildings [134]. Several other building materials and components have been tested for containing PCMs, such as walls, windows, roofs, floors, concrete, gypsum, brick, and mortar, the details of which can be found in the literature [134, 135].

5.7.4 Cold Thermal Energy Storage Applications

Applications as simple as making ice, and then using the ice to cool down anything or to keep something cold, are known as cold thermal energy storage (CTES). The cool packs or ice packs that are even available at retail stores are among such applications. One of the common CTES applications that has been around for decades is for preserving and transport of food and medical supplies [115].

Any refrigeration system, including domestic, commercial, and industrial, can also take advantage of CTES to increase its efficiency [116] by reducing the losses, which is another inherent specification of TES systems. In refrigeration cycles, a CTES unit is normally located around the evaporator [136]. CTES slurries can also be employed as a secondary loop, which receives cold energy and releases it at another point and at any time as required [116].

Space cooling is another well-known CTES application. A popular CTES-based space cooling application is about using the off-peak electricity [115] to generate some cooling effect and storing it in an appropriate PCM for later use, i.e., to minimise the electricity consumption for space cooling at peak time. Apart from the savings offered by this approach, it also helps with shaving the peak of the demand at large scale and removing the pressure from the grid, especially during hot seasons. During such periods, largely due to the high space cooling demand, there is a peak demand for electricity during mid-day and afternoon, putting an immense pressure on the grid. During off-peak hours, an excess amount of electricity can be generated without overloading the grid. This excess amount of energy can be used to charge up CTES systems, which can then help with space cooling during the peak hours to reduce the pressure on the grid. Doing so, during the off-peak hours, the excess electricity can be used to produce cold water or ice. Then this cold water or ice is stored in a CTES unit, which is later employed as a cold supply of an air conditioning system [115]. For example, this concept of CTES-based air conditioning systems have been taken advantage of in the US, China, Korea, Malaysia, Thailand, and Canada [115].

Some applications take advantage of TES systems to store electricity through a process. An application of such that employs both cold and thermal energy storage is called pumped thermal energy storage. Figure 5.13 shows the process diagram of such an arrangement.

A pumped thermal energy storage system employs an SHS approach, and its working fluid is a gas, i.e., usually argon. During the charging phase, the system acts like a refrigeration cycle and consumes electricity to transfer thermal energy from cold storage to hot storage. This way, both storages get charged simultaneously. Then, whenever needed, the cycle is reversed and through the transfer of heat from the hot storage to the cold storage, electricity is produced by a generator located inside the expander [137].

There are other CTES applications, such as liquid air energy storage, superconducting flywheel energy storage, waste cold energy recovery, and boil-off gas re-liquefaction of a liquefied natural gas (LNG) terminal, the details of which can be found in reference [116].

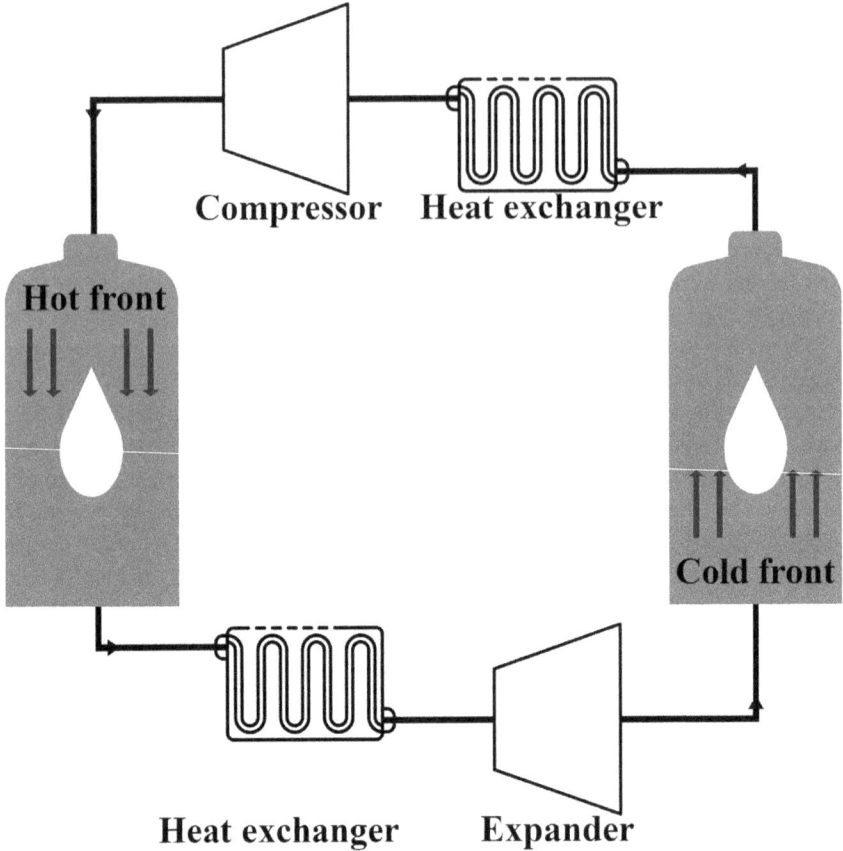

FIGURE 5.13 The process diagram of a pumped thermal energy storage system, recreated from [137], [with permission of reuse from Elsevier].

5.8 CONCLUSIONS

While not being a recent technology, thermal energy storage has recently drawn significant attention due to the important roles they play in thermal energy systems. Particularly, the emergence of renewable energy technologies, especially solar and geothermal, has made TES systems even more popular than before. These green and free sources of energy are intermittent, and in many cases, they cannot meet the energy demand in real-time; that is where TES solutions can play effective roles. TES methods can help with reducing heat losses, addressing the mismatches between the supply and demand of energy, and maintaining a desired temperature depending on applications.

Currently, SHS types of TES, specifically water-based ones, are widely employed worldwide and in various small- and large-scale applications. A common example of SHS is the hot water tanks. Aquifers (large scale) and boreholes are other common water-based SHS approaches that were discussed in this chapter.

In addition to SHS, LHS and TCHS methods have also been developed for use in TES applications. The TCHS method usually offers the highest energy storage density, meaning it can store the highest amount of energy per unit of mass. LHS also generally has a higher energy storage density than what is offered by the SHS method. However, none of the LHS and TCHS methods has proven to be as successful as SHS in a wide range of real-world applications. The two former methods, specifically TCHS, and the technologies associated with them are still in the research and development phase with rooms for their further improvement. The main reason for SHS being successful is that its related technologies are not complicated and have been developing for a long time; they mostly use water as the SHS material, which is readily available; moreover, generally they cost much less than the other two methods.

The high costs and the complexities associated with the technologies around LHS and TCHS have been great barriers to the success of these TES methods. However, solid justifications support the need for further development of these two methods. Firstly, they offer high energy storage densities while being more efficient options than the SHS approach. Secondly, water shortage has already become a concern for some regions, and it will become a more critical issue globally with population growth and industrial developments, which will put further stress on water resources. Considering that, in the future there may be a need to replace water-based SHS systems with other solutions, and that is where LHS and TCHS can come into play. It is noteworthy that these two approaches may require stronger financial support from governments and decision makers in order to unleash their full potentials.

REFERENCES

1. Hirvonen, J., et al., *Effect of apartment building energy renovation on hourly power demand.* International Journal of Sustainable Energy, 2019. **38**(10): p. 918. https://doi.org/10.1080/14786451.2019.1613992

2. De Cian, E., E. Lanzi, and R. Roson, *Seasonal temperature variations and energy demand.* Climatic Change, 2012. **116**(3–4): p. 805. https://doi.org/10.1007/s10584-012-0514-5

3. Jie, B., T. Tsuji, and K. Uchida, *Analysis and modelling regarding frequency regulation of power systems and power supply–demand-control based on penetration of renewable energy sources.* The Journal of Engineering, 2017. **2017**(13): p. 1824. https://doi.org/10.1049/joe.2017.0646

4. Shabani, B., J. Andrews, and S. Badwal, *Fuel cell heat recovery, electrical load management, and the economics of solar-hydrogen systems.* International Journal of Power and Energy Systems, 2010. **30**(4). https://doi.org/10.2316/Journal.203.2010.4.203-4842

5. Assaf, J. and B. Shabani, *Experimental study of a novel hybrid solar-thermal/PV-hydrogen system: Towards 100% renewable heat and power supply to standalone applications.* Energy, 2018. **157**: p. 862. https://doi.org/10.1016/j.energy.2018.05.125

6. Assaf, J. and B. Shabani, *Economic analysis and assessment of a standalone solar-hydrogen combined heat and power system integrated with solar-thermal collectors.* International Journal of Hydrogen Energy, 2016. **41**(41): p. 18389. https://doi.org/10.1016/j.ijhydene.2016.08.117

7. Shabani, B. and J. Andrews, *Hydrogen and Fuel Cells.* In: Sharma, A., Kar, S. (eds) *Energy Sustainability Through Green Energy. Green Energy and Technology,* 2015. Springer, New Delhi. https://doi.org/10.1007/978-81-322-2337-5_17

8. Alva, G., Y. Lin, and G. Fang, *An overview of thermal Energy storage systems.* Energy, 2018. **144**: p. 341. https://doi.org/10.1016/j.energy.2017.12.037

9. Ravi Kumar, K., N.V.V. Krishna Chaitanya, and N. Sendhil Kumar, *Solar thermal energy technologies and its applications for process heating and power generation – A review.* Journal of Cleaner Production, 2021. **282**: p. 125296. https://doi.org/10.1016/j.jclepro.2020.125296

10. Osterman, E. and U. Stritih, *Review on compression heat pump systems with thermal energy storage for heating and cooling of buildings.* Journal of Energy Storage, 2021. **39**: p. 102569. https://doi.org/10.1016/j.est.2021.102569

11. Yousefi, H., M. Aramesh, and B. Shabani, *Design parameters of a double-slope solar still: Modelling, sensitivity analysis, and optimization.* Energies, 2021. **14**(2): p. 480. https://doi.org/10.3390/en14020480

12. Aramesh, M., et al., *A review of recent advances in solar cooking technology.* Renewable Energy, 2019. **140**: p. 419. https://doi.org/10.1016/j.renene.2019.03.021

13. El-Mesery, H.S., et al., *Recent developments in solar drying technology of food and agricultural products: A review.* Renewable and Sustainable Energy Reviews, 2022. **157**: p. 112070. https://doi.org/10.1016/j.rser.2021.112070

14. Aftab, W., et al., *Phase change material-integrated latent heat storage systems for sustainable energy solutions.* Energy & Environmental Science, 2021. **14**(8): p. 4268. https://doi.org/10.1039/d1ee00527h

15. Zou, Y., L. Bo, and Z. Li, *Recent progress in human body energy harvesting for smart bio-electronic system.* Fundamental Research, 2021. **1**(3): p. 364. https://doi.org/10.1016/j.fmre.2021.05.002

16. Guelpa, E. and V. Verda, *Thermal energy storage in district heating and cooling systems: A review.* Applied Energy, 2019. **252**: p. 113474. https://doi.org/10.1016/j.apenergy.2019.113474

17. Percival, W.H. and M. Tsou, *Power from thermal energy storage systems*, in SAE Technical Paper Series. 1962. p. Start Page. https://doi.org/10.4271/620394

18. Garg, H.P., S.C. Mullick, and A.K. Bhargava, *Solar thermal energy storage.* Dordrecht; Boston, MA: Kluwer Academic, 1985.

19. Zhang, S., et al., *Component-dependent thermal properties of molten salt eutectics for solar thermal energy storage: Experiments, molecular simulation and applications.* Applied Thermal Engineering, 2022. **209**: p. 118333. https://doi.org/10.1016/j.applthermaleng.2022.118333

20. Gencel, O., et al., *Glass fiber reinforced gypsum composites with microencapsulated PCM as novel building thermal energy storage material.* Construction and Building Materials, 2022. **340**: p. 127788. https://doi.org/10.1016/j.conbuildmat.2022.127788

21. Shabani, B. and M. Biju, *Theoretical modelling methods for thermal management of batteries.* Energies, 2015. **8**(9): p. 10153. https://doi.org/10.3390/en80910153

22. Nguyen, H.Q. and B. Shabani, *Proton exchange membrane fuel cells heat recovery opportunities for combined heating/cooling and power applications.* Energy Conversion and Management, 2020. **204**: p. 112328. https://doi.org/10.1016/j.enconman.2019.112328

23. Nguyen, H.Q. and B. Shabani, *Review of metal hydride hydrogen storage thermal management for use in the fuel cell systems.* International Journal of Hydrogen Energy, 2021. **46**(62): p. 31699. https://doi.org/10.1016/j.ijhydene.2021.07.057

24. Nguyen, H.Q. and B. Shabani, *Thermal management of metal hydride hydrogen storage using phase change materials for standalone solar hydrogen systems: An energy/exergy investigation.* International Journal of Hydrogen Energy, 2022. **47**(3): p. 1735. https://doi.org/10.1016/j.ijhydene.2021.10.129

25. Nguyen, H.Q., et al., *An experimental study of employing organic phase change material for thermal management of metal hydride hydrogen storage.* Journal of Energy Storage, 2022. **55**: p. 105457. https://doi.org/10.1016/j.est.2022.105457

26. Ben Romdhane, S., et al., *A review on thermal energy storage using phase change materials in passive building applications*. Journal of Building Engineering, 2020. **32**: p. 101563. https://doi.org/10.1016/j.jobe.2020.101563

27. Alva, G., et al., *Thermal energy storage materials and systems for solar energy applications*. Renewable and Sustainable Energy Reviews, 2017. **68**: p. 693. https://doi.org/10.1016/j.rser.2016.10.021

28. Caraballo, A., et al., *Molten salts for sensible thermal energy storage: A review and an energy performance analysis*. Energies, 2021. **14**(4): p. 1197. https://doi.org/10.3390/en14041197

29. Sarbu, I. and C. Sebarchievici, *A comprehensive review of thermal energy storage*. Sustainability, 2018. **10**(1): p. 191. https://doi.org/10.3390/su10010191

30. Zhang, H., et al., *Thermal energy storage: Recent developments and practical aspects*. Progress in Energy and Combustion Science, 2016. **53**: p. 1. https://doi.org/10.1016/j.pecs.2015.10.003

31. Miró, L., et al., *Experimental characterization of a solid industrial by-product as material for high temperature sensible thermal energy storage (TES)*. Applied Energy, 2014. **113**: p. 1261. https://doi.org/10.1016/j.apenergy.2013.08.082

32. Reed, A.L., et al., *Solar district heating with underground thermal energy storage: Pathways to commercial viability in North America*. Renewable Energy, 2018. **126**: p. 1. https://doi.org/10.1016/j.renene.2018.03.019

33. Sanner, B., et al., *Current status of ground source heat pumps and underground thermal energy storage in Europe*. Geothermics, 2003. **32**(4–6): p. 579. https://doi.org/10.1016/s0375-6505(03)00060-9

34. Suresh, C. and R.P. Saini, *Review on solar thermal energy storage technologies and their geometrical configurations*. International Journal of Energy Research, 2020. **44**(6): p. 4163. https://doi.org/10.1002/er.5143

35. Dahash, A., et al., *Advances in seasonal thermal energy storage for solar district heating applications: A critical review on large-scale hot-water tank and pit thermal energy storage systems*. Applied Energy, 2019. **239**: p. 296. https://doi.org/10.1016/j.apenergy.2019.01.189

36. Lee, K.S.(2013). *Underground thermal energy storage*. In: *Underground Thermal Energy Storage. Green Energy and Technology*, 2013. Springer, London. https://doi.org/10.1007/978-1-4471-4273-7_2

37. Chang, C., et al., *Investigation on transient cooling process in a water heat storage tank with inclined sidewalls*. Energy Procedia, 2017. **142**: p. 142. https://doi.org/10.1016/j.egypro.2017.12.023

38. Novo, A.V., et al., *Review of seasonal heat storage in large basins: Water tanks and gravel–water pits*. Applied Energy, 2010. **87**(2): p. 390. https://doi.org/10.1016/j.apenergy.2009.06.033

39. Nordell, B., Snijders, A., Stiles, L., 2021. *The use of aquifers as thermal energy storage systems*, in *Advances in thermal energy storage systems*. Elsevier, pp. 111–138, 2021. https://doi.org/10.1016/b978-0-12-819885-8.00005-x

40. Gao, L., et al., *A review on system performance studies of aquifer thermal energy storage*. Energy Procedia, 2017. **142**: p. 3537. https://doi.org/10.1016/j.egypro.2017.12.242

41. Pasquinelli, L., et al., *The feasibility of high-temperature aquifer thermal energy storage in Denmark: The Gassum Formation in the Stenlille structure*. Bulletin of the Geological Society of Denmark, 2020. **68**: p. 133. https://doi.org/10.37570/bgsd-2020-68-06

42. Gao, L., et al., *Thermal performance of medium-to-high-temperature aquifer thermal energy storage systems*. Applied Thermal Engineering, 2019. **146**: p. 898. https://doi.org/10.1016/j.applthermaleng.2018.09.104

43. Reuss, M. *The use of borehole thermal energy storage systems,* in: *Advances in Thermal Energy Storage Systems,* 2021. Elsevier, pp. 139–171. https://doi.org/10.1016/B978-0-12-819885-8.00006-1

44. Liu, J., et al., *Influencing factors analysis and operation optimization for the long-term performance of medium-deep borehole heat exchanger coupled ground source heat pump system.* Energy and Buildings, 2020. **226**: p. 110385. https://doi.org/10.1016/j.enbuild.2020.110385

45. Gehlin, S. *Borehole thermal energy storage,* in: *Advances in Ground-Source Heat Pump Systems,* 2016. Elsevier, pp. 295–327. https://doi.org/10.1016/B978-0-08-100311-4.00011-X

46. Catolico, N., S. Ge, and J.S. McCartney, *Numerical modeling of a soil-borehole thermal energy storage system.* Vadose Zone Journal, 2016. **15**(1): p. 1. https://doi.org/10.2136/vzj2015.05.0078

47. Park, D., et al., *The effect of aspect ratio on the thermal stratification and heat loss in rock caverns for underground thermal energy storage.* International Journal of Rock Mechanics and Mining Sciences, 2013. **64**: p. 201. https://doi.org/10.1016/j.ijrmms.2013.09.004

48. Park, J.-W., et al., *Analysis on heat transfer and heat loss characteristics of rock cavern thermal energy storage.* Engineering Geology, 2014. **181**: p. 142. https://doi.org/10.1016/j.enggeo.2014.07.006

49. Saxena, A., et al., *A thermodynamic review on solar ponds.* Solar Energy, 2022. **242**: p. 335. https://doi.org/10.1016/j.solener.2022.07.016

50. Leblanc, J., et al., *Heat extraction methods from salinity-gradient solar ponds and introduction of a novel system of heat extraction for improved efficiency.* Solar Energy, 2011. **85**(12): p. 3103. https://doi.org/10.1016/j.solener.2010.06.005

51. Aramesh, M., et al., *A detailed investigation of the walls shading effect on the performance of solar ponds.* Environmental Progress & Sustainable Energy, 2018. **38**(3): p. e13014. https://doi.org/10.1002/ep.13014

52. Aramesh, M., F. Pourfayaz, and A. Kasaeian, *Transient heat extraction modeling method for a rectangular type salt gradient solar pond.* Energy Conversion and Management, 2017. **132**: p. 316. https://doi.org/10.1016/j.enconman.2016.11.036

53. Aramesh, M., F. Pourfayaz, and A. Kasaeian, *Numerical investigation of the nanofluid effects on the heat extraction process of solar ponds in the transient step.* Solar Energy, 2017. **157**: p. 869. https://doi.org/10.1016/j.solener.2017.09.011

54. Aramesh, M., et al., *Energy analysis and shadow modeling of a rectangular type salt gradient solar pond.* Solar Energy, 2017. **146**: p. 161. https://doi.org/10.1016/j.solener.2017.02.026

55. Goswami, Y. *Advances in Solar Energy:* Volume 16. 2017. Routledge. https://doi.org/10.4324/9781315793214

56. Dinçer, I.B. and M. Rosen, *Thermal energy storage: Systems and applications.* 3rd ed. Hoboken, NJ: Wiley, 2022.

57. Kousksou, T., et al., *Energy storage: Applications and challenges.* Solar Energy Materials and Solar Cells, 2014. **120**: p. 59. https://doi.org/10.1016/j.solmat.2013.08.015

58. Su, W., J. Darkwa, and G. Kokogiannakis, *Review of solid–liquid phase change materials and their encapsulation technologies.* Renewable and Sustainable Energy Reviews, 2015. **48**: p. 373. https://doi.org/10.1016/j.rser.2015.04.044

59. Ge, H., et al., *Low melting point liquid metal as a new class of phase change material: An emerging frontier in energy area.* Renewable and Sustainable Energy Reviews, 2013. **21**: p. 331. https://doi.org/10.1016/j.rser.2013.01.008

60. Venkitaraj, K.P. and S. Suresh, *Effects of Al_2O_3, CuO and TiO_2 nanoparticles son thermal, phase transition and crystallization properties of solid-solid phase change material.* Mechanics of Materials, 2019. **128**: p. 64. https://doi.org/10.1016/j.mechmat.2018.10.004

61. Senobar, H., M. Aramesh, and B. Shabani, *Evacuated tube solar thermal collector with enhanced phase change material thermal storage: An experimental study.* Journal of Energy Storage, 2022. **46**: p. 103838. https://doi.org/10.1016/j.est.2021.103838

62. Senobar, H., M. Aramesh, and B. Shabani, *Nanoparticles and metal foams for heat transfer enhancement of phase change materials: A comparative experimental study.* Journal of Energy Storage, 2020. **32**: p. 101911. https://doi.org/10.1016/j.est.2020.101911

63. Aramesh, M. and B. Shabani, *On the integration of phase change materials with evacuated tube solar thermal collectors.* Renewable and Sustainable Energy Reviews, 2020. **132**: p. 110135. https://doi.org/10.1016/j.rser.2020.110135

64. Zarei, M.J., et al., *The effects of fin parameters on the solidification of PCMs in a fin-enhanced thermal energy storage system.* Energies, 2020. **13**(1): p. 198. https://doi.org/10.3390/en13010198

65. Singh, R., S. Sadeghi, and B. Shabani, *Thermal conductivity enhancement of phase change materials for low-temperature thermal energy storage applications.* Energies, 2018. **12**(1): p. 75. https://doi.org/10.3390/en12010075

66. Aramesh, M. and B. Shabani, *Metal foams application to enhance the thermal performance of phase change materials: A review of experimental studies to understand the mechanisms.* Journal of Energy Storage, 2022. **50**: p. 104650. https://doi.org/10.1016/j.est.2022.104650

67. Papadimitratos, A., et al., *Evacuated tube solar collectors integrated with phase change materials.* Solar Energy, 2016. **129**: p. 10. https://doi.org/10.1016/j.solener.2015.12.040

68. Xiao, Q., et al., *Thermal conductivity enhancement of hydrated salt phase change materials employing copper foam as the supporting material.* Solar Energy Materials and Solar Cells, 2019. **199**: p. 91. https://doi.org/10.1016/j.solmat.2019.04.020

69. Li, X., et al., *Preparation of stearic acid/modified expanded vermiculite composite phase change material with simultaneously enhanced thermal conductivity and latent heat.* Solar Energy Materials and Solar Cells, 2016. **155**: p. 9. https://doi.org/10.1016/j.solmat.2016.04.057

70. Singh, P., et al., *A comprehensive review on development of eutectic organic phase change materials and their composites for low and medium range thermal energy storage applications.* Solar Energy Materials and Solar Cells, 2021. **223**: p. 110955. https://doi.org/10.1016/j.solmat.2020.110955

71. Junaid, M.F., et al., *Inorganic phase change materials in thermal energy storage: A review on perspectives and technological advances in building applications.* Energy and Buildings, 2021. **252**: p. 111443. https://doi.org/10.1016/j.enbuild.2021.111443

72. Trigui, A., *Techniques for the thermal analysis of PCM*, in In (Ed.), *Phase change materials – technology and applications* [Working Title]. London: IntechOpen, 2022. https://doi.org/10.5772/intechopen.105935

73. Park, Y.-H., et al., *Measurement of thermal conductivity of Li_2TiO_3 pebble bed by laser flash method.* Fusion Engineering and Design, 2019. **146**: p. 950. https://doi.org/10.1016/j.fusengdes.2019.01.122

74. Cabeza, L.F., *Advances in thermal energy storage systems: Methods and applications.* 2nd ed. Woodhead Publishing Series in Energy. Duxford; Cambridge, MA: Woodhead Publishing, 2020.

75. Liu, M., M.C. Lin, and C. Wang, *Enhancements of thermal conductivities with Cu, CuO, and carbon nanotube nanofluids and application of MWNT/water nanofluid on a water chiller system.* Nanoscale Research Letters, 2011. **6**(1). https://doi.org/10.1186/1556-276x-6-297

76. Rostami, S., et al., *A review of melting and freezing processes of PCM/nano-PCM and their application in Energy storage.* Energy, 2020. **211**: p. 118698. https://doi.org/10.1016/j.energy.2020.118698

77. Leong, K.Y., M.R. Abdul Rahman, and B.A. Gurunathan, *Nano-enhanced phase change materials: A review of thermo-physical properties, applications and challenges.* Journal of Energy Storage, 2019. **21**: p. 18. https://doi.org/10.1016/j.est.2018.11.008

78. Song, Z., et al., *Expanded graphite for thermal conductivity and reliability enhancement and supercooling decrease of $MgCl_2 \cdot 6H_2O$ phase change material.* Materials Research Bulletin, 2018. **102**: p. 203. https://doi.org/10.1016/j.materresbull.2018.02.024

79. Wu, Z.G. and C.Y. Zhao, *Experimental investigations of porous materials in high temperature thermal energy storage systems.* Solar Energy, 2011. **85**(7): p. 1371. https://doi.org/10.1016/j.solener.2011.03.021

80. Qu, Y., et al., *Experimental study on thermal conductivity of paraffin-based shape-stabilized phase change material with hybrid carbon nano-additives.* Renewable Energy, 2020. **146**: p. 2637. https://doi.org/10.1016/j.renene.2019.08.098

81. Sheik, M.A., et al., *Enhancement of heat transfer in PEG 1000 using nano-phase change material for thermal energy storage.* Arabian Journal for Science and Engineering, 2022. https://doi.org/10.1007/s13369-022-06810-9

82. Ghasemi, K., S. Tasnim, and S. Mahmud, *PCM, nano/microencapsulation and slurries: A review of fundamentals, categories, fabrication, numerical models and applications.* Sustainable Energy Technologies and Assessments, 2022. **52**: p. 102084. https://doi.org/10.1016/j.seta.2022.102084

83. Shchukina, E.M., et al., *Nanoencapsulation of phase change materials for advanced thermal energy storage systems.* Chemical Society Reviews, 2018. **47**(11): p. 4156. https://doi.org/10.1039/c8cs00099a

84. Jurkowska, M. and I. Szczygieł, *Review on properties of microencapsulated phase change materials slurries (mPCMS).* Applied Thermal Engineering, 2016. **98**: p. 365. https://doi.org/10.1016/j.applthermaleng.2015.12.051

85. Rajabifar, B., *Enhancement of the performance of a double layered microchannel heat-sink using PCM slurry and nanofluid coolants*, in International Journal of Heat and Mass Transfer, 2015. https://doi.org/10.1016/j.ijheatmasstransfer.2015.05.007

86. Karaipekli, A., T. Erdoğan, and S. Barlak, *The stability and thermophysical properties of a thermal fluid containing surface-functionalized nanoencapsulated PCM.* Thermochimica Acta, 2019. **682**: p. 178406. https://doi.org/10.1016/j.tca.2019.178406

87. Zauner, C., et al., *Experimental characterization and simulation of a fin-tube latent heat storage using high density polyethylene as PCM.* Applied Energy, 2016. **179**: p. 237. https://doi.org/10.1016/j.apenergy.2016.06.138

88. Bhouri, M., et al., *Honeycomb metallic structure for improving heat exchange in hydrogen storage system.* International Journal of Hydrogen Energy, 2011. **36**(11): p. 6723. https://doi.org/10.1016/j.ijhydene.2011.02.092

89. Xiao, X., P. Zhang, and M. Li, *Preparation and thermal characterization of paraffin/metal foam composite phase change material.* Applied Energy, 2013. **112**: p. 1357. https://doi.org/10.1016/j.apenergy.2013.04.050

90. Hassan, F., et al., *Recent advancements in latent heat phase change materials and their applications for thermal energy storage and buildings: A state of the art review.* Sustainable Energy Technologies and Assessments, 2022. **49**: p. 101646. https://doi.org/10.1016/j.seta.2021.101646

91. Al-Maghalseh, M. and K. Mahkamov, *Methods of heat transfer intensification in PCM thermal storage systems: Review paper.* Renewable and Sustainable Energy Reviews, 2018. **92**: p. 62. https://doi.org/10.1016/j.rser.2018.04.064

92. Aramesh, M. and B. Shabani, *Metal foam-phase change material composites for thermal energy storage: A review of performance parameters.* Renewable and Sustainable Energy Reviews, 2022. **155**: p. 111919. https://doi.org/10.1016/j.rser.2021.111919

93. Kalapala, L. and J.K. Devanuri, *Influence of operational and design parameters on the performance of a PCM based heat exchanger for thermal energy Storage – A review.* Journal of Energy Storage, 2018. **20**: p. 497. https://doi.org/10.1016/j.est.2018.10.024

94. Li, Y., S. Liu, Y. Zhang, *Experimental Study of the Heat Transfer Performance of PCMs Within Metal Finned Containers.* In: *Dincer, I., Midilli, A., Kucuk, H. (eds) Progress in Sustainable Energy Technologies Vol II.* 2014 Springer, Cham. https://doi.org/10.1007/978-3-319-07977-6_44

95. Hasse, C., et al., *Realization, test and modelling of honeycomb wallboards containing a phase change material.* Energy and Buildings, 2011. **43**(1): p. 232. https://doi.org/10.1016/j.enbuild.2010.09.017

96. Duan, J., Y. Xiong, and D. Yang, *Melting behavior of phase change material in honeycomb structures with different geometrical cores.* Energies, 2019. **12**(15): p. 2920. https://doi.org/10.3390/en12152920

97. Farid, M.M., et al., *A review on phase change energy storage: materials and applications.* Energy Conversion and Management, 2004. **45**(9–10): p. 1597. https://doi.org/10.1016/j.enconman.2003.09.015

98. Mahmoud, S., et al., *Experimental investigation of inserts configurations and PCM type on the thermal performance of PCM based heat sinks.* Applied Energy, 2013. **112**: p. 1349. https://doi.org/10.1016/j.apenergy.2013.04.059

99. Wang, Z., et al., *Paraffin and paraffin/aluminum foam composite phase change material heat storage experimental study based on thermal management of Li-ion battery.* Applied Thermal Engineering, 2015. **78**: p. 428. https://doi.org/10.1016/j.applthermaleng.2015.01.009

100. Zheng, H., et al., *Thermal performance of copper foam/paraffin composite phase change material.* Energy Conversion and Management, 2018. **157**: p. 372. https://doi.org/10.1016/j.enconman.2017.12.023

101. Ranut, P., *On the effective thermal conductivity of aluminum metal foams: Review and improvement of the available empirical and analytical models.* Applied Thermal Engineering, 2016. **101**: p. 496. https://doi.org/10.1016/j.applthermaleng.2015.09.094

102. Faghri, A., *Review and advances in heat pipe science and technology.* Journal of Heat Transfer, 2012. **134**(12). https://doi.org/10.1115/1.4007407

103. Jouhara, H., et al., *Heat pipe based systems – Advances and applications.* Energy, 2017. **128**: p. 729. https://doi.org/10.1016/j.energy.2017.04.028

104. El-Nasr, A.A. and S.M. El-Haggar, *Effective thermal conductivity of Heat pipes.* Heat and Mass Transfer, 1996. **32**(1–2): p. 97. https://doi.org/10.1007/s002310050097

105. Frazzica, A., V. Brancato, and V. Palomba, *Sorption systems for thermal energy storage,* in: *Advances in Thermal Energy Storage Systems.* 2021. Elsevier, pp. 425–451. https://doi.org/10.1016/B978-0-12-819885-8.00014-0

106. Block, T. and M. Schmücker, *Metal oxides for thermochemical Energy storage: A comparison of several metal oxide systems.* Solar Energy, 2016. **126**: p. 195. https://doi.org/10.1016/j.solener.2015.12.032

107. Yu, N., R.Z. Wang, and L.W. Wang, *Sorption thermal storage for solar Energy.* Progress in Energy and Combustion Science, 2013. **39**(5): p. 489. https://doi.org/10.1016/j.pecs.2013.05.004

108. Sunku Prasad, J., et al., *A critical review of high-temperature reversible thermochemical energy storage systems.* Applied Energy, 2019. **254**: p. 113733. https://doi.org/10.1016/j.apenergy.2019.113733

109. Desai, F., et al., *Thermochemical energy storage system for cooling and process heating applications: A review.* Energy Conversion and Management, 2021. **229**: p. 113617. https://doi.org/10.1016/j.enconman.2020.113617

110. Abedin, A.H., *A critical review of thermochemical energy storage systems*. The Open Renewable Energy Journal, 2011. **4**(1): p. 42. https://doi.org/10.2174/18763 87101004010042

111. Chen, X., et al., *State of the art on the high-temperature thermochemical energy storage systems*. Energy Conversion and Management, 2018. **177**: p. 792. https://doi.org/10.1016/j.enconman.2018.10.011

112. Tetuko, A., B. Shabani, and J. Andrews, *Passive fuel cell heat recovery using heat pipes to enhance metal hydride canisters hydrogen discharge rate: An experimental simulation*. Energies, 2018. **11**(4): p. 915. https://doi.org/10.3390/en11040915

113. Omrani, R., H.Q. Nguyen, and B. Shabani, *Thermal coupling of an open-cathode proton exchange membrane fuel cell with metal hydride canisters: An experimental study*. International Journal of Hydrogen Energy, 2020. **45**(53): p. 28940. https://doi.org/10.1016/j.ijhydene.2020.07.122

114. Tetuko, A.P., et al., *Study of a thermal bridging approach using heat pipes for simultaneous fuel cell cooling and metal hydride hydrogen discharge rate enhancement*. Journal of Power Sources, 2018. **397**: p. 177. https://doi.org/10.1016/j.jpowsour.2018.07.030

115. Saito, A., *Recent advances in research on cold thermal energy storage*. International Journal of Refrigeration, 2002. **25**(2): p. 177. https://doi.org/10.1016/s0140-7007(01)00078-0

116. Yang, L., et al., *A comprehensive review on sub-zero temperature cold thermal energy storage materials, technologies, and applications: State of the art and recent developments*. Applied Energy, 2021. **288**: p. 116555. https://doi.org/10.1016/j.apenergy.2021.116555

117. Chen, S.-L., et al., *An experimental investigation of cold storage in an encapsulated thermal storage tank*. Experimental Thermal and Fluid Science, 2000. **23**(3–4): p. 133. https://doi.org/10.1016/s0894-1777(00)00045-5

118. Hauer, A. and Z. Bayern. *Storage technology issues and opportunities*. in *Committee on Energy Research and Technology (International Energy Agency), International Low-Carbon Energy Technology Platform, Strategic and Cross-Cutting Workshop "Energy Storage–Issues and Opportunities*. 2011. https://www.iea.org/events/workshop-on-energy-storage-issues-and-opportunities

119. Kalair, A., et al., *Role of energy storage systems in energy transition from fossil fuels to renewables*. Energy Storage, 2020. **3**(1). https://doi.org/10.1002/est2.135

120. Sarbu, I. *Thermal Energy Storage*. In: *Advances in Building Services Engineering.*, 2021. Springer, Cham. https://doi.org/10.1007/978-3-030-64781-0_7

121. Yang, T., et al., *Seasonal thermal energy storage: A techno-economic literature review*. Renewable and Sustainable Energy Reviews, 2021. **139**: p. 110732. https://doi.org/10.1016/j.rser.2021.110732

122. Mazlan, M., et al., *Thermal efficiency analysis of the phase change material (PCM) microcapsules*. Sustainable Energy Technologies and Assessments, 2021. **48**: p. 101557. https://doi.org/10.1016/j.seta.2021.101557

123. Teamah, H.M. and M. Teamah, *Integration of phase change material in flat plate solar water collector: A state of the art, opportunities, and challenges*. Journal of Energy Storage, 2022. **54**: p. 105357. https://doi.org/10.1016/j.est.2022.105357

124. Nawsud, Z.A., et al., *A comprehensive review on the use of nano-fluids and nano-PCM in parabolic trough solar collectors (PTC)*. Sustainable Energy Technologies and Assessments, 2022. **51**: p. 101889. https://doi.org/10.1016/j.seta.2021.101889

125. Kalidasan, B., et al., *Phase change materials integrated solar thermal energy systems: Global trends and current practices in experimental approaches*. Journal of Energy Storage, 2020. **27**: p. 101118. https://doi.org/10.1016/j.est.2019.101118

126. Omara, A.A.M., et al., *Phase change materials (PCMs) for improving solar still productivity: A review*. Journal of Thermal Analysis and Calorimetry, 2019. **139**(3): p. 1585. https://doi.org/10.1007/s10973-019-08645-3

127. Garcia Noxpanco, M., J. Wilkins, and S. Riffat, *A review of the recent development of photovoltaic/thermal (PV/T) systems and their applications.* Future Cities and Environment, 2020. **6**(1). https://doi.org/10.5334/fce.97

128. Ebrahimi, M., M. Aramesh, and Y. Khanjari, *Innovative ANP model to prioritization of PV/T systems based on cost and efficiency approaches: With a case study for Asia.* Renewable Energy, 2018. **117**: p. 434. https://doi.org/10.1016/j.renene.2017.10.098

129. Nielsen, J.E., P.A. Sørensen. *Renewable district heating and cooling technologies with and without seasonal storage,* in: *Renewable Heating and Cooling,* 2016. Elsevier, pp. 197–220. https://doi.org/10.1016/B978-1-78242-213-6.00009-6

130. Tulus, V., et al., *Enhanced thermal energy supply via central solar heating plants with seasonal storage: A multi-objective optimization approach.* Applied Energy, 2016. **181**: p. 549. https://doi.org/10.1016/j.apenergy.2016.08.037

131. Michaelides, E.E., *Thermal storage for district cooling – implications for renewable energy transition.* Energies, 2021. **14**(21): p. 7317. https://doi.org/10.3390/en14217317

132. Luo, J., et al., *Battery thermal management systems (BTMs) based on phase change material (PCM): A comprehensive review.* Chemical Engineering Journal, 2022. **430**: p. 132741. https://doi.org/10.1016/j.cej.2021.132741

133. Tyagi, V.V. and D. Buddhi, *PCM thermal storage in buildings: A state of art.* Renewable and Sustainable Energy Reviews, 2007. **11**(6): p. 1146. https://doi.org/10.1016/j.rser.2005.10.002

134. Singh Rathore, P.K., S.K. Shukla, and N.K. Gupta, *Potential of microencapsulated PCM for energy savings in buildings: A critical review.* Sustainable Cities and Society, 2020. **53**: p. 101884. https://doi.org/10.1016/j.scs.2019.101884

135. Lu, S. et al. *A Review of PCM Energy Storage Technology Used in Buildings for the Global Warming Solution.* In: Zhang, X., Dincer, I. (eds) *Energy Solutions to Combat Global Warming. Lecture Notes in Energy, 2017.* vol 33. Springer, Cham. https://doi.org/10.1007/978-3-319-26950-4_31

136. Oró, E., et al., *Energy management and CO_2 mitigation using phase change materials (PCM) for thermal energy storage (TES) in cold storage and transport.* International Journal of Refrigeration, 2014. **42**: p. 26. https://doi.org/10.1016/j.ijrefrig.2014.03.002

137. White, A., G. Parks, and C.N. Markides, *Thermodynamic analysis of pumped thermal electricity storage.* Applied Thermal Engineering, 2013. **53**(2): p. 291. https://doi.org/10.1016/j.applthermaleng.2012.03.030

Part II

Thermal Energy
Advancements in Applications

6 Thermal Energy Storage for Solar Water Heating, Cooking, and Ponds

Navendu Misra, Saurabh Pandey,
Amritanshu Shukla, and Atul Sharma

6.1 INTRODUCTION

Sunlight energy has been utilized by a wide range of thermal energy storage (TES) systems, which have continuously evolved over the years into a wide range of fascinating products having amazing applications. Solar cookers, solar heaters, solar dryers, and a packed bed solar TES system are some of the products developed with the help of various TES systems developed. A basic categorization of the TES system is done with reference to the approach through which energy is stored in the system, and there are broadly three ways (Figure 6.1), which are specific heat-based, latent heat-based, and thermochemical-based.

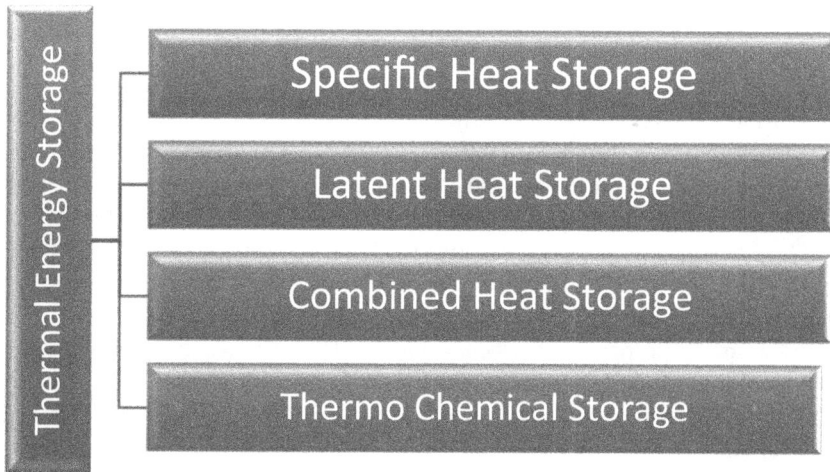

FIGURE 6.1 Types of thermal energy storage systems.

DOI: 10.1201/9781003345558-8

6.1.1 Properties of TES Materials

There are various thermochemical properties of these TES materials, which qualify them for use in different industrial conditions:

1. Melting point: The temperature at which the solid starts to melt in the form of drops is called the melting point.
2. Density: Mass per unit volume is defined as density. The amount of phase change material (PCM) used can be calculated based on the specific gravity of the material.
3. Latent heat of fusion: The heat emitted during the conversion of solid to liquid is called the latent heat of fusion.
4. Specific heat (Cp): Heat required to change a unit temperature for a unit mass is called sensible heat.
5. Thermal conductivity: This is the measurement permissibility of conducting heat through the material.
6. Supercooling: The level of cooling without freezing below the freezing point is called supercooling, so during discharging of heat, the PCMs should freeze comprehensively near their freezing point.
7. Cost and availability: The cost per unit mass should be low so that it can be used widely, and moreover it should be freely available for use.
8. Thermal stability: The materials when subjected to higher temperatures should not crumble and should be operable at a high operating range of temperature.
9. Chemical stability: The TES life depends a lot on the quality of the material being used. Generally, materials with high chemical stability last long in terms of their life and increase the overall life of a TES.

6.1.2 Specific Heat Source (SHS)–Based TES

The sensible heat–based TES stores heat energy in form of temperature difference without any change in the phase of the stored material. The heat stored is the function of the specific heat of the storage material, its mass, and the difference of temperature. Most of the specific heat–based materials available are quite inexpensive with good heat-conducting property, non-toxicity, higher density, and stability. Broadly, the TES systems with specific heat can be classified with respect to the storage medium, as shown in Figure 6.2.

Based on the properties, the behaviour of SHS varies, and accordingly their usages and merits/demerits have been defined. Table 6.1 has detailed various SHS materials with merits and demerits, and it is noteworthy that none of the SHS qualify for all the desired properties, and there is optimization whenever a selection is made.

Further, the thermochemical properties of specific heat–based TES have been consolidated by Sarbu [2] along two different categories with solid-liquid interface and solid-Solid interface.

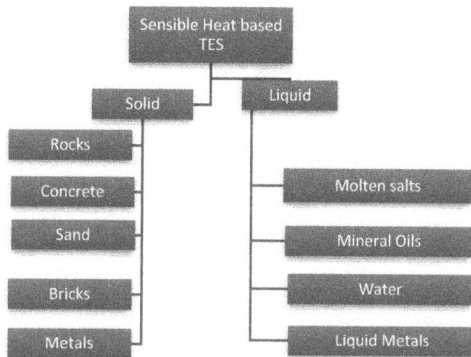

FIGURE 6.2 Types of sensible heat–based thermal storage medium.

TABLE 6.1

Experiment Setup for Solar Pond with Collectors

S. No.	Storage Medium	Operating Temperature (C°)	Merits	Demerits
1	Water	25–90	• High specific heat • Medium thermal conductivity • Simultaneous charging and discharging possible • Non-poisonous and easily available	• High vapour pressure • Corrosiveness
2	Mineral oil	Up to 400	• Less vapour pressure • Don't freeze	• Costly • Less specific heat • Less thermal conductivity
3	Molten salts	Up to 400 (Liquid)	• Higher density • Low vapour pressure	• Lower thermal conductivity • Oxidizing and corrosiveness • Freezing risk
4	Liquid melts	Above 300	• Higher thermal conductivity • Minimum vapour pressure	• Higher in cost
5	Rocks	Up to 1000	• Can perform as heat exchanger and heat storage medium • Non-toxic and easy to obtain • Mechanical resistance higher as compared with thermal cycling	• Poor thermal conductivity • Requires higher mass flowrates to store energy • Pressure drop is a major problem

(Continued)

TABLE 6.1 (*Continued*)
Experiment Setup for Solar Pond with Collectors

S. No.	Storage Medium	Operating Temperature (C°)	Merits	Demerits
6	Concrete	Up to 1000	• Medium thermal conductivity • Great mechanical strength • No container needed	• Cracks form at higher temperatures during repeated cycles • Low specific heats
7	Sand	Up to 550	• Direct use in solar receiver, cheap and non-poisonous, and simply obtainable	• Low density • Low specific heat • Lower thermal conductivity
8	Brick	Up to 400	• Non-corrosive • Non-poisonous and simply obtainable • low-cost	• Lower thermal conductivity and energy density
9	Metals	Up to 1400	• Higher thermal conductivity • Higher energy density	• High cost • Increase in weight

Source: Reprinted/adapted with permission from John Wiley and Sons by Charmala and Saini [1].

6.2 LATENT HEAT–BASED (LHS) TES

The LHS stores heat energy without any variation in temperature of stored material. The sum of heat stored is a function of the latent heat of the storage material and its mass. A latent heat storage system does not undergo temperature rise, so it is safe for operations, and with high-density material use, the size of PCM mass for TES can be reduced significantly. A low size of container material can reduce the cost of fabrication, which means LHS can be a cost-effective solution for industries requiring high volumes of such products. The disadvantage associated with the LHS is low conductivity, which leads to poor thermal transient during cyclic charging cycles. The phase change during charging and discharging cycles leads to the solidification of the phase boundary, which is the main reason behind poor heat transfer. The heat flows via convection between solid and liquid boundaries and the solidified layer results in an increase in thermal resistance. A list of latent heat materials is presented in Figure 6.3.

A latent energy source has the advantage of becoming a higher energy source with lower temperature variance between their storage and retrieval and features such as high energy density, flexible temperature range, and optimal utilization of storage material. Materials such as KNO_3, $NaNo_3$, KOH, and $MgCl_2$ are used in cascade types of heat storage. Reversible chemical storage systems are the inter-metallic compounds made by alloying of different metals by milling or melting. Melting point and enthalpy of PCM materials have been presented in Figure 6.4 by Javadi et al. [3].

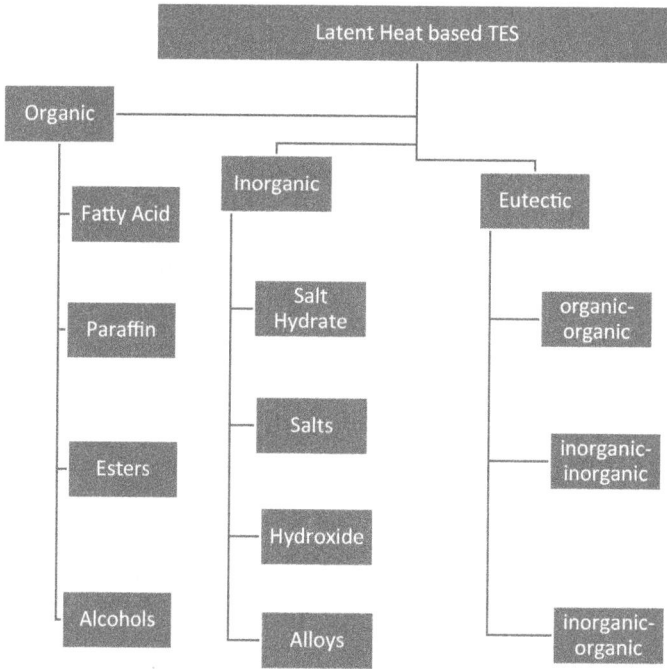

FIGURE 6.3   Type of latent heat–based thermal storage material.

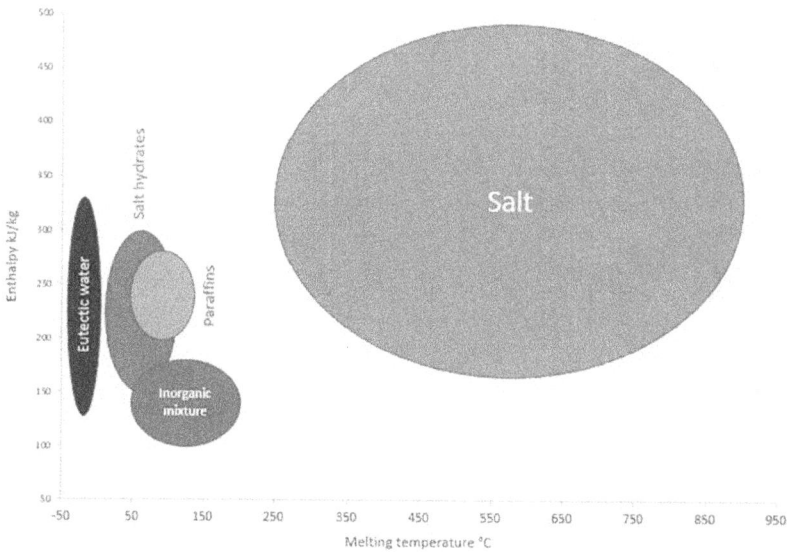

FIGURE 6.4   Melting point and enthalpy of PCM.
Source: Reprinted/adapted with permission from Elsevier, Ref. by Javadi et al. [3].

Melting temperature, enthalpy of fusion, and thermal conductivity are the major properties that help in defining the usage of any application. The melting temperature of organic PCM is generally low, so it can be used preferably for applications near room temperature, but it has low heat conductivity. Similarly, inorganic PCM has a comparatively high melting temperature but relatively better thermal conductivity. Enthalpy plays a major role by defining how much less material of PCM would be required for a desired heat transfer. More mass is required for less enthalpy PCM and vice versa. So the PCM is generally used based on the kind of end conditions and heat transfer environment. Table 6.2 has mentioned various properties of PCM material.

TABLE 6.2
PCM Used for Latent Heat

PCM	Type	Melting Point (°C)	Heat of Fusion (KJ/kg)	Thermal Conductivity (W/m-K)
Paraffin wax (C_{13}-C_{18})	Organic	32	251	0.214
Polyglycerol E600	Organic	22	127.2	0.189
Vinyl stearate	Organic	29	122	0.25
Butyl stearate	Organic	19	140	0.21
1-Dodecanol	Organic	26	200	0.169
n-octadecane	Organic	28	243.5	0.148
Palmitic acid	Organic	57.8	185.4	0.162
Capric acid	Organic	32	152.7	0.153
Caprylic acid	Organic	16	148	0.149
Propyl palmitate	Organic	10	186	N.A
KNO_3/$NaNO_3$	Inorganic	220	100.7	0.56
$CaCl_2.6H_2O$	Inorganic hydrates	29	187	0.53
$LiNO_3$/KNO_3/$NaNO_3$	Inorganic eutectic mixtures	121	310	0.52
(a) KNO_3-$LiNO_3$ (b) KNO_3/$NaNO_3$/$LiNO_3$ (c) Eutectic mixture	Inorganic eutectic mixtures	124	155	0.58
$LiNO_3$/$CaNO_3$& solar salts $NaNO_3$ and KNO_3	Inorganic	130	276	0.56
KNO_3/$NanO_3$(30%/70%)	Inorganic non-eutectic mixture	260	305	0.54
Capric acid (65.12%) &Lauric acid (34.88%)	Organic eutectic mixture	19.67	126	0.21
Polymethylmethacrylate (PMMA)/Capric-Stearic acid mixture	Organic eutectic mixture	21.37	116.25	0.15

Source: Reprinted/adapted with permission from Elsevier. Ref. by Dinker et al. [4].

Dinker et al. [4] studied a comparison of various PCMs with their properties, and it can be construed that the selection of PCM is a trade-off due to associated merits and demerits with respect to end use. A PCM with low thermal conductivity and large variation in volume cannot be used for a small area and quick heat removal; however, a eutectic with higher enthalpy per unit volume and good thermal conductivity can be used. Generally, paraffin is preferred because of its high enthalpy and high heat stability and is consistent with all the metals, but precaution is taken due to its flammable nature. Salt hydrate has high corrosion issues along with supercooling, and otherwise is one of the best low-cost available segments of material for use. Metals have high conductivity and good enthalpy of fusion, but due to higher cost and weight, they are not used prominently. Fatty acids have quite a high cost and flammability at higher temperatures, but due to sharp phase transformation, they are used in solar passive heating appliances. Eutectics have a wide range of use as they are a mix of different organic and inorganic ingredients with properties of both shared as per ratio. Charmala and Saini [1] studied certain latent heat materials for their major thermophysical properties and consolidated features with merits and demerits; they are presented in Table 6.3. Now towards the improvement of features and overcoming the demerits,

TABLE 6.3
Merits and Demerits of Latent Heat Materials

	Paraffin	Non-Paraffin (fatty acid)	Salt Hydrates	Metallic	Eutectics
Phase change temperature	-12 to 71°C	7.8–187°C	11–120°C	30–96°C	4–93°C
Latent heat of fusion	190–260 kJ/kg	130–250 kJ/kg	100–200 kJ/kg	25–90 kJ/kg	100–230 kJ/kg
Features	• Phase transition temperature and enthalpy of fusion depends on chain length • Common PCM	• Broad range of melting point and enthalpy of fusion	• Alloy of water and inorganic salt	• High weight and high cost so less used	• Two or more materials mix • Melting and freezing without phase segregation
Cost	Expensive	Twice as costly as paraffin	Low	High	High
Merits	• No phase segregation and supercooling • Thermodynamically stable • High enthalpy of vaporization • Can be used with metal containers	• Sharper phase transformation	• Easily obtainable • High density • High thermal conductivity • Change in volume is less than that of others • Sharper melting point	• High enthalpy of fusion • High thermal conductivity	• High enthalpy of fusion per unit volume • High thermal conductivity

(Continued)

TABLE 6.3 (*Continued*)
Merits and Demerits of Latent Heat Materials

	Paraffin	Non-Paraffin (fatty acid)	Salt Hydrates	Metallic	Eutectics
Demerit	• Lower thermal conductivity • Do not have well-defined melting temperatures • Flammable • Larger variation in volume	• Mildly corrosive • Flammable at high temperature	• The problem of supercooling • Corrosion to the metallic container	• Lower enthalpy of fusion • Lower specific heat	• Low enthalpy of fusion per unit weight

Source: Reprinted/adapted with permission from John Wiley and Sons, Ref. by Charmala et al. [1].

various measures are taken. One of the measures is the mixing of specific property triggering components such as nanofibers, eutectics, and other PCM. This method has been widely explored and experimented for getting desired properties.

The selection of PCM depends on various features and properties such as high density, low vapor change, and low volume change during phase change. These properties help in reducing the mass of PCM required for any application. Similarly, the PCM selected should be stable, non-toxic, and non-flammable so that it can be used for high-temperature applications and can be safe for the environment. PCM should have high latent heat along with a melting temperature near that of the ambiance of end use. Further, there should be the least super cooling and crystallization so as to ensure proper phase change during charging and discharging cycles. Also, last but not least, the PCM should be cost-effective and easily available for anyone to use. It is always challenging to find a PCM that meets all criteria and specifications, and therefore optimization is done with respect to certain property that is preferably required for the end use.

The main challenges seen generally in using PCM are low thermal conductivity, phase separation, supercooling, and flammability. While going through charging and discharging cycles, the homogeneity of PCM should be maintained at the temperature of use. So generally while selecting any PCM it is ensured that phase change occurs in a well-reversible manner and there is no separation of phases. Thermal conductivity is a main challenge that is mitigated by the use of many additives in the form of graphite, metals, and carbon-based nanomaterials. Graphite can be added in the form of graphite flakes, graphite powder, and expanded graphite. Similarly, nanomaterials can be added for metal oxides, silver nanowires, carbon-based, and simple metallic. Carbon-based nanomaterials are preferred because of their lower weight. Form-stable PCM composites can also be formed wherein the form of the PCM does not change at the melting temperature. Various types of additives have been shown for different carbon-based nanomaterials, as shown by Olabi et al. [5] in Figure 6.5.

Although many additives enhance thermal conductivity performance in a PCM, such as metal oxides, carbon-based nanomaterials have the specific advantage of less weight, which reduces the overall mass of PCM. These carbon nanomaterials have been classified from zero dimension to three dimensions, which is shown in Figure 6.5.

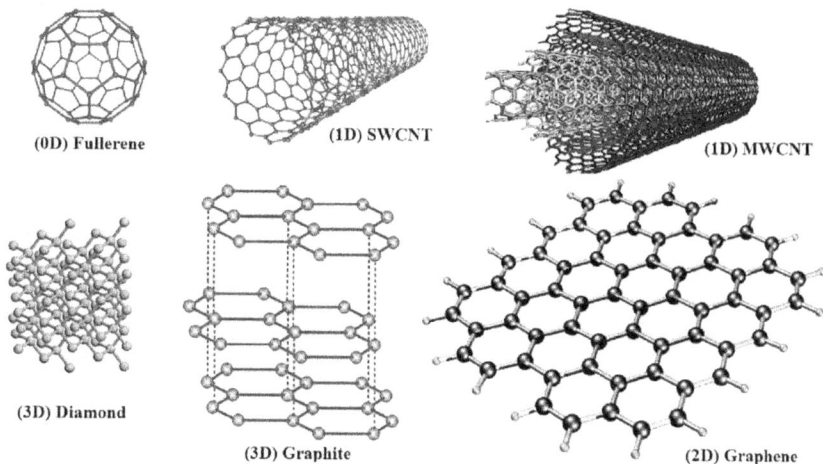

(0D) Fullerene

(1D) SWCNT

(1D) MWCNT

(3D) Diamond

(3D) Graphite

(2D) Graphene

FIGURE 6.5   Different carbon nanomaterials used as an additive in PCMs.

Source: Reprinted/adapted with permission from Elsevier, Ref. by Olabi et al. [5].

Guruprashad et al. [6] explained the technique wherein the PCM is encapsulated at the microscopic level, and it is surrounded by filler, enabling higher heat transfer and reinforcement to the matrix called microencapsulation. Comparatively, manufacturing microencapsulated PCM is an expensive process and requires precise manufacturing controls. There is wide use of the microencapsulation technique, and it is prominently used in buildings and industries. Various types of polymers and monomers are added to the matrix of PCM by either mixing in colloidal mixtures or emulsification of droplets with surfactants or condensation reaction; however, broadly the microencapsulation techniques studied are as follows:

- Concertation
- Suspension polymerization
- Emulsion polymerization
- Polycondensation
- Polyaddition

Furthermore, thermal transmission can be increased through various methods, and one of the most prominent methods is geometric modifications. Fins are added to the main body of the vessel for transferring heat from PCM to the container. There are different ways of attaching a fin to a body; it can be inside and outside or on the cover of the vessel. Dinker et al. [4] deliberated the performance of fins attached to a tube containing heat-carrying fluid passing into PCM, Figure 6.6 (e), and found that better heat transfer can be achieved with the fins.

A combination of SHS and LHS has been studied and designed, wherein both approaches for thermal energy systems have been utilized. As an example, a model studied by Ahmed et al. [7] has been detailed in Figure 6.7. This system contains

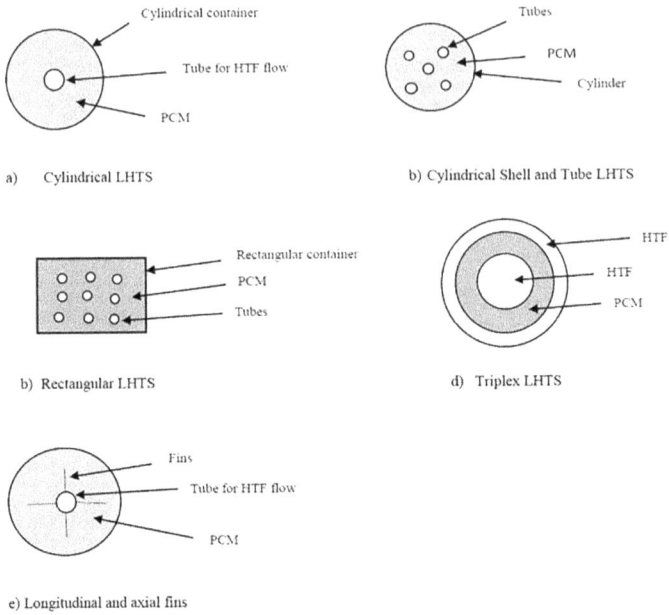

FIGURE 6.6   Different ways of attachment of fins.

Source: Reprinted/adapted with permission from Elsevier, Ref. by Dinker et al. [4].

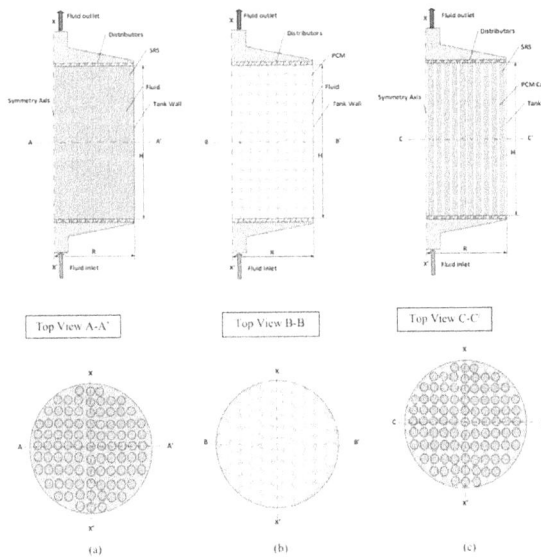

FIGURE 6.7   Schematic of three types of TES: (a) sensible heat, (b) latent heat, and (c) combined system.

Source: Reprinted/adapted with permission from Elsevier, Ref. by Ahmed et al. [7].

PCM encapsulated between cylindrical structures containing sensible heat material combined in a cylinder. The hybrid usage of such systems can be applied widely, to gain advantages of both latent and sensible heat storage materials.

6.3 THERMOCHEMICAL SYSTEMS

Thermochemical systems (TCS) are widely used thermal storage systems due to the broad advantages associated with them. These materials have high energy density in a compact volume added with no heat loss, and further can be stored for longer periods. However, the major disadvantage of the TCS is the high price of capital involved. Various research and studies have been conducted across the world recently. Various facets involved in the study of TCS has been defined by Pardo et al. [8] in Figure 6.8.

Various TCS at medium and high temperatures have been mentioned in Figure 6.9 by Pardo et al. [8]

Thermochemical energy storage materials are classified as insorption and chemical reactions type. The sorption accounts wherein absorbent is absorbed in a liquid or solid during the reaction, so at least two components are needed for starting sorption. The chemical reaction is a high-temperature phenomenon as compared to sorption, but the basic principle remains the same. The pairs tested in one of the studies by Prieto et al. [9] in prototype under actual conditions have been mentioned in Figure 6.10.

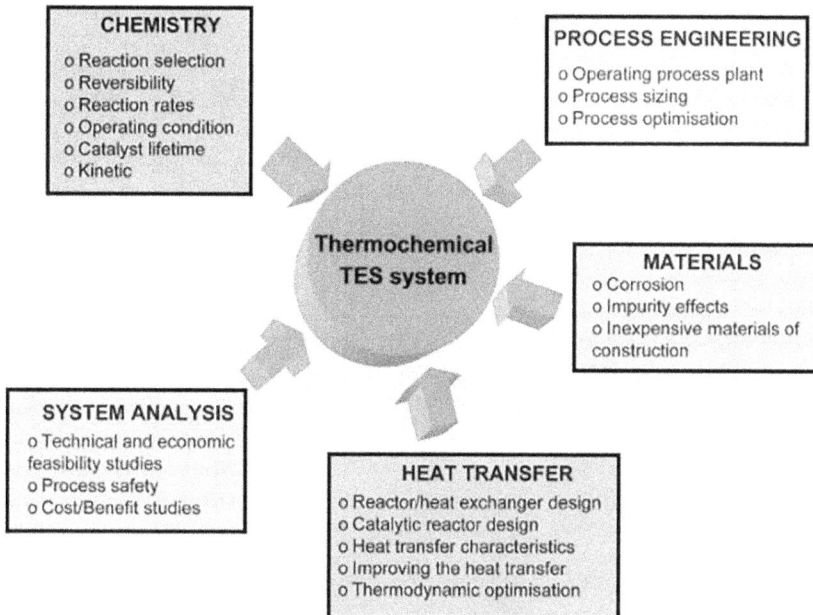

CHEMISTRY
o Reaction selection
o Reversibility
o Reaction rates
o Operating condition
o Catalyst lifetime
o Kinetic

PROCESS ENGINEERING
o Operating process plant
o Process sizing
o Process optimisation

Thermochemical TES system

MATERIALS
o Corrosion
o Impurity effects
o Inexpensive materials of construction

SYSTEM ANALYSIS
o Technical and economic feasibility studies
o Process safety
o Cost/Benefit studies

HEAT TRANSFER
o Reactor/heat exchanger design
o Catalytic reactor design
o Heat transfer characteristics
o Improving the heat transfer
o Thermodynamic optimisation

FIGURE 6.8   Various facets of a thermochemical TES system.

Source: Reprinted/adapted with permission from Elsevier, Ref. by Pardo et al. [8].

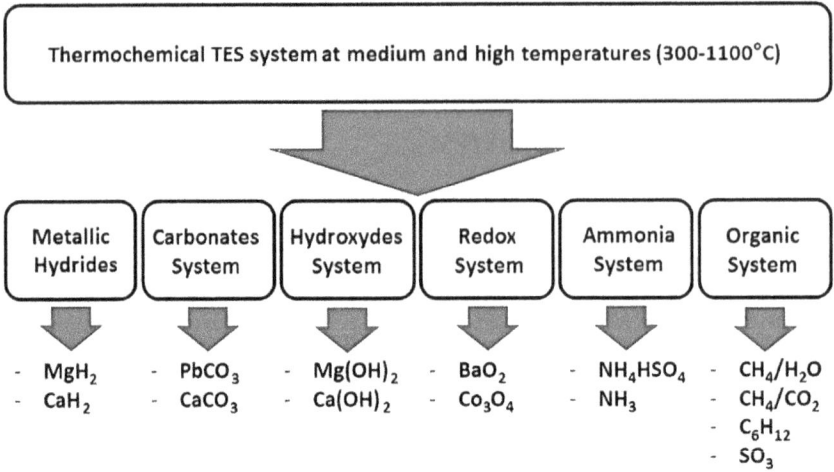

FIGURE 6.9 Various TCS at medium and high temperatures.

Source: Reprinted/adapted with permission from Elsevier, Ref. by Pardo et al. [8].

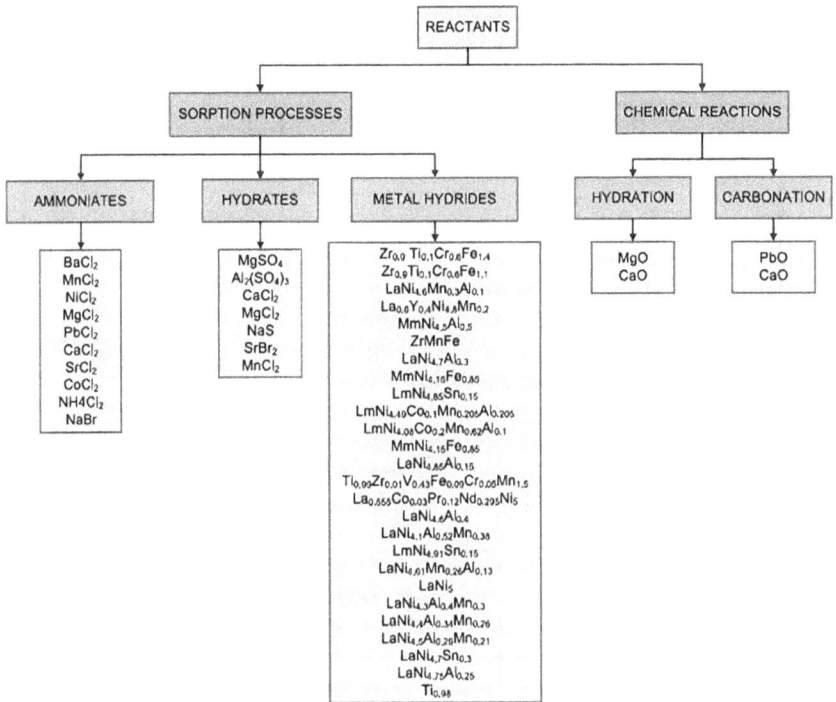

FIGURE 6.10 Pairs tested under actual conditions.

Source: Reprinted/adapted with permission from Elsevier, Ref. by Prieto et al. [9].

6.4 APPLICATION OF TES IN SOLAR WATER HEATING SYSTEMS (SWHS)

A solar water heater (SWH) is a unique system that utilizes solar thermal radiation to heat water for domestic or industrial applications. Unlike conventional heaters that rely on electric current or some fossil fuel, SWH use solar radiation to heat water directly or indirectly. The components of a typical SWH consist of panels, an overhead tank, and a heat exchanger. These panels, also known as collectors, are generally commissioned on the roof and collect solar energy, which is transferred to water in the storage tank through a heat exchanger. Heated water is kept in the tank and can be used as needed. SWH are eco-friendly, cost-effective, and can significantly reduce energy costs compared to traditional water heating systems.

Types of SWH are direct and indirect.

1. Direct SWH: In this type, water is directly heated by the sun's energy through solar panels. The water circulates through the panels and is heated as it flows, then transferred into a tank.
2. Indirect SWH: This type uses a heat exchanger for the transmission of heat from the solar panels to water in the storage tank. A heat-transmission fluid is intermediate, which transfers absorbed heat from the panel to water in the tank.

Additionally, there are also two subtypes of SWH: active and passive.

1. Active SWH: A pump is used to circulate the water or heat-transfer fluid between the panels and the tank.
2. Passive SWH: This type relies on natural circulation, using gravity and thermal convection, to circulate the water or heat-transfer fluid.

Both direct and indirect SWH can be active or passive, with their advantages and disadvantages. The SWH is chosen based on factors such as climate, water temperature requirements, and the user's budget and preferences.

Hohne et al. [10] reviewed SWH technologies and provided highlights against various types of heating technologies. An example of an SWHS along with a flat-plate solar collector is shown in Figure 6.11.

Hohne et al. [10] also made a detailed technoeconomic comparison of various types of technologies being used in SWHS, which has been elaborated in Table 6.4.

Various experiments have been conducted for choosing an optimum PCM for application in SWHS. Kee et al. [11] presented the studies taken up with different types of PCM and the quantity used along with the performance measured.

Wei et al. [12] presented a SWHS to arrest variation of solar radiation during daytime by use of PCM. The oscillating tubes were used in the experiment, and performance was evaluated for a case with and without PCM. The diagram of the experiment is shown in Figure 6.12.

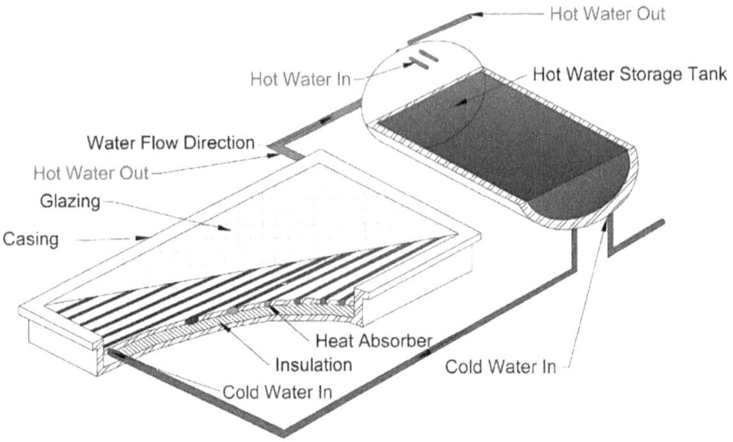

FIGURE 6.11   Solar water heater with flat-plate collector.

Source: Reprinted/adapted with permission from Elsevier, Ref. by Hohne et al. [10].

TABLE 6.4
Technoeconomic Comparison of Solar Water Heater

Technology	Approximate Initial Investment per Thermal kW(USD)	Average Life Expectancy (Years)	Payback Period (Years)
Electric storage tank water heater (WH)	60–90	10	—
Electric tankless WH	184–199	5–8	5–7
SWH	165–273	8–20	3–6
Heat-pump WH	180–282	5–8	5–7
Gas-fired tankless WH	63–88	6–8	5–6
Oil-fired WH	500–2000	10–15	6–14
Biomass WH	1350–2300	10–12	9–15

Source: Reprinted/adapted with permission from Elsevier, Ref. by Hohne et al. [10].

TABLE 6.5
PCM Quality

Category	PCM (Melting Point)	Quantity	Performance
Water storage tank	Paraffin and stearic acid (49–53)	3 kg in 150 L water tank	Water temperature within the PCM melting range for an additional 6–12 hours on average compared to when it isn't PCM.

	Paraffin wax (52)	38 Kg in 101.4 L water tank	Water at 30°C for 11 hours longer than without PCM
	NaOAc.3H$_2$O graphite (57.31–60.75)	~57 L in 105.8 L water tank	Water at 50°C for 6 hours longer than without PCM (simulation result)
	Lauric acid (57–60)	Not available	Water at 43°C for three hours (simulation result)
	Paraffin wax (56.3)	49.4 L (0.039 kg)	Continually emits heat for five hours after dark
	Paraffin wax (55)	5.2 L in 9.5 L water tank	Increases the time to provide hot water by up to 25% when solar radiation is approximated by water at 80°C constant temperature
Solar collector	Paraffin wax (46.70)	42.4 Kg in 1.248 m² solar collector	Collector plate temperature above 40°C for 4 hours after the afternoon's reduction in solar radiation
	Paraffin wax nanocomposite with 1.0 wt% nano-Cu particles (58.4–59.6)	28 kg for 1 m³ solar collector	Water at a temperature of at least 50°C for one hour longer than without PCM
	Ba(OH)$_2$.8H$_2$O (81.83)	14.2 L for 1.272 m³ solar collector	Water tank temperature rising for two hours after the afternoon's reduction in solar radiation
	Dual PCM Tritriacontane (72 °C) and erythritol (118°C)	6 kg of Tritriacontane and 4.2 kg of erythritol for 0.947 m² solar collector	Water at a temperature above 40°C for two hours longer than without PCM

Source: Reprinted/adapted with permission from Elsevier, Ref. by Kee et al. [11].

FIGURE 6.12   Water heating system with oscillating heat pipe.

Source: Reprinted/adapted with permission from Elsevier, Ref. by Wei et al [12].

6.5 APPLICATION OF TES IN SOLAR COOKER

Mawire et al. [13], in 2008, used an oil-pebble bed for thermal energy storage and predicted its performance by calculating energy and exergy balance using two different charging methods. The variation in the study has been made based on constant and variable flowrates, which are done for controlling charging temperature. It was found that for low solar radiation, the rates of exergy and energy are higher when the constant-flowrate method is used (Figure 6.13).

An experiment on the use of several ionic liquids (ILs) was carried out by D. Bhatt et al. [14] in 2012. Seven ILs based on tetrabutyl ammonium cation [N+4444] were used for application in inorganic anions. The efficiency of various ILs being used as PCM was determined by using thermophysical characteristics. The ILs containing bromide, iodate, and nitrate performed more efficiently than the thiocyanate, bromate, and hexafluorophosphate moieties. Himanshu et al. [15] used 16 different combinations of sand, iron grits, stone pebbles, iron balls, and acetamide for TES, along with a parabolic dish collector.

The experiment setup consists of two concentric vessels of TES, wherein a solar cooker was positioned, and different combinations of materials were stored in between in different experiments. The energy stored by the outer TES material was transferred to the inner pot, due to which cooking during off-sunshine hours was successful.

Gagandeep et al. [16] used a parabolic trough collector to focus solar radiation on an absorber tube with thermal energy storage, and the collected heat was relocated through a thermo siphon to the solar cooker. The working fluids used separately were water and thermal oil (engine oil), and acetanilide was used as the TES material in

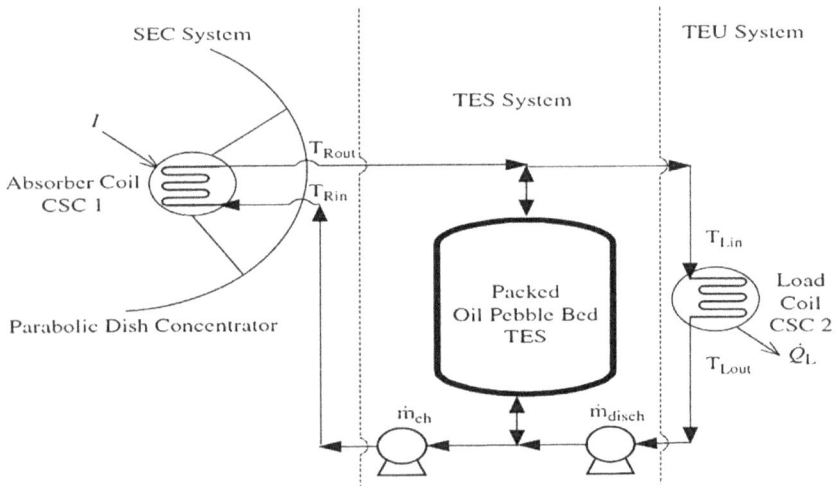

FIGURE 6.13   A schematic diagram of an indirect solar TES and cooking system.

Source: Reprinted/adapted with permission from Elsevier, Ref. by Mawire et al. [13].

the solar cooker. Two times cooking was achieved, and the rate of evening cooking was noticed more with respect to noon cooking. Also by using thermal oil, the heat storage increased by 19.45–30.38% as compared with water. Harvinder et al. [17] compared the performance of two dissimilar heat transfer fluids (HTF) used in solar cookers and the influence of gate valves on the discharge process of PCM. The evacuated tube collector transferred heat by the thermosiphon process through connecting pipes, and the working fluid used was water and thermal oil (engine oil). TES used was commercial-grade acetanilide in the solar cooker.

Various studies for modeling heat transfer through PCM have been carried out, along with a mathematical solution of the enthalpy and exergy. Tarwidia et al. [18] model heat transfer using heat conduction equations in the cylindrical domain by using various PCMs. The best performance among PCMs was found with erythritol during the first 2.5 hours of thermal performance, and magnesium chloride hexahydrate showed maintaining temperature in the range of 110–116.7°C. Results achieved with erythritol were better as compared to others, as it can cook for 3.5 hours under a load of 10 kg of water until it reached 100°C.

Nayak et al. [19] studied a joint case of a developed in-house water heating system with evacuated tubes, solar collector, and storage tank. The use of solar cookers for evening cooking along with water heating systems has been signified by this study. The PCM utilized was stearic acid and acetanilide. The results achieved in the study were cooker efficiency of 30% and collector efficiency of 60–65%, which is satisfactory in terms of establishing the use of solar cookers for evening food.

Use of multiple reflectors has been shown by Talbi et al. [20] in the experimentation of a solar cooker of box type with three reflectors. Salt was used in the base of the cooker as the thermal storage material to improve the efficiency of the solar cooker. The temperature achieved was 140°C, power 225.49W, and efficiency of 93%. This experiment successfully established the usage of a simple box-type construction solar cooker with good efficiency for cooking.

In one of the experiments conducted by Sanjeev et al. [21], water is used as a working fluid in an evacuated tube collector with PCM. Heat given by water is stored in acetanilide during sunshine and is transferred to the cooking vessel entirely throughout the day up till late evening time. To enhance performance, a reflector is placed beneath the evacuated tube collector, which also gave better results as compared to without a reflector. The maximum increase in the temperature of the working fluid, PCM, and cooking vessel with reflector is 18.3%, 20.4%, and 20.8%, respectively.

Babu et al. [22] conducted an experiment wherein a storage tank was used for the collection of heated fluid from a heated working fluid, wherein solar radiations were focused from the parabolic trough collector (PTC) with the manual tracking mechanism (TM). Steric acid is used as the PCM in the storage tank to act as a reservoir so that constant heat is produced even in the case of irregular sunshine. Two working fluids were used, and it was found that engine oil has greater energy extracted than water.

In the latest work, Papede and Patil [23] also addressed the challenge of two meals being cooked by solar cookers, i.e., during the daytime and off-sunshine hours. They developed a system that works during the day as well as at nighttime by using a parabolic collector for storing heat for cooking during the nighttime and during daytime cooking under the impact of available sunshine. It was found that using binary salt as the PCM improved the performance of cooking, and it provided a huge backup for cooking during off-sunshine hours. It was observed by the experiment that two food materials could be cooked simultaneously by this method.

6.6 APPLICATION OF TES IN SOLAR POND

A solar pond is a unique type of artificial pond that uses solar energy to store and distribute heat. It is essentially a large, shallow pond designed to harness the sun's energy and store it in the form of thermal energy in the water. The pond consists of three layers:

1. Upper Convective Zone (UCZ): a shallow upper layer that absorbs the sun's energy and heats up. It protects the solar pond from disturbances outside such as wind, waves, dust, etc.
2. Non-Convective Zone (NCZ): this is a middle layer with a higher salt concentration that acts as a heat barrier for the lower layer. There is no convection transition in this layer.
3. Lower Convective Zone (LCZ): This is a deep bottom layer with the highest salt concentration that serves as the heat storage area.

Ines et al. [24] presented a study on salt gradients in a solar pond. A typical solar pond has been presented in Figure 6.14, where all three zones have been represented, along with the angle of incident radiation (θ_i) and (θ_r) angle of reflected radiation.

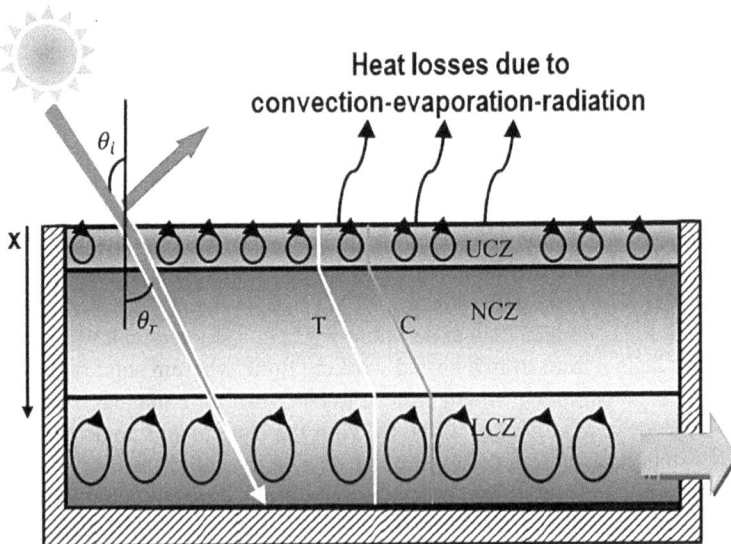

FIGURE 6.14   Solar pond typical schematic view.

Source: Reprinted/adapted with permission from Elsevier, Ref. by Ines et al. [24].

Solar ponds are typically used for industrial and agricultural applications, such as heating greenhouses, drying crops, and generating electricity. They can also be used for recreational purposes, such as swimming and fishing.

One of the main advantages of solar ponds is their low cost, as they do not require any fuel or energy input other than the sun's energy. They are also environmentally friendly, as they do not produce any greenhouse gases or other harmful pollutants.

Solar ponds are typically constructed in areas with high levels of solar insolation, such as desert regions, and are most effective in warm, sunny climates. The cost of construction and maintenance can vary depending on the size and complexity of the pond, but it is generally considered a cost-effective option compared to other renewable energy sources.

Overall, solar ponds offer a unique and environmentally friendly solution for harnessing and storing solar energy and have the potential to be a valuable source of renewable energy in areas with high levels of solar insolation.

Alcaraz et al. [25] studied increasing heat in solar ponds by using solar thermal collectors and found that the efficiency of a solar pond can be increased significantly. Figure 6.15 shows the experiment setup.

There are two main types of solar ponds: salt gradient solar ponds (SGSP) and shallow thermal solar ponds (STSP):

1. Salt Gradient Solar Ponds (SGSP): These deep ponds use a stratified salt solution to trap the sun's energy and store it as heat. The bottom layer of

FIGURE 6.15   Experiment setup for solar pond with collectors.

Source: Reprinted/adapted with permission from Elsevier, Ref. by Alcaraz et al. [25].

the pond has the highest salt concentration and acts as the heat storage area. The middle layer has a slightly lower salt concentration and acts as a heat barrier, while the top layer has the lowest salt concentration and is exposed to the sun's energy.

2. Shallow Thermal Solar Ponds (STSP): These shallow ponds use a shallow layer of water and a black plastic sheet to trap the sun's energy and store it as heat. The black plastic sheet absorbs the sun's energy and heats the water, which can then be used for various applications.

Both types of solar ponds have their advantages and disadvantages, and the choice of which type to use depends on factors such as the climate, the size and location of the pond, and the specific application for which the heat is intended. Regardless of the type, solar ponds are a unique and effective way to harness and store solar energy and have the potential to be a valuable source of renewable energy in many parts of the world.

6.7 CONCLUSION

Each system has some merits and demerits, but thermochemical energy storage systems are the typically used TES, due to their high energy density and long storage time. However, the cost is one of the factors that is optimized while selecting TES for some specific end use. Phase change materials are one of the prospective domains in which lot of studies are being carried out to provide a low-cost, reliable solution that can address the thermal energy storage challenge.

REFERENCES

1. Charmala, S.; Saini, R. (2020). Review on solar thermal energy storage technologies and their geometrical configurations. *International Journal of Energy Research*. https://doi.org/10.1002/er.5143
2. Sarbu, I. (2018). A comprehensive review of thermal energy storage. *Sustainability*, *10*, 191. https://doi.org/10.3390/su10010191.
3. Javadi, F.S.; Metselaar, H.S.C.; Ganesan, P. (2020). Performance improvement of solar thermal systems integrated with phase change materials (PCM), a review. *Solar Energy*, *206*, 330–352. https://doi.org/10.1016/j.solener.2020.05.106
4. Dinker, A.; Agarwal, M.; Agarwal, G.D. (2015). Heat storage materials, geometry and applications: A review. *Journal of the Energy Institute*, S1743967115301215. https://doi.org/10.1016/j.joei.2015.10.002
5. Olabi, A.G.; Force, T.W.; Elsaid, K.; Sayed, E.T.; Ramadan, M.; Atiqure Rahman, S.M.; Abdelkareem, M.A. (2021). Recent progress on carbon-based nanomaterial for phase change materials: Prospects and challenges. *Thermal Science and Engineering Progress*. https://doi.org/10.1016/j.tsep.2021.100920
6. Alva, G.; Liu, L.; Huang, X.; Fang, G. (2017). Thermal energy storage materials and systems for solar energy applications. *Renewable and Sustainable Energy Reviews, 68*, 693–706. https://doi.org/10.1016/j.rser.2016.10.021
7. Ahmed, N.; Elfeky, K.E.; Lu, L.; Wang, Q.W. (2019). Thermal and economic evaluation of thermocline combined sensible-latent heat thermal energy storage system for medium temperature applications. *Energy Conversion and Management, 189*, 14–23. https://doi.org/10.1016/j.enconman.2019.03.040

8. Pardo, P.; Deydier, A.; Anxionnaz-Minvielle, Z.; Rougé, S.; Cabassud, M.; Cognet, P. (2014). A review on high temperature thermochemical heat energy storage. *Renewable and Sustainable Energy Reviews, 32*, 591–610. https://doi.org/10.1016/j.rser.2013.12.014

9. Prieto, C.; Cooper, P.; Fernández, A.I.; Cabeza, L.F. (2016). Review of technology: Thermochemical energy storage for concentrated solar power plants. *Renewable and Sustainable Energy Reviews, 60*, 909–929. https://doi.org/10.1016/j.rser.2015.12.364

10. Hohne, P.A.; Kusakana, K.; Numbi, B.P. (2019). A review of water heating technologies: An application to the South African context. *Energy Reports, 5*, 1–19. https://doi.org/10.1016/j.egyr.2018.10.013.

11. Yiing Kee, S.; Munusamy, Y.; Seng Ong, K. (2017). Review of solar water heaters incorporating solid-liquid organic phase change materials as thermal storage. *Applied Thermal Engineering.* https://doi.org/10.1016/j.applthermaleng.2017.12.032

12. Wu, W.; Dai, S.; Liu, Z.; Dou, Y.; Hua, J.; Li, M.; Wang, X.; Wang, X. (2018). Experimental study on the performance of a novel solar water heating system with and without PCM. *Solar Energy, 171*, 604–612. https://doi.org/10.1016/j.solener.2018.07.005

13. Mawire, A.; McPherson, M.; van den Heetkamp, R.R.J. (2008). Simulated energy and exergy analyses of the charging of an oil–pebble bed thermal energy storage system for a solar cooker. *92*(12), 1668–1676. https://doi.org/10.1016/j.solmat.2008.07.019

14. Bhatt, V.D.; Gohil, K. (2014). Performance evaluation of solar cooker using some [N+4444] based ionic liquids as thermal energy storage materials, 0976–3961. *International Association of Advanced Materials.* https://aml.iaamonline.org/article_14361_d99c09abc17b95b9021e5022ba4034b9.pdf; https://doi.org/10.5185/amlett.2012.9420

15. Agrawal, H.; Yadav, V.; Kumar, Y.; Yadav, A. (2014). Comparison of experimental data for sensible and latent heat storage materials for late-evening cooking based on a dish-type solar cooker. *International Journal of Energy for a Clean Environment, 15*(1), 47–72. https://doi.org/10.1615/interjenercleanenv.2015013791

16. Saini, G.; Singh, H.; Saini, K.; Yadav, A. (2015). Experimental investigation of the solar cooker during sunshine and off-sunshine hours using the thermal energy storage unit based on a parabolic trough collector. *International Journal of Ambient Energy*, 1–12. https://doi.org/10.1080/01430750.2015.1023836

17. Singh, H.; Gagandeep; Saini, K.; Yadav, A. (2015). Experimental comparison of different heat transfer fluid for thermal performance of a solar cooker based on evacuated tube collector. *Environment, Development and Sustainability, 17*(3), 497–511. https://doi.org/10.1007/s10668-014-9556-3

18. Tarwidi, D.; Murdiansyah, D.T.; Ginanjar, N. (2016). Performance evaluation of various phase change materials for thermal energy storage of a solar cooker via numerical simulation. *International Journal of Renewable Energy Development, 5*(3), 199. https://doi.org/10.14710/ijred.5.3.199-210

19. Nayak, N.; Abu Jarir, H.; Al Ghassani, H. (2016). Solar cooker study under Oman conditions for late evening cooking using stearic acid and acetanilide as PCM materials. *Journal of Solar Energy, 2016*, 1–6. https://doi.org/10.1155/2016/2305875

20. Talbi, S.; Kassmi, K.; Lamkaddem, A.; Malek, R. (2018). Design and realization of a box type solar cooker with thermal storage dedicated to the rural regions of the oriental district. *Journal of Materials and Environmental Science, 9*(4), 1266–1284. https://doi.org/10.26872/jmes.2017.9.4.137

21. Kumar, S.; Kumar, A.; Yadav, A. (2018). Experimental investigation of a solar cooker based on evacuated tube collector with phase change thermal storage unit in Indian climatic conditions. *International Journal of Renewable Energy Technology, 9*(3), 310. https://doi.org/10.1504/IJRET.2018.093007.

22. Babu Sasi Kumar, S.; Chinna Pandian, M. (2019). Experimental analysis of a solar cooker with a parabolic trough enhanced with PCM based thermal storage. *IOP Conference Series: Materials Science and Engineering, 574*, 012019. https://doi.org/10.1088/1757-899X/574/1/012019

23. Papade, C.V.; Kanase-Patil, A.B. (2022). *Binary salt phase change material for concentrated solar cooker: storage and usages*. Taylor and Francis, pp. 6698–6708. https://doi.org/10.1080/15567036.2022.2096724

24. Ines, M.; Paolo, P.; Roberto, F.; Mohamed, S. (2019). Experimental studies on the effect of using phase change material in a salinity-gradient solar pond under a solar simulator. *Solar Energy, 186*, 335–346. https://doi.org/10.1016/j.solener.2019.05.011

25. Alcaraz, A.; Montalà, M.; Valderrama, C.; Cortina, J.L.; Akbarzadeh, A.; Farran, A. (2018). Increasing the storage capacity of a solar pond by using solar thermal collectors: Heat extraction and heat supply processes using in-pond heat exchangers. *Solar Energy, 171*, 112–121. https://doi.org/10.1016/j.solener.2018.06.061

7 Solar Thermal Energy for Drying and Other Small Scales of Industrial Applications

Saurabh Pandey, Abhishek Anand, Amritanshu Shukla, and Atul Sharma

7.1 INTRODUCTION

Cooking food for survival is one of the first applications of fire as thermal energy after its discovery, which was a milestone in human evolution. The availability of thermal energy precedes the existence of living beings and is easily available in nature. There is a requirement for minimum ambient temperature for living beings to survive, which is why there is a strong need for thermal energy in our lives. The available thermal energy from the sun fulfills the requirement of favorable thermal conditions that are needed for survival on Earth, and this solar thermal energy varies at different locations on Earth. In the last two to three decades, the global population has increased rapidly, and energy consumption per capita has also increased, except in 2009 and 2020, following the global financial crisis and coronavirus pandemic. According to the Statista report published in 2021, the global primary energy consumption in 2019 is 581.5 exajoules, of which fossil-based or conventional fuels constitute a major part, but their reserves are limited, small, and non-renewable (Korachagaon and Bapat, 2012). From primary supply to final consumption, a large amount of energy is ruined, and a huge amount of greenhouse gases are emitted. The increase in the level of greenhouse gases due to the excessive use of fossil fuels to fulfill energy needs contributes to harmful environmental impacts such as climate warming. The major consequences of climate warming are the melting of ice in the Arctic and Antarctic regions as a result of sea level increase. Therefore, the urge is to move toward a clean and unconventional energy stream, conserve maximum energy, avoid geopolitical issues that occur in conventional energy reserves, and create economic instabilities.

Among all unconventional energy sources, solar energy has received increasing attention because it is abundant, free, clean, and does not pollute the environment. Behind the excess core of the sun is the reason that involves a fusion reaction, where two hydrogen atoms fuse and continuously release a stupendous amount of energy toward the blue planet. The sun emits energy in the form of electromagnetic waves, commonly known as radiation, which is composed of photons and temperature. This form of energy proclaims both magnetic and electrical waves that propagate in a packet of energy called photons. A study estimates solar energy's global potential of 1,600 to 49,800 exajoules is much higher than the energy consumed by the world population (global primary energy consumption in 2019 is 581.5 exajoules). This excited difference forced us to focus on solar energy, which can completely fulfill the world's total energy demand, and it is currently estimated that it continuously provides approximately 4 billion years. Except for some parts of the world, solar radiation falls on each corner with sufficient intensity; however, it varies with latitude, time of day, and season. The intensity of solar radiation that falls on the earth is approximately 1380 W.m^{-2}. That is sufficient to satisfy the need for thermal energy that uses solar radiation in the form of heat for low-temperature and high-temperature applications, such as drying, hot water supply, and harvesting electricity through indirect conversion techniques, such as concentrated solar power by using reflector sterilization, pasteurization, washing, cooling, dyeing, bleaching, degreasing, distillation, and electricity generation through concentrated solar power plants and photovoltaic cells. To overcome these shortcomings and achieve a clean and green global energy goal, it is necessary to intensify research on the efficacy and economic ways to capture, seize, and mutate solar energy. Sometimes, owing to unavailability during consumption hours and its intermittent nature, there is a necessity for energy storage systems, such as thermal energy systems in which phase change materials (PCMs) play a vital role and battery-based energy storage systems. A miniature or major industrial energy ecosystem involves a power supply (electric power, direct heat, gases, cold fluids, and steam), production plants, recovery systems, and cooling systems, as shown in Figure 7.1. Generally, industrial production processes require thermal energy that is dissipated in the production plant, and recovery and re-cooling systems are an integral part of the plant. This chapter focused on solar thermal energy for drying and other small-scale industrial applications.

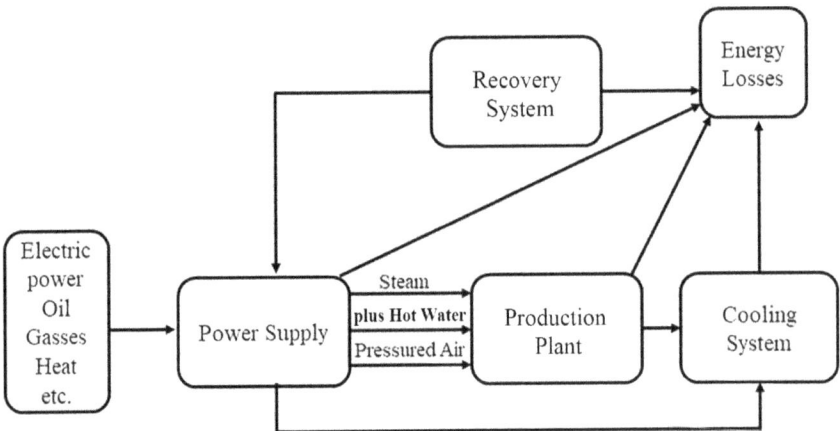

FIGURE 7.1   Industrial energy ecosystem (Schnitzer, Brunner and Gwehenberger, 2007).

7.2 ASSORTMENT OF SOLAR THERMAL ENERGY CHANNELS

There are many methods for categorizing solar thermal energy systems. Depending on the operating temperature, solar thermal systems are divided into low-temperature (between 30–120°C), medium-temperature (between 130–400°C), and high-temperature (greater than 400°C) systems. Flat-plate collectors, evacuated tubular collectors, solar ponds (Kalogirou, 2003), and solar chimneys are examples of low-temperature solar thermal systems with efficiencies in the range of 15–40%, while linear Fresnel reflectors (LFR) and parabolic trough collectors are examples of medium-temperature solar thermal systems with efficiencies of 50–60% (Sharma *et al.*, 2015). High-temperature systems with efficiencies of 60–80% include a central receiver (CR) and parabolic dish collector (PDC) (Zheng, 2017). Based on the axis tracking and concentration nature, they are divided into single-axis tracking, two-axis tracking, concentration modulating systems, and non-concentrating or stationary collectors. Solar radiation is focused on a focal point by concentrating collectors, and concentrated solar radiation is collected by placing an absorber at the concentrator's focal point. For drying and other small-scale industrial applications, the operating temperature lies in the low-temperature range (30–120°C), for which flat-plate collectors and evacuated tube collectors (ETC) are used as solar thermal systems (Sobiski and Swierczyna, 2009). Further discussion focuses solely on drying and small-scale industrial applications of solar thermal energy (Ravi Kumar, Krishna Chaitanya and Sendhil Kumar, 2021).

7.3 APPLICATIONS

7.3.1 Drying

It has been observed that most of the developing countries of the world are unable to meet their food supply needs. One of the major reasons for this is population growth, which directly affects the food balance. To correct this problem, attention must be given to the type and amount of food that spoils as a result of subpar processing procedures and a lack of storage facilities. Whether it is a household, or a small- or large-scale enterprise, energy is a significant issue that needs to be addressed. The primary method of food preservation is sun drying, although conventional open drying has a number of drawbacks. Many scientists and researchers have worked to identify better alternatives in recent years to lessen this effect. This has resulted in the invention of solar dryers for many types (as shown in Figure 7.2) of domestic, small, and large-scale industries, and they are constantly being improved. Solar dryers maintain accurate drying conditions, which are challenging tasks in open sunlight, such as maintaining accurate conditions of temperature, velocity, and humidity (Ndukwu *et al.*, 2020). Indirect solar dryers and direct solar dryers are alternatives to controlled sun dryers. This helps produce high-quality goods along with drying and dust-free goods in less time. These can be forced convection or natural convection (passive or active mode). Passive dryers have an overall efficiency that ranges between 20–40% depending on the material, moisture content, air temperature, velocity, and humidity (Patil and Gawande, 2016). A thermal energy storage system, such as latent heat or

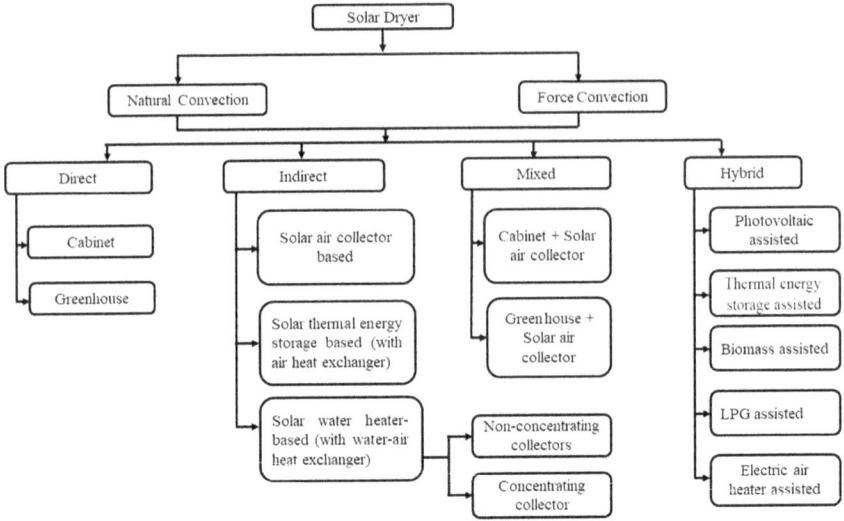

FIGURE 7.2 Types of solar dryer (Agrawal and Sarviya, 2016).

a sensible heat storage unit, can be added to a solar dryer to continue the drying process, even when the sun's rays do not fall on Earth (Kumar and Singh, 2020).

The International Energy Outlook of the US Energy Information Administration (EIA), published in 2019, states that 4% of the heat produced is used for greenhouse heating, 46% for space and water heating, and the remaining 50% is used by industry (Goldstein, Gounaridis and Newell, 2020). Owing to the increase in demand, energy consumption in the industrial, transportation, residential, and commercial sectors is increasing continuously. It uses renewable energy commercially to power electricity, but its role is relatively high compared with different sources (Monthly Energy Review, US EIA, 2020). According to research conducted by various groups and the International Energy Agency (IEA) in 2018, a significant portion of total energy consumed by industries falls into different temperature ranges. Approximately 11.5% of industries require temperatures below 150°C, while 8.5% require temperatures in the range of 150–400°C. The rest of the industries require temperatures that exceed 400°C (Hrudey, 2012). This trend in energy use is essentially the same across all industries worldwide. This section discusses the many small-scale businesses and their need for thermal energy for solar drying.

7.3.1.1 Agricultural and Food Industry Applications

Moisture is a major factor in food spoilage caused by microorganisms and bacteria. Drying the food material reduces the mass of the material, thereby reducing the cost of transportation as well as reducing spoilage and the growth and reproduction of microbes (Chandramohan, 2016). Agricultural products require a significant quantity of heat to prevent dehydration, which solar thermal energy may provide. Utilizing solar energy also contributes to a reduction in the use of conventional energy sources. For the thermal analysis of solar dryers, many researchers have proposed their perspectives on different types of dryers to improve efficiency through mathematical, analytical, and experimental studies (El-Sebaii and Shalaby, 2013; Lamidi *et al.*, 2019).

Black turmeric's drying kinetics were investigated by utilizing a mixed-mode forced convection solar drier connected to a thermal energy storage device (Lakshmi *et al.*, 2018). Results from open sun drying were contrasted with those from this study. Two sets of 200 g thin layers of black turmeric were employed in the experiment; one set was dried using a solar dryer and the other using direct sun drying. Then 35 kg of paraffin wax was utilized as the thermal energy storage material, and a shell and tube heat exchanger was linked to the solar dryer to support the thermal energy storage system for drying the samples even at night. During the night, the air temperature at the exit from the system was observed to be 6–10°C higher than the surrounding air temperature. The turmeric dried from an initial moisture content of 73.4% to 8.5% (Wb) in 1110.00 minutes (18.50 h) in the solar dryer and 2790.00 minutes (46.50 h) in the open sun. The estimated energy required for drying the product was 5.21 kWh per kg of moisture. The efficiency of the collector was determined to be 25.6%, while the overall drying efficiency was found to be 12%. It was concluded that sun-dried turmeric had better color, phenolic content, and flavonoid content compared to fresh turmeric (Baniasadi, Ranjbar and Boostanipour, 2017; Atalay, 2019; Atalay, Çoban and Kıncay, 2017). Similar studies have been conducted on various fruits and vegetables, including bitter gourd, red pepper, garlic, orange apricot, and apple, using solar dryers (Ndukwu *et al.*, 2017; (Zachariah, Maatallah and Modi, 2021; Shringi, Kothari and Panwar, 2014). The data collected from these studies included drying capacity, initial and final moisture content, dry air temperature, drying duration, energy storage materials used, the mass of water evaporated, and energy efficiency, as listed in Table 7.1.

TABLE 7.1

Solar Drying Experimental Analysis Data of Different Vegetables

Product	Orange Slices	Garlic clove	Bitter ground	Red Chili	Chili	Apricot Slices	Black Turmeric
Capacity (Kg)	10	—	2.5	1	40	—	15
Initial moisture (%)	93.5	55	92	72.2	72.8	86	73.4
Final moisture (%)	10.76	6.5	4	10.1	9.1	25	8.5
Drying air temperature (°C)	55.4	39–69	48	41.83	50.4	65	65

(Continued)

TABLE 7.1 (*Continued*)
Solar Drying Experimental Analysis Data of Different Vegetables

Product	Orange Slices	Garlic clove	Bitter ground	Red Chili	Chili	Apricot Slices	Black Turmeric
Drying time (h)	7.2	8	10	36.5	24	1.55	18.5
Thermal energy storage material	Packed bed (Pebbles)	PCM	PVC caps, Al pipes, and paraffin wax	NaCl	Gravel	Paraffin wax	Paraffin wax
Evaporated water (kg of water/kWh)	0.505	0.54–1.05	0.16	0.602	0.870	0.933	0.192
Drying efficiency (%)	34.4	—	18.6	11.89	21	10.7	12
Exergy efficiency (%)	63.34	—	67–88	—	—	—	—
Reference	(Atalay, 2019)	(Shringi, Kothari and Panwar, 2014)	(Zachariah, Maatallah and Modi, 2021)	(Ndukwu *et al.*, 2017)	(Mohanraj and Chandrasekar, 2009)	(Baniasadi, Ranjbar and Boostanipour, 2017)	(Lakshmi *et al.*, 2018)

7.3.1.2 Rubber Industry

Rubber is an elastic material used in the production of a variety of products. Either artificial synthetic methods or natural latex from plants is used to create rubber. Drying of raw material is an important process in rubber manufacturing from raw material to the final product, in which it is important to maintain the quality of rubber sheet made from raw material after drying. According to Breymayer (Breymayer *et al.*, 1993), improper drying of raw rubber material produces low-grade rubber sheets, which account for around 80% of the entire production and must be sold at competitive pricing. The two most common methods for drying are smoke drying and the direct hot air-drying process. Moist rubber sheets are dried by creating hot air using open solar drying or using conventional energy sources in the air drying method. In an example of a smoke-drying technique, smoke is generated by burning biomass and firewood in a smokehouse to dry rubber sheets. Rubber usually requires a temperature between 45–60°C to dry, which can be easily achieved using solar energy. As a secondary heat source, Bremeyer built a dryer integrated with solar air collectors and conventional firewood/biomass smokehouses, in which the solar air collectors recirculated the dryer exhaust air to provide a temperature between 45–60°C (Tanwanichkul, Thepa and Rordprapat, 2013). The rubber sheets with a total weight of 0.32 tonne were dried using the solar dryer, and finally it was found that the moisture content of the sheets reduced from 60% to 0.5%, thereby reducing the fuel consumption. The need has decreased. Compared to the solar dryer used alone, it took five days for the biomass to attain a moisture content of 0.5% in the dryer. However, the technology devised is very cost-efficient as the cost would be high if commercial electricity is used for a week (instead of sun-drying). The natural rubber was dried using an indirect solar dryer with the help of an electric air heater, which produced high-quality rubber as the end result. The moisture contents of the rubber sheet were reduced from 60% to 0.5% (wb) in 8.8 hours at 100°C (Pratoto, Daguenet and Zeghmati, 1998).

The greenhouse solar dryer was found to be more effective than traditional open solar drying methods in drying rubber sheets. The dryer could maintain temperatures between 32–55°C, allowing for efficient drying of 0.75 tonnes of rubber sheets without affecting their quality. Over the course of five days, the dryer was able to reduce the moisture content of the rubber sheets from 24–30% to 0.4–3.9% (Janjai et al., 2015).

Traditionally, drying rubber sheets using natural and open solar methods can take up to 20 days, leading to significant changes in the color of the final product. Extended drying times and high moisture levels can also result in the growth of fungi, further spoiling the color. These factors often result in decreased prices for the product, causing discomfort for rubber producers. Solar drying can address these problems and allow for the rubber sheets to be dried in a timely manner, preserving their freshness and attracting customers. However, more research and experimentation is needed to optimize the drying process for different types of rubber sheets and to make the process accessible to industries and farmers (Lingayat, Balijepalli and Chandramohan, 2021).

7.3.1.3 Tea Industry

One of the most widely consumed beverages is tea, which is mostly produced in China and India. One of the primary small- and large-scale enterprises is the tea sector. Withering, drying, grading, and packing are a few of the energy-intensive processes involved in the manufacture of tea. These actions collectively consume 15% electrical energy and 85% thermal energy (Sharma et al., 2019). Solar energy provides the thermal energy required for the drying stage of the tea-making process. Palaniappan and Subramaniam examined the financial feasibility of a rooftop solar dryer for drying tea (Palaniappan and Subramanian, 1998). Coal and firewood were offered as additional thermal energy sources for the dryer, and a collector space of 212 m² was created, where the use of solar energy reduced the annual fuel usage in the dryer by 25%. In this category of tea drying, some researchers have performed experimental explication of solar-assisted drying systems for drying herbal teas (Sopian et al., 2000), in which the solar collector has V-shaped grooves to generate sufficient turbulence and increase the rate of heat transmission at a flow rate of 15.1 m³/min, a 20 m² collector area can produce hot air with temperatures in excess of 50°C. An auxiliary heating system was used when the air temperature was below 50°C. The herb leaves were successfully dried from 87% (wb) moisture content to 54% (wb) moisture content in 12 hours.

To dry black tea, Pu and Tripathi created and examined vacuum-assisted sun dryers (Pou and Tripathy, 2020). Input factors such as vacuum level and loading rate were taken into account in connection to output factors such as color, odor, drying time, and energy use. With the least amount of energy and drying time, the system was calibrated to create a wine with the greatest possible color and fragrance index. The calculated values of liquor color, aroma index, drying time, and energy consumption were 20.08, 11.65, 4.44, 140.66, and 21450.7 kJ, respectively, at an ideal vacuum level of 570.71 mmHg and a loading rate of 0.96 kg m^2.

After chamomile leaves (an herbal tea) were dried in a solar dryer with an electric heater acting as a backup heat source, as well as a solar collector, reflector, water-air heat exchanger, and solar-heated water storage tank, the drying kinetics were evaluated (Amer, Gottschalk and Hossain, 2018). During the day, heat energy was stored in water, which was used as a thermal energy storage medium. The volumes of the two drying chambers were between 32 kg and 35 kg. The required drying temperature was controlled using a temperature controller. Compared with open sun drying, the system proved helpful in reducing the drying time of chamomile from 75% to 6% (wb MC), and the Midilli model was used to fit the chamomile drying kinetics.

Previous studies on drying tea leaves produced the following main conclusions: Vacuum-assisted sun dryers help produce high-quality black tea by providing the proper vacuum level and loading rate (Lingayat, Balijepalli and Chandramohan, 2021). An exergo-economic analysis must be done to calculate the energy and exergy losses from the dryer. Additionally, the expenditures related to the drying method and payback length can be recognized because they are significant keywords for the tea sector. A solar dryer featuring a reflector, a thermal energy storage substance, and two drying chambers might drastically shorten the drying period by up to 20 hours when compared to open sun drying. Small-scale farmers in rural areas can benefit from the commercialization of these dryer types. Preheating the air used in the processing of tea with a solar air heating system reduces the amount of fuel used annually by up to 26% (Palaniappan and Subramanian, 1998; Amer, Gottschalk and Hossain, 2018).

7.3.1.4 Marine Industry

Drying marine products is another key application of solar energy in the marine industry. A solar dryer with a thermal energy storage system was created to dry shrimp. The energy was captured by a water-based thermal energy storage system at the height of the solar day (Murali *et al.*, 2020). To assist the dryer, additional heating was provided by liquefied petroleum gas (LPG): 50 kg of prawns were dried for 6 hours, going from 76.71% (wb) to 15.38% (wb). The LPG water heater system helped with the remaining 26.07% of the heat, while the solar collector supplied 73.93% of it (Murali *et al.*, 2020).

Marine meal is an extremely delicate tissue that requires careful drying. The partial wetness of food damages tissues, rendering it potentially useless. The moisture content of fish is between 15–20% (wb). The dryer uses a large amount of energy during the thorough drying process, making the solar energy insufficient or requiring additional research to obtain the same results. Hybrid dryers can be created in such a case to continuously operate the dryer to prevent microorganism reactions. Most of the time, dry salt was used to dry sea dishes. Higher moisture content was always extracted from the tissues using dry saline. When the muscle is dried at a higher temperature and moisture content, the salt spreads uniformly throughout its geometry because the diffusion coefficient depends on both temperature and moisture content. Other drying information provided in this study includes the fish-to-salt ratio of 3:1, the effective drying time of 6–20 hours, and the air temperature range of 40–60°C. The average collector outlet temperature and drying efficiency for drying seafood are 75°C and 25.42%, respectively. The chamber's inside has an average temperature range of 60–65°C (Rahman, 2006).

7.3.2 Paper Industry

Currently, coal, petroleum fuel, biomass, and electricity constitute the majority of the fuel mix used in the paper industry. It is better to use fewer fossil fuels while meeting the growing energy needs of the paper industry. This can be accomplished by increasing fuel utilization efficiency and replacing fossil fuels with renewable energy sources by utilizing them in appropriate processes and end use. The energy used by the paper industry is mostly thermal and electrical energy. Low and intermediate temperatures (50–250°C) are utilized for heating processes, which account for around 75% of the energy used to make paper (Kalogirou, 2003). Processes including pulping, drying, bleaching, washing (using hot water), and heating boiler feed water use the majority of the process heat. Heat is transferred through a heat-transfer medium, which can be water, steam, air, or thermic oil, depending on the needs of a particular process. Typical paper manufacturing processes for pulp preparation, bleaching, delinking, and paper drying require temperatures of 120–170°C, 120–150°C, 60–90°C, and 90–200°C, respectively (Lauterbach et al., 2012; Sharma et al., 2015). With or without integrating it with conventional fossil fuel–based process heating systems, modern technology makes it possible to efficiently gather solar energy to generate heat at the necessary temperatures (below 250°C). Agro-residues and recycled fiber are used as feedstock in small-scale paper factories, where solar energy for industrial process heat may be appropriate. To assess the potential for solar industrial process heating in paper mills, the following methodology was employed. Data were gathered for several factors after a thorough analysis of the paper industry, including the size and feedstock usage classification of paper mills, the use of cogeneration in paper mills, the annual production of paper (based on raw materials), and the specific thermal energy requirements for process heating. Finally, a few commercially available solar collectors that could generate process heat in the necessary temperature range (50–250°C) were chosen in order to estimate their instantaneous efficiency for the required process heating temperatures of the paper industry based on their performance equations. Different solar collector types were employed for various temperature ranges.

Some examples of the use of solar thermal energy in paper industries are as follows: For drying 12 A4-sized sheets (each weighing 22 g) simultaneously, an indirect solar dryer was created. The temperature at the collector exit ranged from 45–70°C. Over a drying period of 65–75 minutes, the dryer significantly reduced the moisture content of the sheets from 50% to 7% (wb), and the results of an economic analysis revealed that the dryer had a 95-day payback period (Madhavan and Ramachandran, 2015). An experiment showed the ability of thermal energy to process heat and its benefits in the scarcity of carbon dioxide emittance in the paper sectors. Based on the location and raw material usage, the annual process heat requirements are calculated, and the efficiency of commercial systems is assessed (Sharma *et al.*, 2017). The researchers used a convective solar dryer, whose diffusion coefficient varies with temperature and were found to vary between 1.56 and 6.98 m²/s, to determine the drying kinetics and properties of the carob pulp at different drying temperatures between 50–80°C and velocities between 0.18–0.09 m/s. The estimates place the respective power and energy efficiency between 4.23–7.25% and 30.12–80.5% (Search *et al.*, 2004).

7.3.3 TEXTILE INDUSTRY

The manufacturing of yarn, fabric, and its pursuant design or distribution are the main concerns of the textile industry, where natural or synthetic fibers are used as raw materials. The fiber and yarn configurations encompass all types of textile fibers, including natural fibers such as cotton, jute, silk, and wool as well as synthetic/human-made fibers such as polyester, viscose, nylon, acrylic, and polypropylene. Also included are filament yarns, such as partially oriented yarns (Schnitzer, Brunner and Gwehenberger, 2007).

Spinning, weaving, and finishing are the three main steps involved in textile production. The primary energy consumers are spinning and weaving. Operations such as spinning, weaving, bleaching, drying, and curing require energy in the form of steam and/or hot water (Suleman, Dincer and Agelin-Chaab, 2014). Table 7.2 lists the temperatures required for various procedures. The type of products being processed (fiber, yarn, fabric, or cloth), the machine, the specific process, and the stage of the finished product all affect how much energy is required in the textile business. Typically, to produce 1 kg of fabric, 6–8 kg of water vapor, or steam, 0.8–1.5 kWh of energy, and 0.35 m³ of natural gas are required (Mahmood and Harijan, 2012; Kumar, Chaitanya and Kumar, 2021).

Figure 7.3 depicts the relationship between the solar field and different stages of the textile industry. The heat requirement in the textile industry is between 40–120°C.

TABLE 7.2

Temperature Requirements for Different Procedures Involved in the Textile Industry (Kumar, Chaitanya and Kumar, 2021)

Procedure	Temperature (°C)
Yarn Conditioning	105–120
Sizing	80–85
De-sizing	60–90
Bleaching	90–93
Dyeing	48–80
Scouring	90–110
Finishing	90–93

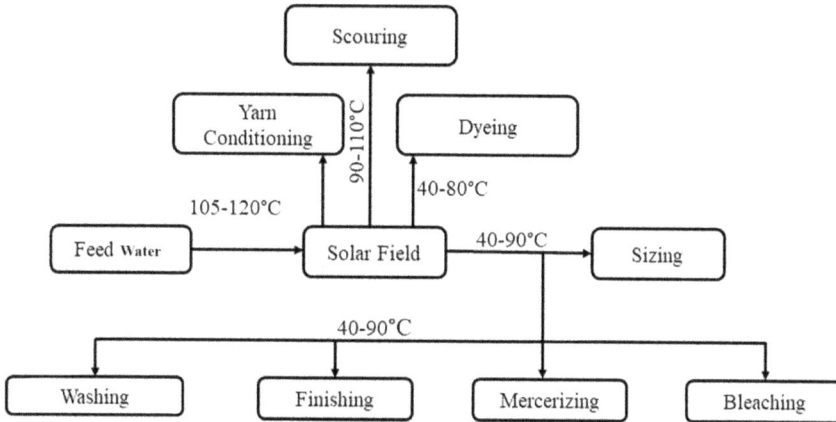

FIGURE 7.3   Relationship between the solar field and different stages of the textile industry.

With the exception of yarn conditioning, most procedures do not require more than 100°C of heat. Evacuated tube collectors can produce working fluid output temperatures of approximately 120°C and can be installed on the building roof (Sabiha *et al.*, 2015). As a result, an evacuated tube collector might be a preferable choice to satisfy the textile industry's need for heat. If there is a significant need for heat and land around the textile business, the linear Fresnel reflector system may also be preferable (Ramos, Ramirez and Beltran, 2014). To lessen the effects of changes in solar insolation, auxiliary heaters and buffer thermal energy storage may be used.

7.3.4 LEATHER INDUSTRY

Leather is made from the skins of cows, pigs, goats, and sheep and is primarily produced in developing nations such as India and China. This includes both semi-finished and finished leather goods. These include gloves, boots, saddles, clothing, and other leather products (Farjana *et al.*, 2018). The following categories can be used to group leather production units:

- Tanneries: Process unfinished skins or hides to create completed and semi-finished leathers.
- Consumer goods production units: Manufacture finished leather into goods such as fashionable footwear, protection gear, clothing, gloves, and other items.
- Integrated units: Process verdant skin and hides to create downstream consumer items, as already indicated (Navarro *et al.*, 2020).

The two main production cycles used in leather businesses involve various chemical and physical processes. Tanning, in which raw hides and skins are converted into finished leather, is the first cycle of leather production, followed by the second cycle of using finished leather to manufacture consumer leather goods. The

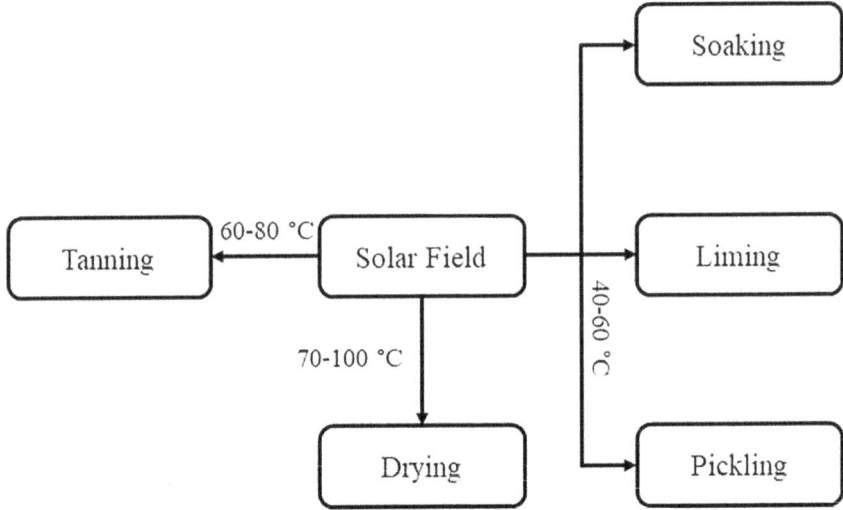

FIGURE 7.4 Integration of leather formation process and solar system.

pre-tanning (beam house), tanning, and finishing processes are all parts of the tanning process. Consumer leather goods are made by marking, cutting, and dressing
the finished leather as well as sewing and stitching (Saravanabhavan *et al.*, 2006),
in which energy plays an important role, with the percentage of energy used for
leather industry processes for tanning, finishing, drying, and remaining heat handling being 13%, 37%, 40%, and 10%, respectively (Ravi Kumar, Krishna Chaitanya
and Sendhil Kumar, 2021). There is a significant demand for warm water supply,
between 60–80°C, because tanning involves numerous chemical and physical processes that require enormous volumes of warm water at key working temperatures.
Furthermore, this sector requires heat energy in the range of 70–100°C for drying
processes to manufacture high-quality leather. Various drying techniques have been
employed with great success, significantly reducing the use of conventional energy.
This demonstrates the substantial potential of solar thermal applications that can be
used at different temperatures for different processes. The leather business requires
heat for pre-tanning (soaking, liming, and pickling), tanning, and post-tanning (drying) activities at temperatures ranging from 40–60°C, 60–80°C, and 70–100°C,
respectively. Figure 7.4 shows how the solar system and leather formation process
are integrated (Maina, Ollengo and Nthiga, 2019).

7.3.5 Foodstuff Processing and Quencher Industry

The food processing sector is a highly fragmented industry that includes sub-
segments for fruits and vegetables, milk and milk products, beer and alcoholic beverages, meat, and poultry, maritime goods, grain processing, packaged or convenience
meals, and packaged drinks, to name just a few. Heat is needed for the majority of food
business processes, including drying, cooking, sterilizing, washing containers, and

TABLE 7.3

Different Food Industry Operations and Their Respective Operating Temperatures (Kalogirou, 2003)

Sector	Procedure	Temperature (°C)
Meat	Washing	60–90
	Sterilization	90–100
	Cooking	60–90
Milk	Pasteurization	60–90
	Sterilization	100–120
	Drying	120–180
	Boiled feed water	60–90
Tinned food	Sterilization	110–120
	Pressurization	60–80
	Cooking	60–90
	Bleaching	60–90
Beverages	Sterilization	60–80
	Pasteurization	60–70
	Malting-steeping	90–110
Rice mill	Paddy cooking	80–120
	Drying	60–70

chilling. Solar thermal collectors can be used to gather this heat energy. Table 7.3 lists some of the most significant food industry processes along with the temperature ranges in which they operate. Dairy, beer, and rice mill industries all have a reasonable demand for thermal energy. Other industries just need a tiny amount of thermal energy to carry out their processes (Müller, Brandmayr and Zörner, 2014).

7.3.6 RICE MILL

By removing the husks from paddy grains, the rice milling process creates polished rice. Rice mill process phases include pre-cleaning, parboiling, husking, husk aspiration, paddy separation, whitening, polishing, length grading, blending, weighing, and bagging (Nabavi-Pelesaraei *et al.*, 2019). Figure 7.5 shows the several processes used in the rice milling sector and the different forms of solar thermal energy used. All contaminants and empty grains were removed from the paddy during the pre-cleaning process at the rice mill, for which the paddy is parboiled, which requires process heat. After parboiling, the husk was removed from the paddy and separated from the brown rice and the unhusked paddy. The unhusked paddy was separated from the rice during the whitening process, and the brown rice barn layer and pathogens were also eliminated. Hot water at 80°C or above was pumped for 6–8 hours to ripen the paddy. Following this procedure, steam was immediately introduced into the paddy for cooking for approximately 10 minutes. In the rice mill sector, the parboiling process can be performed using solar thermal systems (Vijayaraju and Bakthavatsalam, 2020).

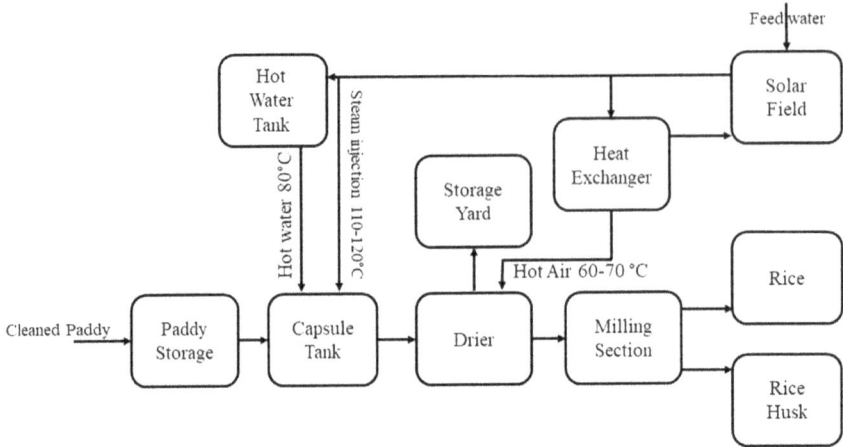

FIGURE 7.5 Flowchart of the rice milling sector and the different forms of solar thermal energy used (Ravi Kumar, Krishna Chaitanya and Sendhil Kumar, 2021).

7.3.7 BEER INDUSTRY

Distinct nations and populations have different drinking habits. A variety of alcoholic beverages fall under the categories of soft liquors, hard liquors, etc. The oldest and most widely used alcoholic beverages in the world are beer and wine. Beer is made by steeping a carbohydrate source in water, which allows it to ferment with yeast. Malting, milling, mashing, lautering, boiling, fermentation, conditioning, filtering, and packaging are processes used in the beer industry (Mauthner *et al.*, 2014). The malting, mashing, and meshing sections of the manufacturing process require heat energy between 60–110°C (Eiholzer *et al.*, 2017), either in the form of hot water or heat, and evacuated tube collectors can be integrated with the beer industry processes, which is more efficient than a flat-plate collector in producing hot water or water vapor, i.e., steam at 120°C. The process used in the brewing industry may not be significantly affected by variations in solar insolation. However, it is possible to reduce the volatility by using supplementary heaters and/or buffer storage (Pino, Lucena and Macho, 2019). A flowchart of the integrated beer-making process with the solar field is shown in Figure 7.6.

7.3.8 MILK INDUSTRY

For extended use of milk and other milk products, dairy plants process raw milk. It involves two basic procedures: heating milk to make it safe and dehydrating dairy goods such as butter, cheese, and dry milk to make them more durable and longer-lasting; and drying, cleaning, evaporation, sterilization, pasteurization, and storage. Fresh milk was placed in the storage tank after cooling to a temperature of 5–7°C (Ravi Kumar, Krishna Chaitanya and Sendhil Kumar, 2021). Milk was clarified or filtered to remove any undesirable organic elements before processing.

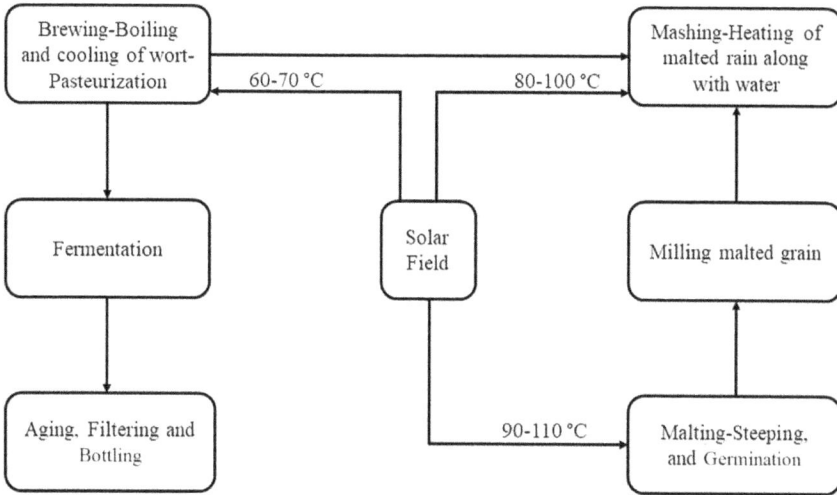

FIGURE 7.6 Integration of the beer-making industry with the solar field (Ravi Kumar, Krishna Chaitanya and Sendhil Kumar, 2021).

Milk that has been previously processed is pasteurized at 71.6°C for 15 seconds (Quijera, Alriols and Labidi, 2011). The pasteurized milk is cooled in cold storage before further packaging and distribution. The industry uses a significant amount of heat energy for milk processing (such as pasteurization, sterilization, spray drying, and evaporation) and electrical energy for refrigeration during milk pre-chilling, chilling of milk after pasteurization, cold storage of packed milk, compressed air requirements for pneumatic milk packaging machines, milk homogenization, and clarification operations. Heat to a temperature of 120°C is necessary for sterilization, and spray drying uses 16% of the overall energy used (Quijera, Alriols and Labidi, 2011). Twenty percent of the energy used by the dairy sector is spent on cooling and maintaining chilly temperatures. Its processes require temperatures below 250°C, but the milk sector needed a large amount of process heat. Preservation methods, which include a variety of chilling processes, also greatly increase the amount of thermal energy used in the sector. The dairy sector needs a lot of heat in addition to cooling, as drying processes are essential for processing finished goods. The combination of the dairy business and solar field is depicted in Figure 7.7. The maximum quality of heat required by the dairy industry is 120°C for chilling, sterilization, and spraying applications. An ETC or LFR system is the optimum consonant for the milk industry, where the linear Fresnel reflector system is preferred over the evacuated tube collector system when extreme heat is required and ample acreage is available (Ramaiah and Shekar, 2018). By using warm water or evaporated forms of water from various parts of the solar field, different processes can be provided with varied grades of heat. To prevent variation during prolonged cloudy or rainy seasons, the solar field may be connected to an auxiliary heat source and/or thermal buffer storage.

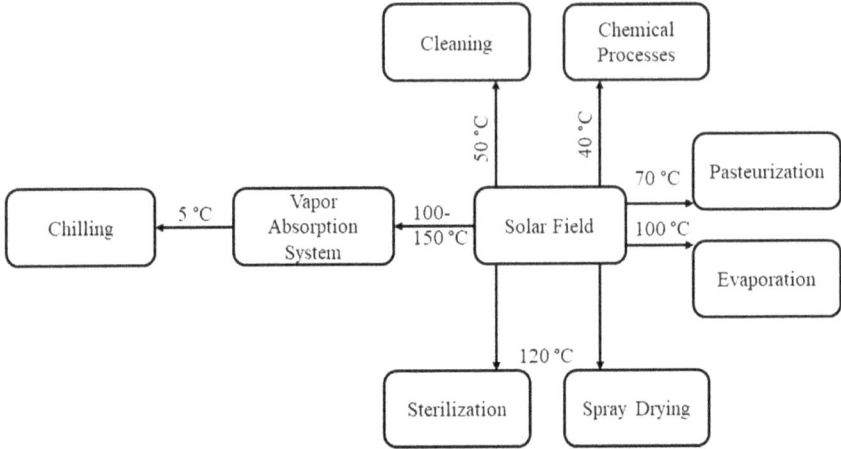

FIGURE 7.7 Combination of the milk industry and solar field (Sharma *et al.* (2017).

7.4 CONCLUSION

This chapter discusses a variety of small-scale industrial applications, showing the use of solar drying in agricultural and food industry applications and rubber, tea, and marine industries, which can be used to reduce people's dependence on conventional energy and provide cheap and clean energy. Solar thermal energy has been used in a variety of small-scale applications, mainly in the paper, textile, leather, food, and beverage industries. For applications that have a temperature of less than 120°C, flat-plate collectors or evacuated tubular collectors are used. In this way, the use of solar thermal energy in small-scale industries can promote the participation of clean and green energy and reduce the dispensation of greenhouse gases in the environment, in addition to improving the income and lifestyle of farmers and small- and middle-class businesspeople. In this way manufacturers and suppliers both make a big contribution to the world.

REFERENCES

Agrawal, A. and Sarviya, R.M. (2016) 'A review of research and development work on solar dryers with heat storage', *International Journal of Sustainable Energy*, 35(6), pp. 583–605. Available at: https://doi.org/10.1080/14786451.2014.930464.
Amer, B.M.A., Gottschalk, K. and Hossain, M.A. (2018) 'Integrated hybrid solar drying system and its drying kinetics of chamomile', *Renewable Energy*, 121, pp. 539–547.
Atalay, H. (2019) 'Performance analysis of a solar dryer integrated with the packed bed thermal energy storage (TES) system', *Energy*, 172, pp. 1037–1052.
Atalay, H., Çoban, M.T. and Kıncay, O. (2017) 'Modeling of the drying process of apple slices: Application with a solar dryer and the thermal energy storage system', *Energy*, 134, pp. 382–391.

Baniasadi, E., Ranjbar, S. and Boostanipour, O. (2017) 'Experimental investigation of the performance of a mixed-mode solar dryer with thermal energy storage', *Renewable Energy*, 112, pp. 143–150.

Breymayer, M. *et al.* (1993) 'Solar-assisted smokehouse for the drying of natural rubber on small-scale Indonesian farms', *Renewable Energy*, 3(8), pp. 831–839.

Chandramohan, V.P. (2016) 'Numerical prediction and analysis of surface transfer coefficients on moist object during heat and mass transfer application', *Heat Transfer Engineering*, 37(1), pp. 53–63.

Eiholzer, T. *et al.* (2017) 'Integration of a solar thermal system in a medium-sized brewery using pinch analysis: Methodology and case study', *Applied Thermal Engineering*, 113, pp. 1558–1568.

El-Sebaii, A.A. and Shalaby, S.M. (2013) 'Experimental investigation of an indirect-mode forced convection solar dryer for drying thymus and mint', *Energy Conversion and Management*, 74, pp. 109–116.

Farjana, S.H. *et al.* (2018) 'Solar process heat in industrial systems–A global review', *Renewable and Sustainable Energy Reviews*, 82, pp. 2270–2286.

Goldstein, B., Gounaridis, D. and Newell, J.P. (2020) 'The carbon footprint of household energy use in the United States', *Proceedings of the National Academy of Sciences*, 117(32), pp. 19122–19130.

Hrudey, S. (2012) 'Environmental and Health Impacts of Canada's Oil Sands Industry R'.

Janjai, S. *et al.* (2015) 'Experimental performance and neural network modeling of a large-scale greenhouse solar dryer for drying natural rubber sheets', *Journal of Control Science and Engineering*, 1(1), pp. 48–53.

Kalogirou, S. (2003) 'The potential of solar industrial process heat applications', *Applied Energy*, 76(4), pp. 337–361.

Korachagaon, I. and Bapat, V.N. (2012) 'General formula for the estimation of global solar radiation on earth's surface around the globe', *Renewable Energy*, 41, pp. 394–400. Available at: https://doi.org/10.1016/j.renene.2011.11.002.

Kumar, P. and Singh, D. (2020) 'Advanced technologies and performance investigations of solar dryers: A review', *Renewable Energy Focus*, 35, pp. 148–158. Available at: https://doi.org/10.1016/j.ref.2020.10.003.

Lakshmi, D.V.N. *et al.* (2018) 'Drying kinetics and quality analysis of black turmeric (Curcuma caesia) drying in a mixed mode forced convection solar dryer integrated with thermal energy storage', *Renewable Energy*, 120, pp. 23–34.

Lamidi, R.O. *et al.* (2019) 'Recent advances in sustainable drying of agricultural produce: A review', *Applied energy*, 233, pp. 367–385.

Lauterbach, C. *et al.* (2012) 'The potential of solar heat for industrial processes in Germany', *Renewable and Sustainable Energy Reviews*, 16(7), pp. 5121–5130.

Lingayat, A., Balijepalli, R. and Chandramohan, V.P. (2021) 'Applications of solar energy based drying technologies in various industries – A review', *Solar Energy*, 229(March), pp. 52–68. Available at: https://doi.org/10.1016/j.solener.2021.05.058.

Madhavan, S. and Ramachandran, P.N. (2015) 'Design, fabrication and testing of a solar paper dryer', *International Research Journal of Engineering and Technology*, 2, pp. 1911–1914.

Mahmood, A. and Harijan, K. (2012) 'Utilizing solar thermal energy in textile processing units', in *Energy, Environment and Sustainable Development*. Springer, pp. 121–130.

Maina, P., Ollengo, M.A., Nthiga, E.W. 2019. Trends in leather processing: A Review. *International Journal of Scientific and Research Publications (IJSRP)* 9, p. 9626. https://doi.org/10.29322/IJSRP.9.12.2019.p9626

Mauthner, F. *et al.* (2014) 'Manufacture of malt and beer with low temperature solar process heat', *Energy Procedia*, 48, pp. 1188–1193.

Mohanraj, M. and Chandrasekar, P. (2009) 'Performance of a forced convection solar drier integrated with gravel as heat storage material for chili drying', *Journal of Engineering Science and Technology*, 4(3), pp. 305–314.

Müller, H., Brandmayr, S. and Zörner, W. (2014) 'Development of an evaluation methodology for the potential of solar-thermal energy use in the food industry', *Energy Procedia*, 48, pp. 1194–1201.

Murali, S. *et al.* (2020) 'Design and performance evaluation of solar-LPG hybrid dryer for drying of shrimps', *Renewable Energy*, 147, pp. 2417–2428.

Nabavi-Pelesaraei, A. *et al.* (2019) 'Comprehensive model of energy, environmental impacts and economic in rice milling factories by coupling adaptive neuro-fuzzy inference system and life cycle assessment', *Journal of Cleaner Production*, 217, pp. 742–756.

Navarro, D. *et al.* (2020) 'Life cycle assessment and leather production', *Journal of Leather Science and Engineering*, 2(1), pp. 1–13.

Ndukwu, M.C. *et al.* (2017) 'Energy and exergy analysis of a solar dryer integrated with sodium sulfate decahydrate and sodium chloride as thermal storage medium', *Renewable Energy*, 113, pp. 1182–1192.

Ndukwu, M.C. *et al.* (2020) 'Development of a low-cost wind-powered active solar dryer integrated with glycerol as thermal storage', *Renewable Energy*, 154, pp. 553–568.

Palaniappan, C. and Subramanian, S. V. (1998) 'Economics of solar air pre-heating in south Indian tea factories: A case study', *Solar Energy*, 63(1), pp. 31–37.

Patil, R. and Gawande, R. (2016) 'A review on solar tunnel greenhouse drying system', *Renewable and Sustainable Energy Reviews*, 56, pp. 196–214.

Pino, A., Lucena, F.J.P., Macho, J.G. (2019). Economic Analysis for Solar Energy Integration in a Microbrewery, in: *2019 International Conference on Smart Energy Systems and Technologies (SEST). IEEE*, pp. 1–6. https://doi.org/10.1109/SEST.2019.8849128

Pou, K.R. and Tripathy, P.P. (2020) 'Process optimization of vacuum-assisted solar drying of Crush, Tear and Curl (CTC) black tea', *Journal of Biosystems Engineering*, 45(1), pp. 24–32.

Pratoto, A., Daguenet, M. and Zeghmati, B. (1998) 'A simplified technique for sizing solar-assisted fixed-bed batch dryers: Application to granulated natural rubber', *Energy Conversion and Management*, 39(9), pp. 963–971.

Quijera, J.A., Alriols, M.G. and Labidi, J. (2011) 'Integration of a solar thermal system in a dairy process', *Renewable Energy*, 36(6), pp. 1843–1853.

Ramaiah, R., Shashi Shekar, K.S. (2018). Solar Thermal Energy Utilization for Medium Temperature Industrial Process Heat Applications – A Review. *IOP Conference Series: Materials Science and Engineering*, 376, 012035. https://doi.org/10.1088/1757-899X/376/1/012035

Ramos, C., Ramirez, R. and Beltran, J. (2014) 'Potential assessment in Mexico for solar process heat applications in food and textile industries', *Energy Procedia*, 49, pp. 1879–1884.

Ravi Kumar, K., Krishna Chaitanya, N.V.V. and Sendhil Kumar, N. (2021) 'Solar thermal energy technologies and its applications for process heating and power generation – A review', *Journal of Cleaner Production*, 282, p. 125296. Available at: https://doi.org/10.1016/j.jclepro.2020.125296.

Sabiha, M.A. *et al.* (2015) 'Progress and latest developments of evacuated tube solar collectors', *Renewable and Sustainable Energy Reviews*, 51, pp. 1038–1054.

Saravanabhavan, S. *et al.* (2006) 'Reversing the conventional leather processing sequence for cleaner leather production', *Environmental Science & Technology*, 40(3), pp. 1069–1075.

Schnitzer, H., Brunner, C. and Gwehenberger, G. (2007) 'Minimizing greenhouse gas emissions through the application of solar thermal energy in industrial processes', *Journal of Cleaner Production*, 15(13–14), pp. 1271–1286. Available at: https://doi.org/10.1016/j.jclepro.2006.07.023.

Search, H. *et al.* (2004) 'A general solution of unsteady stokes equations', 229. Available at: https://doi.org/10.1016/j.

Shafiur Rahman, M. (2006). Drying of Fish and Seafood, in: *Handbook of Industrial Drying*, Third Edition. CRC Press. https://doi.org/10.1201/9781420017618.ch22

Sharma, A. *et al.* (2019) 'Study of energy management in a tea processing industry in Assam, India', in *AIP Conference Proceedings*. AIP Publishing LLC, p. 20012.

Sharma, A.K. *et al.* (2015) *Potential of Solar Energy Utilization for Process Heating in Paper Industry in India: A Preliminary Assessment, Energy Procedia*. Elsevier B.V. Available at: https://doi.org/10.1016/j.egypro.2015.11.486.

Sharma, A.K. *et al.* (2017) 'Potential of solar industrial process heating in dairy industry in India and consequent carbon mitigation', *Journal of Cleaner Production*, 140, pp. 714–724. Available at: https://doi.org/10.1016/j.jclepro.2016.07.157.2

Shringi, V., Kothari, S. and Panwar, N.L. (2014) 'Experimental investigation of drying of garlic clove in solar dryer using phase change material as energy storage', *Journal of Thermal Analysis and Calorimetry*, 118, pp. 533–539. Available at: https://doi.org/10.1007/s10973-014-3991-0.

Sobiski, P. and Swierczyna, R. (2009) 'Capture and containment ventilation rates for commercial kitchen appliances measured during RP-1362', *ASHRAE Transactions*, 115, p. 161.

Sopian, K. *et al.* (2000) 'Experimental studies on a solar assisted drying system for herbal tea', in *World Renewable Energy Congress VI*. Elsevier, pp. 1139–1142.

Suleman, F., Dincer, I. and Agelin-Chaab, M. (2014) 'Energy and exergy analyses of an integrated solar heat pump system', *Applied Thermal Engineering*, 73(1), pp. 559–566.

Tanwanichkul, B., Thepa, S. and Rordprapat, W. (2013) 'Thermal modeling of the forced convection Sandwich Greenhouse drying system for rubber sheets', *Energy Conversion and Management*, 74, pp. 511–523.

Vijayaraju, K. and Bakthavatsalam, A.K. (2020) 'Adoption of Integrated Solar System with Thermic fluid working medium for Parboiling Rice Mills in Tamilnadu', in *IOP Conference Series: Earth and Environmental Science*. IOP Publishing, p. 12049.

Zachariah, R., Maatallah, T. and Modi, A. (2021) 'Environmental and economic analysis of a photovoltaic assisted mixed mode solar dryer with thermal energy storage and exhaust air recirculation', *International Journal of Energy Research*, 45(2), pp. 1879–1891.

Zheng, H. (2017) 'Solar concentrating directly to drive desalination technologies', in *Solar Energy Desalination Technology*. Available at: https://doi.org/10.1016/b978-0-12-805411-6.00009-9.

8 Waste Heat Recovery from Industry Through Thermal Energy Storage Mediums

*Abhishek Anand, Pragya Gupta,
Muhamad Mansor, Amritanshu Shukla, and
Atul Sharma*

8.1 INTRODUCTION

Energy consumption is an essential driver for the economic development of any country. It is a vital parameter that determines the level of production of any region. The International Energy Agency (IEA) report suggests that the rise in prime power consumption, more than 49%, has caused massive carbon emissions in the last 20 years. The significant chunks of this energy are derived from the burning of fossil-based resources, which is responsible for the degradation of the environment. Renewable energy sources have to play a significant position in solving the current environmental problems by providing clean and economic power without impairing the current pace of economic growth.

The statistics have shown that the industrial sectors are the major consumer of energy worldwide. The industrial division comprises 37% of worldwide power consumption. As a consequence of large-scale industrial activities, industrial waste heat (IWH) generation is a prime concern. About 33% of this energy is wasted as heat without any proper usage. IWH is released into the environment through hot industrial devices, exhaust gases, cooling units, etc. This heat is mainly low-temperature heat (100–400°C). There is a great prospect of recovering this IWH and reusing it in different applications. This strategy of recovery and reuse will ensure lesser wastage of energy and promote energy efficiency, lesser environmental impacts, and low production cost.

For the IWH recovery, two types of technologies can be usually thought of, i.e., active and passive, which are described in Figure 8.1. In the active IWH, the recovered energy is transformed into other energy forms or at a higher temperature. In contrast, passive IWH relies upon using the recovered energy at the same or a lower temperature. Thermal energy storage (TES) can be enthusiastically exploited to rescue waste heat. In this regard, sensible heat, latent heat, and thermochemical energy deposits can be extensively explored and studied for the IWH option. Phase change materials (PCMs) can be most credible for rescuing, storing, and reusing unused heat. PCMs have an advantage over sensible heat storage materials in that they can store an enormous amount of heat energy over a narrow band of the temperature range. Based on the heat requirement site, IWH can be divided into on-site and off-site. When the

DOI: 10.1201/9781003345558-10

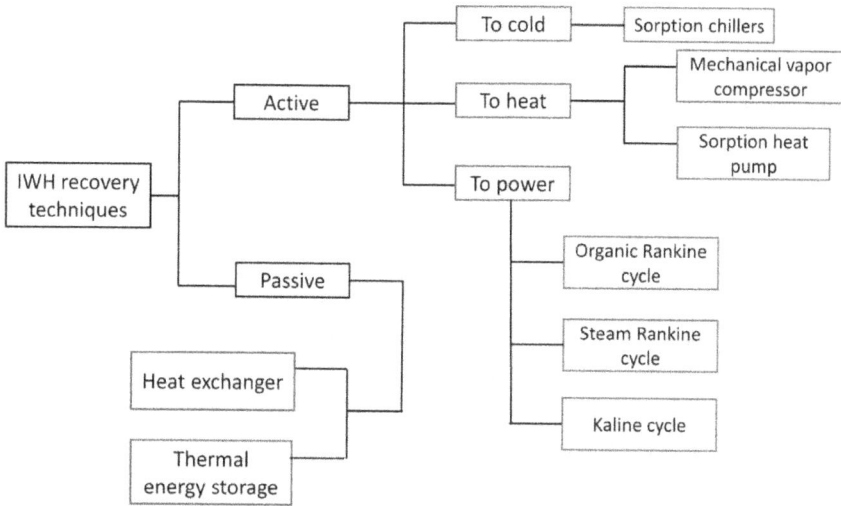

FIGURE 8.1   IWH recovery technique.

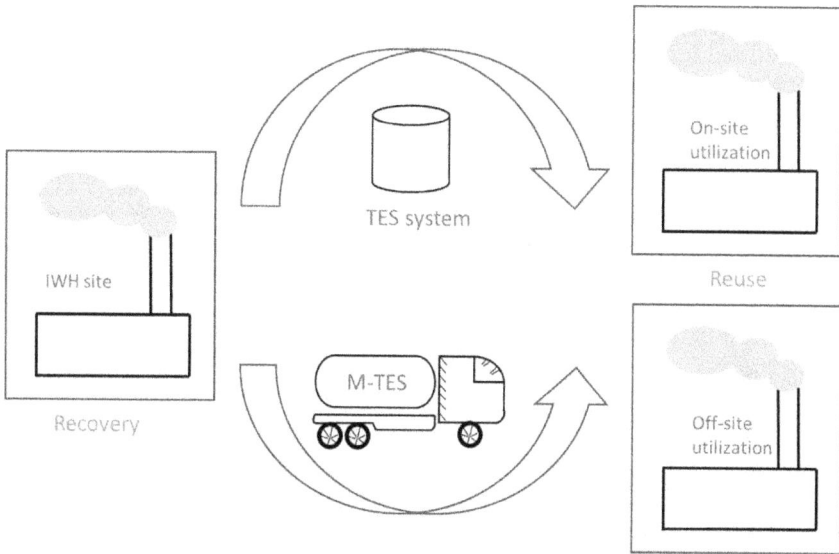

FIGURE 8.2   On-site and off-site heat recovery and reuse.

heat recovery and demand are located in the same spot, it is called on-site. Off-site
is considered when the heat recovery and demand sites are located at some distance
from each other. The on-site and off-site heat reuse has been schematically shown
in Figure 8.2. In the off-site case, the IWH is transported from the recovery to the
demand site using a mobile-TES system (also known as M-TES).

The present chapter focuses on IWH recovery, storage, and utilization using the TES, particularly the PCM in the industrial sector. In the first part, on-site reuse is considered and divided into manufacturing industries and different industrial exercises. The manufacturing industries cover heavy industries such as metallic and non-metallic, food processing, paper and pulp, and chemical industries. The incineration plants, vehicle engines, and power has been covered under other industrial activities.

8.2 ON-SITE WASTE HEAT RECOVERY AND REUSE

On-site waste heat retrieval explains how to decrease the final product's total cost by reducing the energy utilized by the system and suppressing carbon dioxide emissions. In specific sectors, between 15–50% of the total energy is misplaced as it wastes heat from exhausted gas, convection cooling, and other forms. Waste heat is dissipated, which results in energy waste, water use, increased greenhouse gasses, and environmental contamination. Reusing waste heat is a practical approach to these problems that has advantages for the environment and the economy and is of interest to researchers.

8.2.1 Manufacturing Industry

The technology can potentially recycle excess heat from the exhaust of the manufacturing industry. It significantly induces overall efficiency by reusing vapour, which can be utilized for heating purposes in the production unit or for the power system. The technology recovers excessive energy during processing without any other cost. Although a significant amount of energy is wasted during manufacturing, it sometimes cannot be used properly because of its low temperature and poor quality. The dynamic use of low-temperature excessive energy can be successfully transformed into mechanical or electrical energy. In order to outline the methods and develop capabilities and acceptability for manufacturing applications, an evaluation of several waste heat recovery systems has been done in this study (Singh and Pedersen, 2016).

8.2.1.1 Manufacture of Essential Metals

The refining or smelting of ore can be exploited for ferrous and non-ferrous metals. Various on-site waste heat recovery researchers evaluated the foundry and other metallurgical techniques. An elevated temperature was designed for unused heat recovery from the foundry furnace of the Swedish pump manufacturing unit. The recovered heat was considered for space heating application, which benefited in the minimization of energy consumption. The predicted unexploited heat was around 10 GW h/y, which can be recovered using a water-based chiller system and accumulated in borehole storage through the heat pump. The projection of heat recovery from the foundry furnace was 3800 MWh yearly. The usage of waste heat is advantageous in the reduction of 1500 tons per year of carbon dioxide. The electric arc heating device reduces power use and carbon dioxide discharge, comparable to continuous steel manufacturing using a blast furnace. Although electric power compensates for most of the energy, direct fuel combustion is also responsible for a significant percentage of thermal energy. Throughout the latest research, outlines are available to enhance energy efficiency and lower carbon dioxide emissions in an electric arc furnace. We create potential process designs based on these technologies, including the switch to Oxyfuel slag preheating, onsite use of the CO_2-rich gasification process and off-gas heat, and recovery of residual heat produced by the high-temperature gas channel of an electric arc furnace to processed boilers. The power system model is used in a brief study to calculate the possibilities for efficiency gains, carbon dioxide reduction, and refuse heat usage. In a design, scientists examine how buffer limits, carbon dioxide and excess heat usage ratios, and energy and CO_2 emission savings relate to one another. The monetary analysis produces the ideal system layout for various framework conditions (Dock and Kienberger, 2022).

The heat pump combined with exhaust energy has been considered an efficient and advanced option to prevent global warming to 1.5°C by 2050. The excess energy from steel is recovered using a novel rotational heat pump technique for room heating. Higher energy efficiency, as well as heating capacity, were attained using centrifugal compressors with paralleled connected system topology. According to simulations, a temperature increase of far more than 30°C will result in a coefficient of performance greater than 6. To verify the results obtained, a 10 MW-sized centrifugal pump was set up. It is verified in China's Angang Lingshan steel manufacturing company. The tested heating capacity and coefficient of performance were 6.67 and 9.67 MW, respectively, with wastewater inlet temperatures of 32.5°C and heated water exit temperatures of 62.5°C. In stimulating situations, the system's capacity outweighed the heating load. The method demonstrated benefits over traditional techniques during the commercial and conservational evaluation (Hu et al., 2022).

In another report, the scientists investigated how to retrieve discontinuous waste heat from a foundry for storage in a 10 MWh thermal energy storage system. It is a packed bed with a direct heat exchanger at 300°C to make it the dual thermal storage system. They aim to employ the energy stored as a heating intake for various heating strategies and rooms as high as 100°C.

The researchers have also concentrated on using aluminium melt in sand molds as thermal energy storage to capture latent heat during solidification. It involved setting up a sand mold cavity with aluminium melt. The melt was near the cavity as it could soak heat released at the time of solidification. The casted pattern was then released from the mold and moved via a conveyor to an insulation container, where it was supported to heat the scrap via conduction. The energy usage declined by using this scrap in the furnace as fuel. Around 6.4% of the unused energy was thought to be recoverable and usable using the technique.

8.2.1.2 Manufacturing of Non-Metallic Mineral Products

The classification belongs to recycling unused energy discarded by non-metallic industries, including thermoelectric power generators. The process analyzes the possibility of power generation due to excess heat produced by silicon casting using inline measurements and calculations. During silicon casting, a thermoelectric generator based on bismuth-tellurium modules was taken into operation. The power was calculated at the highest temperature differential. Here, the forecasting has been done for the produced power to accelerate beyond the measured statistics. It was also predicted that the output power could double the heat transfer coefficient at the cooler side by bringing the generator nearby the heat source. The generator design can be altered to produce more power using minimum thermoelectric material. It offers recommendations for planning thermoelectric generators to exploit power generation by utilizing unused energy released by silicon during casting (Børset *et al.*, 2017).

High temperatures and energy-intensive processes are required to produce silicon. As a necessary component of silicon furnace operation, abrupt releases of high-temperature gases to the outlet result in significant oscillations in the unused heat recovery procedure. Because of this, unused heat recuperation is complex, which often limits the generated steam and diminishes system efficiency. It is evaluated for a silicon processing facility in Norway to retrofit thermal power deposit to the current unused heat recapture system to prevent this restriction and boost system efficiency. It is discovered that a vapour accumulator similar to current bypasses, which doesn't interrupt the current unused heat recapture system, is most suited for the situation. The analysis employs actual plant parameters to verify a dynamic simulation model developed in Modelica. Financial analyses are employed to specify the configuration for the steam accumulator, and a dynamic simulation model is used to assess how well they function. According to the analytical model, stored volumes under 10 m³ can generate annual returns of about 23 k€, and the economic efficiencies' approximations are appropriate. Based on federal spending, capital expenses, eliminating transit and on-site setup, are roughly 120 k€, with payback periods extending between three to six years. The payback periods extend from seven to ten years when transportation and on-site setup are considered. The plant operators thought it was feasible to include this size of storage. Consequently, the data would be used to develop a business plan with a more thorough financial estimate (Rohde *et al.*, 2022).

The graphite furnace frequently used to produce high-purity graphite ultimately releases a lot of excess heat that includes being recovered significantly to improve energy efficiency. The reliability of a graphite furnace for heat transfer was numerically examined during electrical heating and simple cooling periods. The temperature-dependent characteristics with heat transfer factors were employed with on-site observations to forecast overall waste heat recovery capability. The results demonstrate that the fossil fuel can exceed 368°C without igniting the volatiles in the coal, while the centre temperature could exceed 3000°C during graphitization manufacturing. Regarding total power input and heat output from the furnace, the heat absorption from the upper surface of coal is responsible for 48.5% and 66.9%, respectively. The capability for unused heat recovery from the graphitization heater is explored. The results can be used as a guide for the performance and technological advancement of graphitization furnace heat recovery (Lan *et al.*, 2022).

8.2.1.3 Paper and Pulp Industry

With a production of 41.8% metric tons of wood pulp per year, and around 22% of the average global production, Europe ranks second in producing and consuming paper and board. Sulfate pulp (which accounts for 60% of complete output), mechanical and sub-pulp (32% of overall manufacturing), and sulfite pulp (5% of overall output) are the three significant categories of wood pulp used for papermaking. Sweden, Finland, Germany, and Portugal are Europe's four largest pulp providers. However, Sweden and Finland have more than 57% of the entire output, whereas Italy, Germany, France, and the UK are the four most significant customers. Just 0.7 metric tons of pulp is produced in Italy annually, most of which is made using mechanical and semi-chemical pulping techniques. The remainder is made using sulfate pulping. Germany, Finland, Sweden, and Italy are the top four paper manufacturers for different types of paper, such as graphic, sanitary, household, and packaging paper, with annual production totals of 22.8, 13.1, 11.7, and 9.1 metric tons for respective countries. Other countries followed were Germany with 13.1 metric tons, Finland with 11.7 metric tons, and Italy with 9.1 metric tons.

Regarding size, more than 50% of the paper industry in Italy has an annual paper production of less than a 25-kilo ton, and many of them are private firms. Despite global growth in pulp and paper sector production from 2000 to 2018, the European trend remained downward in 2019, with a decrease of roughly 3%. Due to international uncertainty and business pressures, the EU paper and board output fell in 2019 along with the EU economy, in contrast to a marked increase in international pulp production (+0.8%) brought on by export sector requirements and investments in additional capacity.

Because of the high demands in the pulp and paper industry, energy costs significantly impact the industry's sustainability. Different fuels are used depending on where the paper and pulp industries are situated. For instance, Italian pulp and paper industries using natural gas accounts for 95% and oil for 5% of the energy. Regarding energy consumption, Italy's pulp and paper industry used roughly 2.40 Gm^3 of natural gas and 7.00 TWh of electric energy in 2019. For energy production, cogeneration facilities are present in 85 out of 153 Italian pulp and paper mills; in this study, 50 were examined (Cioccolanti *et al.*, 2021).

Their combined natural gas and power use in 2019 was around 0.45 Gm^3 and 1.46 GWh, which are roughly 19% and 2.1% of the natural gas along with electricity needs of the total Italian pulp and paper business, respectively. With a total installed capacity of 613 MW, or roughly 13% of the total installed capacity in Italy's pulp and paper industry, the CHP factories of the examined paper industries have received particular attention. Their electricity generation capacity varies from 1–105 MW. The analysis provides accurate projections on seeking additional low-grade unused heat rescued from CHP plans. The reported paper industries represent an example of the Italian pulp and paper industry.

The late 1970s saw the selection of an American paper and pulp production plant to consider the possible energy conservation from deploying a thermal energy storage system to collect unused heat from internal procedures, as shown in Figure 8.3, when a hog fuel boiler system is running around more significant base

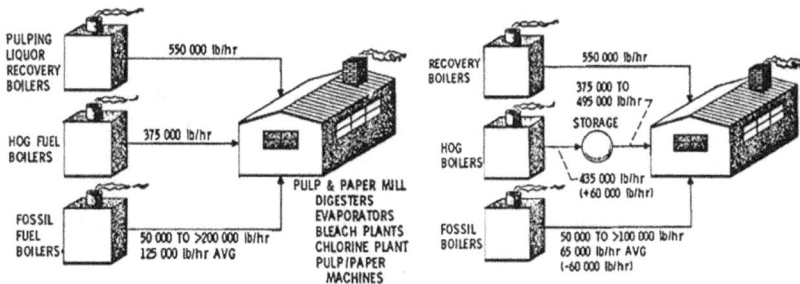

FIGURE 8.3 The role of thermal energy storage in industrial energy conservation (Miró, Gasia and Cabeza, 2016).

load. It is suggested that installing 173 m^3 water storage to collect the extra vapour so it may be used again in the same process when there is a spike in demand. According to the estimates, this results in a 50% reduction in the demand for fossil fuels. Annual fuel oil savings were anticipated from this installation (Miró, Gasia and Cabeza, 2016).

8.2.1.4 Food Processing and Beverage Industry

According to reports, the food industry uses roughly 26% of the energy in the EU overall and ranks as the UK's fourth-largest processing unit power consumer. Low-medium temperature is the most common classification for waste heat generated in food enterprises.

The quantity of unused heat accessible in the food business greatly varies from industry to industry and depends mainly on the procedure. Because different sectors employ different production processes, the only way to accurately estimate the quantity of beneficial waste heat is to undertake an extensive audit of the processes' energy consumption. However, there are widespread prospects for unused heat retrieval in the food business. It is supposed that energy waste ranges from 10–45%, depending on the operation. Theoretically, hot streams of air or water utilized in manufacturing and heating and cooling systems are the primary origins of unused heat.

For instance, the unused heat source in the red meat processing sector can be divided into byproduct rendering, rescue from refrigeration methods, and meat industries. For instance, a slaughterhouse's most energy-intensive operation is skeleton chilling. However, the conversion of byproducts can be a significant energy consumer. Activities such as scalding, singeing, and hair removal have also taken significant energy. The cleaning process that requires a lot of hot water can also be considered a huge power customer. For instance, heat is discharged into the surroundings during hog singeing processes, although power is required to dry off the hog corpses. Unused heat rescue could be heavily utilized to increase production efficiency by providing the power needed for the dehairing and scalding procedures. The singeing procedures generate waste flue gas with a temperature of 800°C. It could be used for boilers and to preheat the water through unused heat retrieval devices such as economizers.

On the other hand, it also has the potential to retrieve heat from overflow hot water using automatically operated scalding chambers. Because it entails processes such as boiling, cooling, smoking, etc., manufacturing processed meat requires more energy than killing animals. Refrigeration and product curing account for most heat loss in food processing. Unused heat contains heat from condensers, wastewater, smoking vents, and cooking exhaust, depending on the type of operation. Once more, rescue from these origins needs to be researched and established in specific instances. Recovery from some waste heat sources, including wastewater and cookstove outlets, could be challenging and extravagant due to the grease and food scraps in the exhaust.

Simultaneously in the poultry industry, where bird meat is prepared and processed, the scalding, cooling, and freezing processes account for energy use and loss. Heat can be recovered when the scalder and chiller overflow occurs by gathering and transferring the energy to the scalder or chiller procedures. The boiler used to process wash water can be preheated using heat recovered from refrigeration condenser systems. A de-superheater is placed between the compressor and condenser to rescue heat between 60–90°C temperatures. It can be used to conduct the process of extracting heat from a refrigerant condenser.

The preliminary origin of waste heat in dairy processing plants is heat rejected by the pasteurization procedure and refrigeration condensers. However, it is also possible that the excess heat obtained from the dryer might be utilized for heating the stored air for the spray dryer. With heat exchangers such as economizers or CO_2 heat pumps, chilled milk can be heated in the regenerator by recovering the heat from the pasteurization and milk cooling operations.

The heat from condensers is utilized for heating culture tanks for various activities, preheating boiler water, and providing hot water for clean-up. The study has shown that the complete energy cost for preparation can be decreased by more than 46% with a reimbursement period of roughly 40 months when using a heat pump with an inner heat exchanger for coordinated heat rescue and water generation.

In a different investigation, heat pumps increase the energy efficiency of a sizable meat processing facility. Heat pumps were employed in this study for maintaining hot water temperature of 65°C by recovering heat from the condensers, showing a potential daily energy savings of up to £530.

The refrigeration technique in which the condenser releases unused heat during freezing operations allows the unused heat to be quickly recovered from a heated refrigerant. However, it has been claimed that wastewater and exhausts are the introductory origins of unused heat in canning operations. Excess heat rescued from fruit and vegetable processing was utilized to heat water, wash cans, cleanse plants, boil water, etc.

Bakeries and biscuit factories generate excess heat from cooking ovens, fryers, pan washers, and boilers. The recovered waste heat can generate hot water for cleaning. Another potential for further usage is heat recovery from cooking oven exhausts. For instance, recovered low-quality excess heat from ovens can be used to make biscuits with a thermoacoustic heat engine.

The technology's engine, which comprises two heat exchangers and a stack of parallel plates housed in a cylindrical case and transforms thermal to acoustic energy, works without mechanical parts. According to the study, producing 1030 W of acoustic energy in thermal engine efficiency of 5.4% is possible by recovering waste heat at 150°C. However, using air preheaters can reduce oven fuel usage by 4%. In one experiment, the primary air reserve was heated to 105°C while the heat was retrieved from the exhaust of an industrial cooking oven. According to this study, raising the prior air temperature can conclude operating cost savings of at least £4,200 (Jouhara *et al.*, 2018).

8.2.1.5 Chemical Industry

A twin-screw, compressor-based, high-temperature heat pump technique was developed to rescue unused heat in the dyeing manufacturing unit. An on-site skein dyeing procedure examined the pump system's performance. The heat pump technology can efficiently use excess heat at various ranges to supply the power needed for the dyeing technique. The system's heating capacity could be precisely adjusted to perform the appropriate dyeing liquid temperature rising rate between 0.6–2.5°C/min throughout various heating operations. The outcomes of the on-site analysis additionally showed that the heat pump could be dependably utilized to raise the temperature to 95°C while maintaining a mean COP of 4.2 at the time of the heating procedure. An economic estimation has shown that a heat pump can keep 47% of running costs compared to conventional steam heating (Wu *et al.*, 2016).

8.2.2 OTHER INDUSTRIAL ACTIVITIES

8.2.2.1 Power Plants

The waste heat produced by renewable power stations will only sometimes be enough to fulfil the needs of a particular area. Then, excess heat from different origins can be blended with sustainable power sources to meet the demand. Different systems with sizable excess heat, such as industrial facilities and process industries, might produce reusable high-grade heat when combined with renewable thermal energy.

In the case of a conventional solar panel–based heat pump, the excess heat from a solar-powered system is collected. In various distributed generating applications, these solar power–incorporated heat pumps can fit or even outperform solar photovoltaics in terms of effectiveness, execution, and expense. The quantity of electricity produced by a Rankine heat pump and photovoltaics of equal extent will be comparable. Still, the heat pump system can provide 4–6 units of usable heat for every unit of energy produced. More research needs to address excess heat from photovoltaic-based power facilities. To recover excess heat, it was typically used for desalination – the fusion of various thermodynamic cycles to produce electricity or water heating applications.

For a hypothetical 5000 kWe power plant directed towards steam production with the parabolic channel, it has investigated the thermoeconomic factors of power generation and water recovery. The cogen power systems have a severe issue since, depending on the season, there may need to be more than just chilling or warming as a product to fulfil regional markets. During various months of the year, these systems must effectively utilize waste heat. They suggested a trigeneration solar heating, cooling, and power generation strategy for a home in a remote area in Western China. Because the spiral screw expander's working temperatures were suitable for Cogen, they used it.

Steam is transferred via two heat exchangers after being separated from the superheated steam produced by the spiral screw in the steam partition. The first heat exchanger releases heat for radiative heat processing and chilling, and the second channels heat for superheated water storage. After that, the solar domain is fed with the fluid via a steam divider. Compared to a typical solar power scheme, the technique included a greater solar power transformation efficiency at 58.0% instead of 10.2%.

Many photovoltaic designs also utilized and explored PCMs. This type of system has a unique design, including a phase change storage connected to a photovoltaic board and a fluid process built into the container to transport power to a heat exchanger. In this design, a PCM-based photovoltaic provides power storage capability, which can benefit solar power applicability because it enables energy generation without direct sunlight.

Increased latent heat of fusion, increased thermal conductivity, chemical stability, non-corrosiveness and toxicity, and temperature variation for the particular design are requirements for using PCMs in photovoltaic. Electrical power increased by 13.6%, according to a thermoelectric analysis of the PVT system with PCM. More PVT opportunities will open up as PCM technology advances, better suited to various environments and applications.

Recovering excess heat and utilizing it for various thermodynamic processes can boost efficiency in geothermal power stations. Adding more power generation stages increases net power output and improves geothermal resource use. Yet, adding phases also results in higher maintenance and investment expenditures. There are potential plans to incorporate these processes to extract more power. It is generally advised to keep the number of stages to two.

The recovery of heat from low-temperature geothermal origins is the primary function in a binary process power system. It makes it possible to exploit geothermal resources that would not otherwise be able to produce energy. The working fluid in this plant is often a low boiling point organic liquid, typically running on the organic Rankine or Kalina cycles. Using a heat exchanger, the liquid takes heat from the primary fluid, expands it, and transforms it into a vapour. Then, a turbine is powered by this vapour to produce energy. After cooling in a condenser, the excess vapour is returned to the cycle. Alkanes, fluoro-alkanes, ethers, and fluoro-ethers are possible working fluids (DeLovato *et al.*, 2019).

8.2.2.2 Incineration Plants

Energy recapture from excess heat converts unwanted products into proper energy, power, or fuel. This method is frequently referred to as waste-to-energy. Waste heat retrieval from incineration facilities is the analytical measure to maximize waste utilization and lower carbon emissions. Avesta (Sweden) constructed a 15,000 m^3 water-filled stone cave in 1981 to initially store excess power generated in a local ignition plant for research. These systems can also meet high powers due to the high injection power.

In an available sorption procedure, portable power storage using zeolite can use industrial debris when a pipeline-based linkage is not economically feasible. Over a year, a plant was built, run, and monitored that used litter incineration plant steam extraction to set and hold with 130°C hot air and an industrial drying technique at the client 7 km from the charging process.

The storage has a capacity of 2.3 MWh, can hold 14 tonnes of zeolite, and employs dryer exhaust air at the discharging status, with a temperature of 60°C and a relative humidity of 0.09 kg/kg. It also keeps 616 kg of carbon dioxide in each process and demonstrates no degradation within the limitations of the measuring tools. The desired power output cannot be achieved due to maldistribution within the dense zeolite bed. When considering small-scale mass production, the direct power expenses can be decreased to 73 €/MWh (Krönauer et al., 2015).

Up to 6000 tonnes of garbage are landfilled daily in Ano Liosia, Athens (Greece), and the dump gas is utilized to power an ICE power system on the landfill zone. With a 23.5 MW installed capacity, the power plant contains 15 ICEs. The increased electrical effectiveness and quick load retort of ICEs make them an excellent choice for power generation. But vehicle exhaust heat still accounts for more than half of the energy content of landfill gas emissions. Compared to ORC recourses, the thermodynamic study of all waste heat recovery cycle options reveals that water/steam processes typically increase electrical system efficacy up to 37%. A pentane cycle in the absence of a thermal lubricant middle trajectory can achieve a station efficiency of roughly 36% with the latter. Due to pentane's intense flammability, using this cycle type could be dangerous. Thus, most industries use pentane cycles to employ thermal petroleum trajectories (Gewald et al., 2012).

Three distinct situations are considered and contrasted to examine the idea's implications. The first scenario involves a plant for old-design waste that uses a fuel-burning warming-up procedure. The second strategy implicates a heat pump plant that uses steam extraction from turbines for water preheating. The third technique involves removing the previous preheating devices from the process and replacing them with power recovered from the incinerator's exhaust gas. The three scenarios are subjected to energy, exergy, and economic studies, and the critical interpretation elements of power systems have been distinguished.

The findings indicate the efficiency of the heat retrieving case, which considers heat and power outcomes. It could achieve 94%. In contrast, facilities, outdated structures, and steam recovery lines can achieve 77% and 82% efficacy, respectively. The proposed solution has the highest energetic efficiency of 81% of the three designs. The development expense will be the minimum, with a significance of less than 11 $/GJ. The impacts of various operational factors on the presented system's effectiveness are evaluated (Alrobaian, 2020).

In this study, it is suggested and technoeconomically assessed to use the hot flue gas exergy of the waste-fired CHP unit to boost the organic Rankine cycle's potential for optimizing the net power production of the hybrid cycle. Also, the system's performance was examined using alternative, ecologically friendly organic working fluids. According to the study's findings, the size of the organic Rankine cycle may be significantly increased by using the same CHP unit's flue gas, which would boost the plant's power output.

So, when the organic Rankine cycle and the primary CHP cycle operate at nominal load, the combined plant's net exergy and energy efficiency values under various operational situations are improved relative to its initial configuration by around 10% and 20%, respectively. The payback period for the parallelization project was reduced by nearly 10% (from 7.4 years to 6.7 years) because of this exergy utilization, making it economically advantageous as well. The substitute organic working fluid does not affect the system's technical and economic performance indices (Arabkoohsar and Nami, 2019).

8.2.2.3 Vehicle Engines

Both transportation and traditional power generation systems require engines as crucial components. In reality, the cooling streams and exhaust gases squander roughly
two-thirds of the energy that goes into an engine. The exhaust gas temperature, for
instance, in a four-stroke diesel engine is roughly 400–500°C while the engine is
operating at total capacity. As these exhaust gases are released into the environment,
it is possible to recover a significant amount of proper heat from them. However, the
engines might not always run continuously in certain circumstances. The recovery
of waste heat and bridging the gap between energy supply and demand is possible in
such circumstances using TES systems.

A water-cooled diesel engine attached to an electrical dynamometer, integrated
with a recovery heat exchanger, and connected to the TES system made up the experimental configuration as shown in Figure 8.4. The stainless steel cylindrical vessel
was part of the TES system, and 55 kg of castor oil and 15 kg of paraffin contained
in 48 cylindrical capsules, respectively, were chosen as the visual and latent TES
materials in the first investigation.

If results were compared to the identical setup without TES technology, fuel consumption might be reduced by up to 15.2%. The thermal behavior of the TES system with paraffin was working at different engine load conditions. Depending on
the load, a direct correlation between charging rate and efficiency was discovered.
Also, this device recovered around 15% of the overall heat. In the first TES tank, 19
kg of d-sorbitol and 15 kg of paraffin were utilized as the TES and HTF materials,
respectively, while 55 kg of castor oil was employed in the second TES tank. Findings demonstrated the viability of the chosen TES and HTF for the intended application, and the advantages of the cascaded mode as total waste heat recovery improved
in comparison to the initial single tank configuration by up to 20%. The hopes for
generating financial gains are already mentioned when deploying TES systems for
automotive applications.

The TES system comprised a stainless steel vessel filled with 40 low-density polyethylene spherical containers, each containing about 100 g of PCM (paraffin). Water
served as the HTF in this system. The energy efficiency of the integrated system

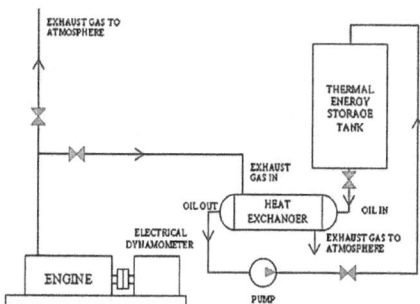

FIGURE 8.4 Schematic diagram of the experimental setup (left); picture of the experimental setup (right) (Miró, Gasia and Cabeza, 2016).

fluctuated between 3.2% and 34.2%, and results showed that roughly 7% of the total exhaust heat was recovered.

For specific automotive engine applications, the benefits show TES addition to a diesel car engine at below-freezing conditions. A combination of an exhaust gas heat recovery system and a latent heat TES accumulator was used in place of the additional heater since, under certain circumstances, the engine often requires additional heating to maintain the desired operating temperature (above 70°C). In a conventional shell and tube structure, 4 kg of a commercial PCM with a melting point of 75°C was chosen for the heat accumulator.

The reaction of the TES accumulator was quicker than the original system, and the slow response wastes were significantly decreased. A tank containing 4.2 kg of xylitol with a heat capacity of 1300 kJ was later proposed for storing waste heat from a hot engine coolant. The heat stored may be used to reheat the engine and the interior of the vehicle swiftly. The researchers built and modeled a TES system for the engine exhaust streams in the context of marine transportation. A 1000 m³ cylindrical thermal oil storage tank was considered for the TES system.

Based on numerical findings, it was possible to cut the boilers' fuel usage by 80%. The numerically proposed method can provide energy in remote places by combining a diesel engine with a CAES unit. The heat exchanger receives the engine's exhaust fumes and the reservoir's compressed air, which are then recycled. A ceramic matrix and an inorganic salt that melted at 210°C made up the suggested PCM. Findings revealed that the integrated system used just 50% as much fuel to meet the exact demand as a single diesel unit (Miró, Gasia and Cabeza, 2016).

According to an analysis of the energy distribution in internal combustion engines, about 65–70% of the energy input is lost through dissipation in cooling circuits (about 30%) and exhaust gases (about 35–40%). In other words, the radiator and exhaust gas system are two significant heat producers for internal combustion engines. A study estimated that recovering 400–500 W of waste thermal energy as electrical energy can reduce carbon dioxide (CO_2) emission values of 6–7 gm/km, underscoring the potential and significance of waste heat recovery for internal combustion engines (Burnete *et al.*, 2022).

Only a portion of the exhaust energy is usable because of irreversibility, ambient conditions, etc. At exhaust temperatures of 600°C and higher, the exhaust exergy is more significant than 30% of its energy value, but it diminishes when the exhaust temperature is lowered. Exhaust exergy is low near the engine's optimal efficiency point but high at low engine loads (across most driving cycle operating ranges). According to a second law analysis, the exhaust exergy is almost as large as brake work in contemporary gasoline automobiles.

So, if a WHR system can be implemented, a lot of exhaust energy can be used to increase system efficiency overall. The exhaust waste heat of an SI engine has been reported to range from 4.6–120 kW depending on operating conditions, which is consistent with this analysis. However, the maximum ideal limit of usable energy is reported to be between 1.7–45 kW (Aghaali and Ångström, 2015).

8.3 OFF-SITE WASTE HEAT RECOVERY AND REUSE

The off-site IWH relies on recovering the waste heat through various industrial processes and transporting it to the desired heat demand site that can be used later, which is located at some distance from the waste heat generation site. For heat transportation, two mediums are generally used, i.e., district heating and cooling (DHC) and mobilized thermal energy storage (M-TES). Due to its economic viability, DHC is an appropriate method for supplying heat to densely inhabited areas with high heat demand, whereas M-TES is suitable for when the investment and maintenance cost of the DHC's infrastructure is high. The research and investigation have shown that M-TES can be feasible for transportation up to 35 km by truck, 200 km by train, and three days by ship.

For an efficient M-TES system, certain requirements should be fulfilled (Miró, Gasia and Cabeza, 2016), for example, the following:

- High energy density to transport the maximum amount of energy in a very small volume
- High charging and discharging capability to ensure compatibility with different IWH sources and heat demands
- Temperature stability
- Appropriate working strategy ensuring storage mass and number of cycles

Guo *et al.* (2013) conducted research to evaluate the viability of shipping IWH from a Swedish utility plant to a steelworks business. Comparisons were made among three different ways to move the IWH: (1) a typical DH solution requiring new underground plumbing, (2) train transportation using PCM, and (3) train transportation using zeolite sorption technology. The sorption option using zeolites (thermochemical materials, often known as TCM) was the most effective in terms of price.

Li *et al.* (2013) looked at the economic costs of employing an M-TES system with a 28 m³ prototype container that uses erythritol as the PCM to send heat to a demand site that is between 10–50 kilometers from the IWH source. The authors also compared it to other ways of supplying heat. The cost was between $30–60 USD/MWh, according to the results, and it was noted that the cost fluctuation was inversely correlated with the heat demand but proportional to the transport distance. In addition, the cost was more responsive to PCM pricing compared to other factors, such as shipping costs.

In a case study from Sweden, Wang et al. (Wang, 2010; Wang *et al.*, 2014) showed how the M-TES technology was used to recover IWH from a combined heat and power (CHP) plant. This IWH was delivered to a DH network that served a small village 20 km from a CHP plant with hot water during the summer and space heating and tap water during the winter. In the CHP plant, four probable IWH sources were found, and a TES material was suggested for each stream based on the streams' exit temperatures. The results revealed that 224 MWh of heat could be delivered each

TABLE 8.1
PCMs with Their Thermophysical Properties

PCM	Type	Melting Temperature (°C)	Thermal Conductivity (W/m·K)	Latent Heat of Fusion (kJ/kg)	Specific Heat (kJ/kg·K)	References
D-mannitol	Organic	165	0.6	341	1.26	(Alva, Lin and Fang, 2018; Du et al., 2021)
Erythritol	Organic	118	0.732 (solid) 0.326 (liquid)	339	1.35 (solid) 2.74 (liquid)	(Yuan et al., 2019; Du et al., 2021)
NaOH	Inorganic	318	—	158	—	(Nomura, Okinaka and Akiyama, 2010; Chiu et al., 2016)
Xylitol	Organic	93	0.52 (solid) 0.36 (liquid)	280	1.27 (solid)	(Diarce et al., 2015; Höhlein, König-Haagen and Brüggemann, 2017)
Magnesium chloride hexahydrate	Inorganic	115	—	167	—	(Lin, Alva and Fang, 2018; Gan and Xiang, 2020)
Sodium acetate trihydrate	Inorganic	58	0.7 (solid)	264	2.79 (solid)	(Xiao et al., 2018; Wang et al., 2019)
Sodium carbonate (17%)-sodium hydroxide (83%	Inorganic-Inorganic	285	—	252	—	(Du et al., 2021)

month, assuming an M-TES container energy efficiency of 85%, and that two TES containers were charged and discharged every day.

Table 8.1 shows the various phase change materials and their thermophysical properties that can be used in M-TES applications.

8.4 MAJOR ISSUES AND CHALLENGES

One of the major concerns of the TES system is the maturity level of the TES. The other barriers include financial barriers, industrial production processes, and commercial confidentiality associated with the sectors. When we talk about the maturity

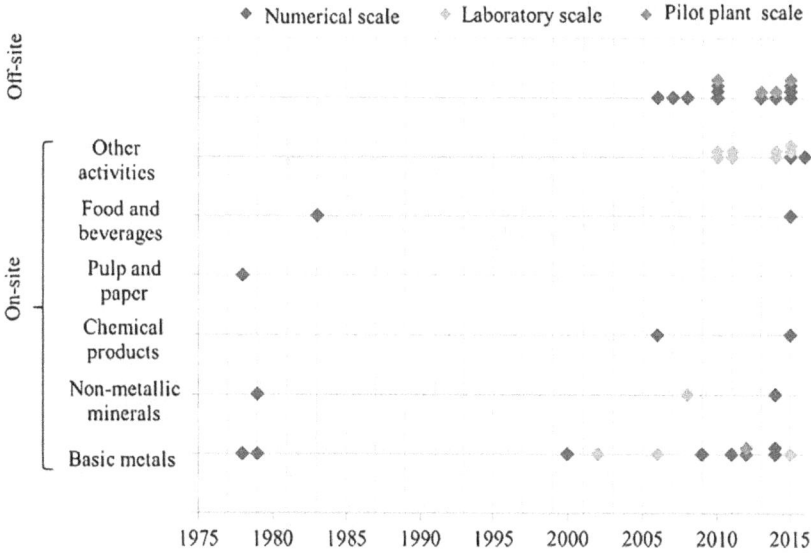

FIGURE 8.5 Developmental level of the projects.

and commercialization level, only underground thermal energy storage (UTES), pit thermal energy storage (PTES), and residential hot water systems with storage have reached the advanced maturity level. The rest are all in the development stage. Figure 8.5 shows that the majority of the projects are numerical scale and very few are pilot and real-scale projects, but in recent years, due to rising environmental awareness, IWH with M-TES is gaining popularity as a result of the large-scale commercialization of TES materials and systems. The major financial and administrative barriers include uncertainty regarding the investing company's economic future and expectations of returns that are excessive.

8.5 CONCLUSION

Industries are the major consumer of energy worldwide. IWH recovery, storage, and reuse would not only help in CO_2 reduction but would also reduce energy costs and make industrial activities less energy intensive and energy efficient. However, due to technological and financial constraints, the potential of IWH recovery has not been fully utilized.

IWH with a TES unit can be beneficial as a thermal management tool in comparison to other technologies, as the IWH source is intermittent and the distance between the energy source and heat demand site vary to a large scale and is not fixed. Thus, it helps in reducing the mismatch between energy demand and supply. TES with on-site heat recovery and reuse is beneficial for many internal processes, and off-site M-TES can efficiently deliver the energy from the source to the heat demand site.

Most of the study reported at present are numerical in nature due to the technical and financial constraints. The TES mediums used are also scarce. There is great potential for TES in IWH in the future due to the increasing commercialization of the TES material and systems.

8.5.1 ACKNOWLEDGMENT

The work was supported by Tenaga Nasional Berhad (TNB) and UNITEN through the BOLD refresh Postdoctoral Fellowships under the project code of J510050002-IC-6 BOLDREFRESH2025-Centre of Excellence.

REFERENCES

Aghaali, H. and Ångström, H.-E. (2015) 'A review of turbocompounding as a waste heat recovery system for internal combustion engines', *Renewable and Sustainable Energy Reviews*, 49, pp. 813–824. Available at: https://doi.org/10.1016/j.rser.2015.04.144.

Alrobaian, A.A. (2020) 'Improving waste incineration CHP plant efficiency by waste heat recovery for feedwater preheating process: energy, exergy, and economic (3E) analysis', *Journal of the Brazilian Society of Mechanical Sciences and Engineering*, 42(8), p. 403. Available at: https://doi.org/10.1007/s40430-020-02460-w.

Alva, G., Lin, Y. and Fang, G. (2018) 'An overview of thermal energy storage systems', *Energy*, 144, pp. 341–378. Available at: https://doi.org/10.1016/j.energy.2017.12.037.

Arabkoohsar, A. and Nami, H. (2019) 'Thermodynamic and economic analyses of a hybrid waste-driven CHP–ORC plant with exhaust heat recovery', *Energy Conversion and Management*, 187(March), pp. 512–522. Available at: https://doi.org/10.1016/j.enconman.2019.03.027.

Børset, M.T. *et al.* (2017) 'Exploring the potential for waste heat recovery during metal casting with thermoelectric generators: On-site experiments and mathematical modeling', *Energy*, 118, pp. 865–875. Available at: https://doi.org/10.1016/j.energy.2016.10.109.

Burnete, N.V. *et al.* (2022) 'Review of thermoelectric generation for internal combustion engine waste heat recovery', *Progress in Energy and Combustion Science*, 91(March), p. 101009. Available at: https://doi.org/10.1016/j.pecs.2022.101009.

Chiu, J.N. *et al.* (2016) 'Industrial surplus heat transportation for use in district heating', *Energy*, 110, pp. 139–147. Available at: https://doi.org/10.1016/j.energy.2016.05.003.

Cioccolanti, L. *et al.* (2021) 'District heating potential in the case of low-grade waste heat recovery from energy intensive industries', *Applied Thermal Engineering*, 191(February), p. 116851. Available at: https://doi.org/10.1016/j.applthermaleng.2021.116851.

DeLovato, N. *et al.* (2019) 'A review of heat recovery applications for solar and geothermal power plants', *Renewable and Sustainable Energy Reviews*, 114(August 2018), p. 109329. Available at: https://doi.org/10.1016/j.rser.2019.109329.

Diarce, G. *et al.* (2015) 'Eutectic mixtures of sugar alcohols for thermal energy storage in the 50–90°C temperature range', *Solar Energy Materials and Solar Cells*, 134, pp. 215–226. Available at: https://doi.org/10.1016/j.solmat.2014.11.050.

Dock, J. and Kienberger, T. (2022) 'Techno-economic case study on Oxyfuel technology implementation in EAF steel mills – Concepts for waste heat recovery and carbon dioxide utilization', *Cleaner Engineering and Technology*, 9(June), p. 100525. Available at: https://doi.org/10.1016/j.clet.2022.100525.

Du, K. *et al.* (2021) 'A state-of-the-art review of the application of phase change materials (PCM) in Mobilized-Thermal Energy Storage (M-TES) for recovering low-temperature industrial waste heat (IWH) for distributed heat supply', *Renewable Energy*, 168, pp. 1040–1057. Available at: https://doi.org/10.1016/j.renene.2020.12.057.

Gan, G. and Xiang, Y. (2020) 'Experimental investigation of a photovoltaic thermal collector with energy storage for power generation, building heating and natural ventilation', *Renewable Energy*, 150, pp. 12–22. Available at: https://doi.org/10.1016/j.renene.2019.12.112.

Gewald, D. *et al.* (2012) 'Waste heat recovery from a landfill gas-fired power plant', *Renewable and Sustainable Energy Reviews*, 16(4), pp. 1779–1789. Available at: https://doi.org/10.1016/j.rser.2012.01.036.

Guo, S. *et al.* (2013) 'Numerical simulation study on optimizing charging process of the direct contact mobilized thermal energy storage', *Applied Energy*, 112, pp. 1416–1423. Available at: https://doi.org/10.1016/j.apenergy.2013.01.020.

Höhlein, S., König-Haagen, A. and Brüggemann, D. (2017) 'Thermophysical characterization of MgCl2·6H2O, Xylitol and erythritol as phase change materials (PCM) for Latent Heat Thermal Energy Storage (LHTES)', *Materials*, 10(4), p. 444. Available at: https://doi.org/10.3390/ma10040444.

Hu, B. *et al.* (2022) 'Ten megawatt scale vapor compression heat pump for low temperature waste heat recovery: Onsite application research', *Energy*, 238, p. 121699. Available at: https://doi.org/10.1016/j.energy.2021.121699.

Jouhara, H. *et al.* (2018) 'Waste heat recovery technologies and applications', *Thermal Science and Engineering Progress*, 6(January), pp. 268–289. Available at: https://doi.org/10.1016/j.tsep.2018.04.017.

Krönauer, A. *et al.* (2015) 'Mobile sorption heat storage in industrial waste heat recovery', *Energy Procedia*, 73, pp. 272–280. Available at: https://doi.org/10.1016/j.egypro.2015.07.688.

Lan, Y. *et al.* (2022) 'Investigation of the waste heat recovery and pollutant emission reduction potential in graphitization furnace', *Energy*, 245, p. 123292. Available at: https://doi.org/10.1016/j.energy.2022.123292.

Li, H. *et al.* (2013) 'Economic assessment of the mobilized thermal energy storage (M-TES) system for distributed heat supply', *Applied Energy*, 104, pp. 178–186. Available at: https://doi.org/10.1016/j.apenergy.2012.11.010.

Lin, Y., Alva, G. and Fang, G. (2018) 'Review on thermal performances and applications of thermal energy storage systems with inorganic phase change materials', *Energy*, 165, pp. 685–708. Available at: https://doi.org/10.1016/j.energy.2018.09.128.

Miró, L., Gasia, J. and Cabeza, L.F. (2016) 'Thermal energy storage (TES) for industrial waste heat (IWH) recovery: A review', *Applied Energy*, 179, pp. 284–301. Available at: https://doi.org/10.1016/j.apenergy.2016.06.147.

Nomura, T., Okinaka, N. and Akiyama, T. (2010) 'Waste heat transportation system, using phase change material (PCM) from steelworks to chemical plant', *Resources, Conservation and Recycling*, 54(11), pp. 1000–1006. Available at: https://doi.org/10.1016/j.resconrec.2010.02.007.

Rohde, D. *et al.* (2022) 'Thermal energy storage for increased waste heat recovery at a silicon production plant in Norway', *Applied Thermal Engineering*, 215(June), p. 118909. Available at: https://doi.org/10.1016/j.applthermaleng.2022.118909.

Singh, D.V. and Pedersen, E. (2016) 'A review of waste heat recovery technologies for maritime applications', *Energy Conversion and Management*, 111(X), pp. 315–328. Available at: https://doi.org/10.1016/j.enconman.2015.12.073.

Wang, W. (2010) *Mobilized Thermal Energy Storage for Heat Recovery for Distributed Heating, School of Sustainable Development of Society and Technology.* Malardalen University, Sweden.

Wang, W. *et al.* (2014) 'Experimental study on the direct/indirect contact energy storage container in mobilized thermal energy system (M-TES)', *Applied Energy*, 119, pp. 181–189. Available at: https://doi.org/10.1016/j.apenergy.2013.12.058.

Wang, Y. *et al.* (2019) 'Preparation and thermal properties of sodium acetate trihydrate as a novel phase change material for energy storage', *Energy*, 167, pp. 269–274. Available at: https://doi.org/10.1016/j.energy.2018.10.164.

Wu, X. *et al.* (2016) 'Performance evaluation of a capacity-regulated high temperature heat pump for waste heat recovery in dyeing industry', *Applied Thermal Engineering*, 93, pp. 1193–1201. Available at: https://doi.org/10.1016/j.applthermaleng.2015.10.075.

Xiao, Q. *et al.* (2018) 'Solar thermal energy storage based on sodium acetate trihydrate phase change hydrogels with excellent light-to-thermal conversion performance', *Energy*, 165, pp. 1240–1247. Available at: https://doi.org/10.1016/j.energy.2018.10.105.

Yuan, M. *et al.* (2019) 'Characterization and stability study of a form-stable erythritol/expanded graphite composite phase change material for thermal energy storage', *Renewable Energy*, 136, pp. 211–222. Available at: https://doi.org/10.1016/j.renene.2018.12.107.

9 Advancement in Thermal Energy Storage for Agricultural Application

Sara Baddadi, Rabeb Ayed, and Salwa Bouadila

9.1 INTRODUCTION

Field agriculture is undergoing a revolutionary transition, spurred by advances in agricultural technology. Protected crops have evolved from simple covered greenhouse structures to high-tech plant factories that optimize plant and human labor productivity. A modern greenhouse works as a system.

Greenhouse technology is strategic to intensify agricultural production and help meet global food demand as it provides an appropriate microclimate for plants to achieve better quality and higher yields. Greenhouses provide sustainable environments for growing crops in almost any location. Local production, increased yields per area, longer harvest periods and better controlled growing environments make greenhouses a feasible solution in disadvantaged areas where heat is not readily available.

Livestock feed worldwide is estimated to double by 2050. In fact, the unavailability of high-quality fodder profoundly affects livestock production and reproduction (Bradley & Marulanda, 2001). Fodder production is influenced by several constraints such as land unavailability, water scarcity, rainfall uncertainty as well as natural disasters. The mentioned limitations of conventional forage cultivation make hydroponics technology an attractive alternative to forage cultivation, especially for livestock. The hydroponic greenhouse is an above-ground cultivation offering a suitable choice to meet the growing needs of farm animals, guaranteeing a significant reduction in labor costs, space and water requirements and ensuring freshness and the good palatability of the fodder (Godde et al., 2021). Hydroponic technology is used to produce larger, healthier and tastier crops on a large scale. As a result, farmers have a great interest in setting up their production units.

Most greenhouse growers combine hydroponic technology with a controlled environment to optimize the greenhouse microclimate and produce the highest quality produce. In a sophisticated structure, everything can be controlled, including temperature, sunlight, humidity and lighting, allowing year-round cultivation. Producers strive to maximize yields and minimize expenses. An auxiliary conditioning device often seems essential to improve the microclimate of the greenhouse. Heating systems of an agricultural greenhouse are basically designed to respond to heat loss. Heating is the main cost involved in greenhouses, and this is usually done by burning

DOI: 10.1201/9781003345558-11

fossil fuels. That is why the heating of greenhouses is considered to be the most energy-intensive function during all of their operations. This huge energy consumption of agricultural greenhouses leads not only to increased operating costs, but also to environmental pollution. The adverse environmental effects of fossil fuels provoke climate change and grow concerns for energy security, making the use of renewable energy sources more urgent than ever.

Thus, to address the aforementioned challenges and improve the sustainability of greenhouses, farmers, engineers and governments must seek innovative solutions to ensure a sustainable supply of food and energy. The combination of renewable energy sources in heating devices has grown rapidly in recent decades. Solar energy is the most abundant source of renewable energy available on earth. Thus, solar systems appear to be the most promising technology that can provide some of the clean energy to meet the huge energy needs of greenhouses.

The solutions must make better use of renewable energy sources and improve energy efficiency. The intermittent characteristics of many renewable energy sources can be offset by using thermal energy storage systems that better match supply and demand. Since the 1970s, thermal energy storage systems have proven to be important tools for increasing energy efficiency compared to conventional energy systems. Thermal energy storage systems provide alternative heating and cooling solutions to reduce electricity and fossil fuel consumption and also replace mechanical cooling devices. Greenhouses need a lot of thermal energy, and a significant part of their cost is heating. Therefore, major benefits can be driven from thermal energy storage.

In this study, two additional heating systems with thermal energy storage were specifically designed and implemented to achieve energy optimization of the microclimate of the hydroponic greenhouse.

9.2 HYDROPONIC AGRICULTURAL GREENHOUSE SYSTEM (HAGS)

9.2.1 DESCRIPTION OF HAGS

The hydroponic agricultural greenhouse system, called HAGS, is a southeast-oriented structure, thermally insulated, whose dimensions are 8 meters by 4 meters by 4 meters. It consists of a galvanized framework covered with polyurethane sandwich panels and has two separate rooms. The first room has two glazed sides, one with single glazing on the north and the other with double glazing on the south. The greenhouse features screens that can be raised or removed to regulate light and solar radiation. For the vegetation cultivation, there are two rows of metallic structure, with five levels spaced 0.5 meters apart from each one. The second room is used for the germination of crops such as barley, maize and wheat fodder.

The HAGS is an under-shelter prototype including all the necessary equipment for ventilation, irrigation, lighting and environmental control. It has an irrigation system to optimize the supply of the nutrient solution composed of water and essential elements to promote plant growth. It is also equipped with a ventilation system of three fans, one on the south and two on the north, to improve air quality. The greenhouse also has a lighting system with heat lamps to supplement weak sunlight in winter. The germination tank is supplied with oxygen through an electric aerator.

The temperature inside the greenhouse is measured with K-type thermocouples. To measure the relative humidity, a HMP155A sensor was protected from the solar radiation and placed in the center at 1.2 meters above the floor, and a second HMP155A sensor was placed outside to register the external temperature and relative humidity. Two Kipp and Zonen pyranometers were also installed to measure global solar irradiation. A CR5000 data logger recorded all of the parameters every 10 minutes.

9.2.2 THERMAL BEHAVIOR OF HAGS

Figure 9.1 exhibits the variation of the air temperature inside the greenhouse (represented by the air temperature at the central position in the greenhouse as being a significant temperature of that inside the greenhouse) accompanied with the ambient temperature and solar radiation variation.

The temperature inside the greenhouse has the same evolution as the ambient temperature and also follows the solar radiation. For daytime temperatures ranging from 14–27°C, the temperature inside the greenhouse varies between 18–34 °C. The latter are constantly higher than the ambient temperature during the day, and it can be seen that the difference between the two temperatures clearly depends on the values of the solar radiation. Indeed, on February 8, a very sunny day when the temperature outside varied between 20–23°C, the temperature inside the greenhouse exceeded 34°C at 1 p.m., while the next day when the solar radiation does not reach 270 W/m^2, the difference between the two indoor and outdoor temperatures is around 3°C. In the early hours of the morning, the values of the indoor and outdoor temperatures approach each other, and the ambient temperature can exceed that inside the greenhouse, which generally varies between 12–16°C.

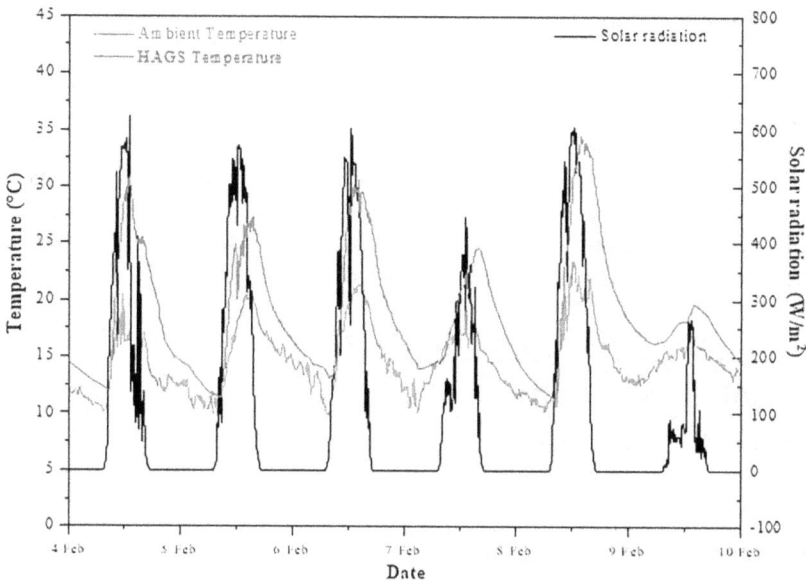

FIGURE 9.1   Evolution of the temperature inside the HAGS.

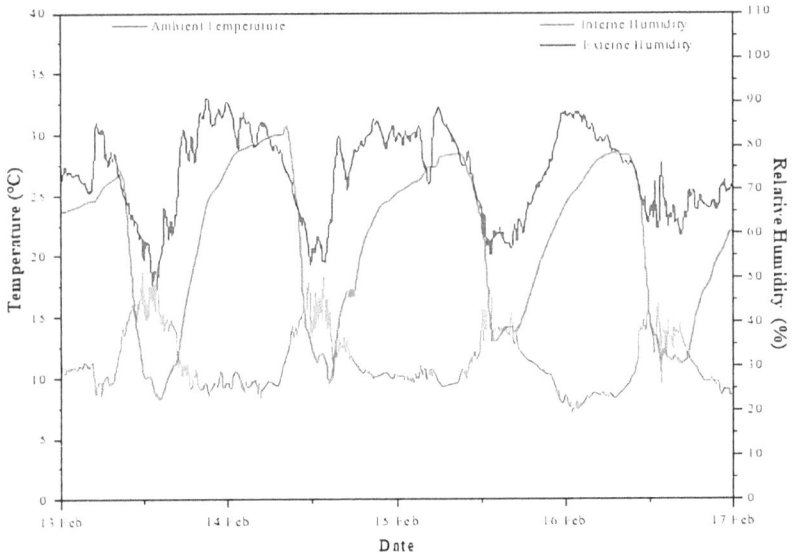

FIGURE 9.2   Evolution of humidity inside and outside the greenhouse.

Humidity is also a climatic parameter that intervenes in the microclimate in hydroponic greenhouses and that then influences the growth of plants. Figure 9.2 shows the variation in indoor and outdoor humidity and the variation in ambient temperature.

The variation of humidity and that of temperature are inversely proportional. Low humidity values accompany the sunniest days when the temperature is very high. We notice that the humidity inside the greenhouse is much more stable than that outside, which is explained by the effect of the insulation of the greenhouse, which plays a very important role in stabilizing humidity and reducing climatic variations. The humidity inside the greenhouse is almost always lower than that outside, and the difference between the two humidity levels can exceed 30%. At 2 p.m., when the temperature is equal to 17°C, low humidity equal to 23% is recorded. From midnight on, the humidity begins to increase until it reaches 80% around 9 a.m. These high values can be justified by the location of the greenhouse very close to the sea.

9.3 SOLAR HEATING SYSTEM WITH SENSIBLE HEAT STORAGE (SHS_SHS)

The first studied system for the heating of the HAGS is a solar heating system with sensible heat storage called SHS_SHS.

9.3.1 DESCRIPTION OF SHS_SHS

The SHS_SHS was specially designed for heating the hydroponic greenhouse as a sensible heat storage system. The principle of the SHS_SHS is as follows (Baddadi et al., 2022):

FIGURE 9.3   The multi-stage exchanger system.

hot water, heated by two vacuum collectors and then stored in a storage tank, is circulated by a circulation pump from a water storage tank to coils distributed in the different floors of the greenhouse, constituting a multi-floor, multi-layer exchanger.

The main idea of the realization of this device is to heat the climate of the hydroponic greenhouse while homogenizing the temperature inside and attenuating the thermal stratification that can exist between the different levels of the culture trays. For this reason, the operating principle of this system is to provide the same heat to the different levels of the greenhouse by a multi-stage exchanger thanks to the hot water, which will circulate there continuously. Figure 9.3 illustrates the exchanger

formed by the multi-layer coils arranged under the culture trays at the different vertical levels. During the sunny period, the absorber tubes capture the solar energy and
the exchanger transfers it in the form of thermal energy to the heat transfer fluid.
The water heated by the sensors is then stored in the tank until needed. Once the
circuit starts to operate, the pump delivers the stored hot water to the water collector
and beyond to the coils to circulate it under all of the culture trays. After crossing
the multi-stage exchanger, the water is then collected in the second collector to be
sucked up and returned to the tank, where it will reward the thermal energy evacuated
and gain other degrees Celsius to do the same again. Thus, the hot water leaves the
tank with the same temperature and circulates in the multi-stage exchanger, generating almost the same amount of heat.

9.3.2 Behavior of SHS_SHS

It is essential to first follow the evolution of the temperature of the water, which
will circulate in the system for heating the HAGS. For this, the water temperatures
respectively at the inlet and outlet of the collectors and that in the storage tank are
presented in Figure 9.4.

In the absence of the sun, the inlet and outlet temperatures are close to around
17°C, while the tank temperature is around 40°C. Then, the more the sunshine and
the ambient temperature increase, the more the water temperatures move away and
record remarkable differences between them. During the day, the temperature at the
outlet of the collectors increases to reach 80°C, exceeding the temperature of the
water at the inlet by around 30°C.

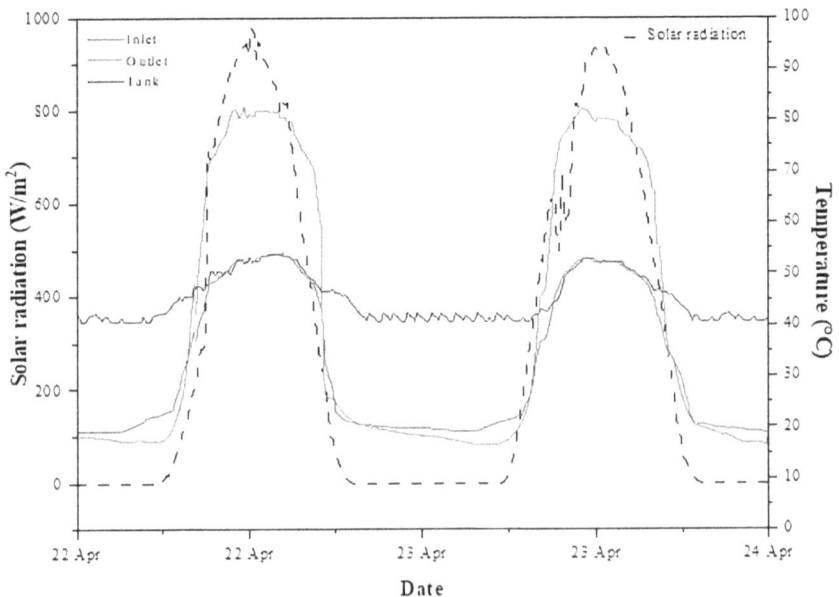

FIGURE 9.4   Variation of water temperatures at the sensor inlet and outlet.

Concerning the temperature of the tank, it does not vary widely since it is relatively isolated, but it obviously follows the variation of that of the output of the collector with a delay, which depends on the sunshine. On sunny days, the temperature of the water in the tank often exceeds 50°C and approaches 40°C at night. This temperature directly influences the climate of the greenhouse since it is almost at this temperature that the water will circulate in the coils of the multi-stage exchanger.

9.3.3 CONTRIBUTION OF SHS_SHS IN HEATING HAGS

To clearly highlight the effect of heating on the behavior of the hydroponic greenhouse, we will present the temperatures at several levels during two separate periods, one without heating and the other after using the SHS_SHS solar installation.

Figure 9.5 and Figure 9.6 represent, respectively, the vertical variation of the temperature inside the hydroponic greenhouse without and with heating for two durations with relatively similar climates.

If we compare the temperatures of the first day of each period, we notice that the maximum temperature T1 (at the first position at the top) is around 34°C, while it is around 39°C after heating. The same applies almost for the entire period when a temperature increase of up to 5°C is observed. If we look at thermal stratification, we notice that during the day, the temperatures of the different levels of the greenhouse

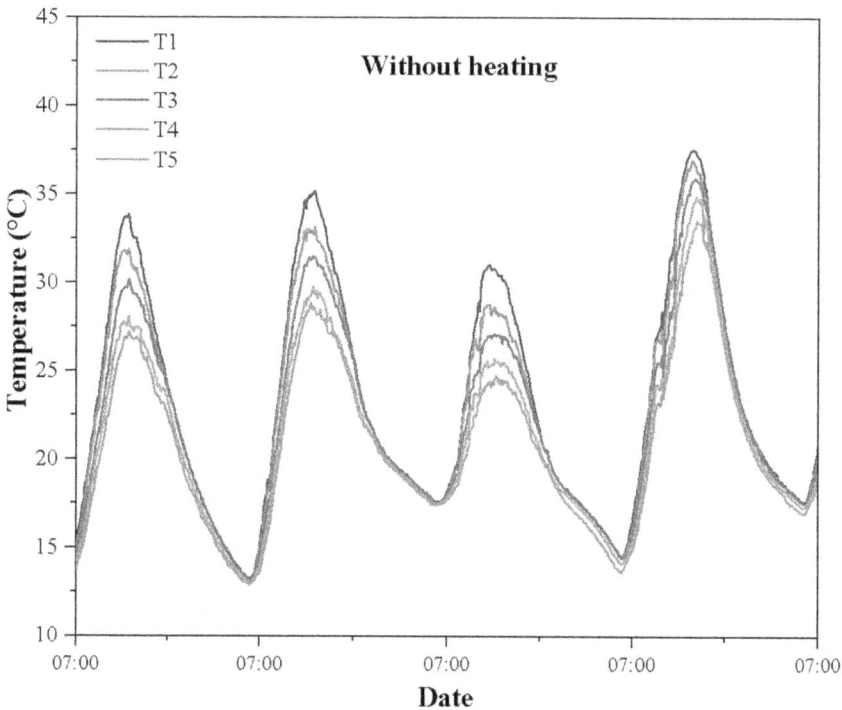

FIGURE 9.5 Vertical temperature stratification without heating.

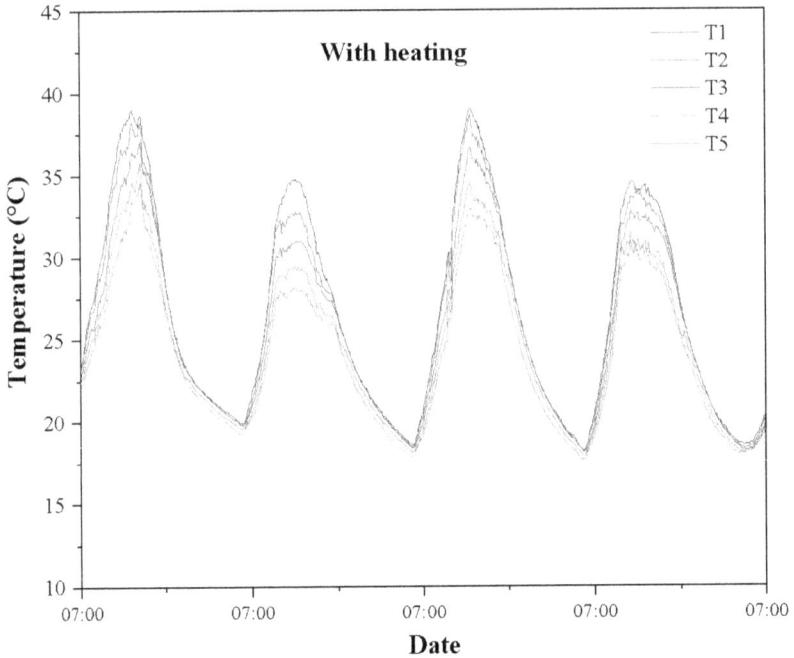

FIGURE 9.6 Vertical temperature stratification with heating.

without heating show a total difference of up to 7°C recorded between 12–14 hours during maximum sunshine. From about 7 p.m. this temperature difference is reduced considerably until the temperatures at the different levels become equal.

9.4 SOLAR HEATING SYSTEM WITH LATENT HEAT STORAGE (SHS_LHS)

The second studied system for the heating of the HAGS is a solar heating system with latent heat storage called SHS_LHS.

9.4.1 Description of SHS_LHS

Before realizing the SHS_LHS, it was sized according to the HAGS requirements (Baddadi et al., 2019; Bouadila et al., 2022). As shown in Figure 9.7, it is singularly formed of two beds, each comprising 156 spherical nodules. These nodules contain PCMs, and they form the absorber of the collector. The 312 nodules are arranged on two metal grids of 12 columns and 13 rows each. These two beds are superimposed in a square box with a side of 1.05 m and a volume of 0.28 m³ and mounted on a metal support facing due south and forming an angle of inclination equal to 30 degrees with respect to the horizontal. Above the box, there is a converging made of aluminum sheet, which presents the link between the sensor and the greenhouse.

FIGURE 9.7 Solar heating system with latent heat storage.

To ensure the greenhouse effect and minimize convective and radiative heat loss from the absorber, a transparent cover glass with a surface area of 1 m² and a thickness of 0.004 m covers the SHS_LHS. Two openings are also present in the design, one opening below the collector allowing the entry of cold air and a second opening located at the top in the upper part of the convergent for the exit of hot air from the collector. The exchange between the penetrating air and the absorber inside the sensor is ensured by an air extractor controlled by a variable speed drive. In order to ensure the thermal insulation of the collector, a 5-cm-thick layer of Armaflex covers the converging as well as the side and rear faces of the collector. This insulation is resistant to heat and humidity, it is flexible and has a quick and easy use.

9.4.2 BEHAVIOR OF SHS_LHS

Figure 9.8 shows that the behavior of the SHS_LHS considerably follows climatic changes. On the sunniest days, the absorber temperature often exceeds 30°C and the outlet temperature is around 40°C. February 20 is characterized by the highest values. For an ambient temperature equal to 24°C and sunshine over 800 W/m², the temperature of the absorber reaches 35°C and the outlet temperature exceeds 43°C around 1 p.m.

The SHS_LHS typically provides above ambient temperature by more than 20°C. We can notice that the absorber is characterized by a particular look. This dissimilarity is explained by the phenomenon of latent thermal storage, which takes place in the nodules of the absorber. Indeed, during the day, the extractors of the collector are closed and the collector captures solar radiation due to its absorber and stores the excess thermal energy in the form of latent thermal energy by the MCP nodules. In

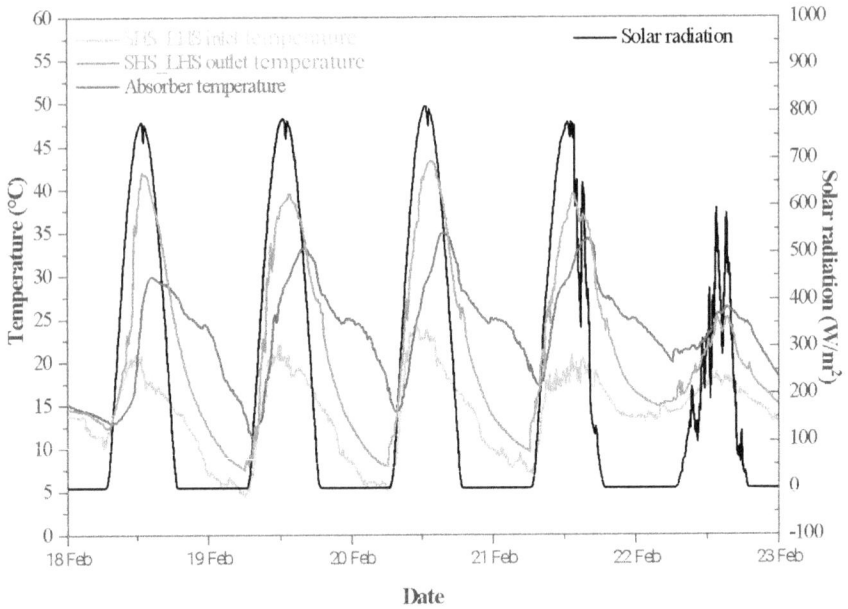

FIGURE 9.8 Evolution of SHS_LHS temperatures.

the evening, in the absence of the sun, the collector extractors open and the energy already stored is restored to ensure the nocturnal heating of the greenhouse, which explains the phase shift seen on the significant curve of the absorber.

9.4.3 Contribution of SHS_SHS in Heating HAGS

Once the temperature inside the hydroponic greenhouse drops, the collector extractors open and the energy already stored by the collector is restored to ensure the heating of the greenhouse. So, to visualize the effect of the SHS_LHS on the climate within the greenhouse, we will expose the air temperature in the middle of the greenhouse accompanied by the ambient temperature and the sunshine for a week when the sensor is triggered at 6 p.m. After exhibiting the different sensor temperatures, Figure 9.9 implements the evolution of the temperature inside the greenhouse after the use of the SHS_LHS.

It is clear that the SHS_LHS increased the temperature of the greenhouse, particularly at night during the operation of the collector. During the day, the temperature is above 32°C and it reaches 37°C at 1:50 p.m. The lowest temperature is recorded on February 28 when the solar radiation did not reach 400 W/m². At night, the temperature inside the hydroponic greenhouse varies from 17–20°C, registering an increase of about 6°C compared to the temperature of the greenhouse without heating. We notice the improvement of the temperature particularly during the night periods, which are characterized by inevitable temperature drops that can damage the growth of plants, especially those that require heat. This contribution of greenhouse climate optimization

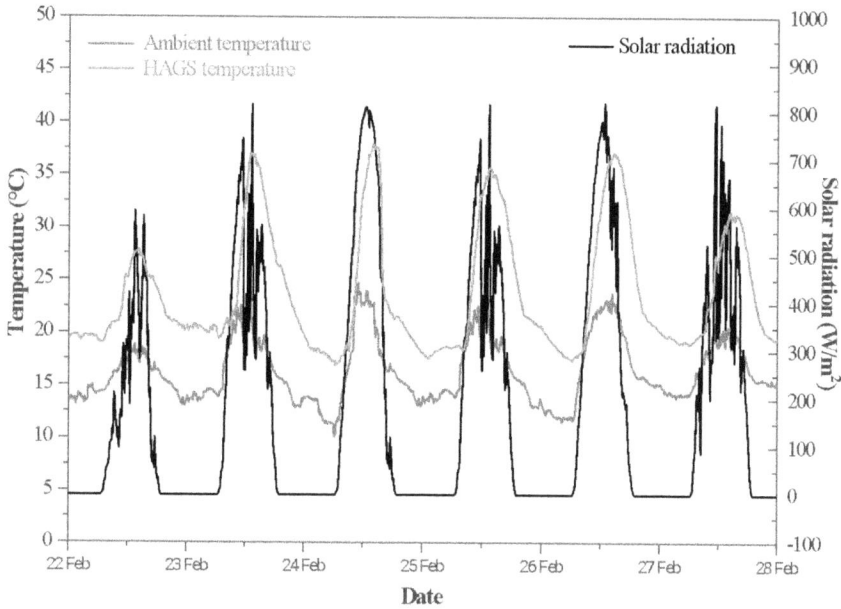

FIGURE 9.9 Evolution of the temperature inside the greenhouse heated by SHS_LHS.

is explained by the role of the solar collector. The energy stored by the MCP has successfully rewarded the energy lost during cold durations and nighttime periods.

9.5 FINDINGS

Two types of solar heating systems with thermal energy storage were investigated to heat a hydroponic agricultural greenhouse system. The first system is a solar heating system with sensible heat storage. The absorber of this collector has played an indispensable role in improving the night climate and mitigating low and severe temperatures in a hydroponic greenhouse. Indeed, the phase change material of the absorber stored the excess thermal energy by latent heat during the daytime during very sunny periods and restored it in the evening and at times when the temperature drops. This restored energy has succeeded in improving the atmosphere within the greenhouse and attenuating the harsh climate that can harm the plants. The second system is a solar heating system with sensible heat storage. The latter managed to raise the temperature and improve the climate in the hydroponic greenhouse without heating. The experimental results show an increase in the temperature prevailing in the greenhouse of 5°C.

REFERENCES

Baddadi, S., Ayed, R., & Bouadila, S. (2022). Harnessing solar energy for homogeneous spatial variability of a greenhouse air temperature: System design and implementation.

The 13th International Renewable Energy Congress IREC, 6–10. https://doi:10.1109/IREC56325.2022.10002011

Baddadi, S., Bouadila, S., & Guizani, A. (2019). Beneficial use of two packed beds of latent storage energy for the heating of a hydroponic greenhouse. Energy Procedia, 162, 156–163. https://doi.org/10.1016/j.egypro.2019.04.017

Bouadila, S., Baddadi, S., Skouri, S., & Ayed, R. (2022). Assessing heating and cooling needs of hydroponic sheltered system in Mediterranean climate: A case study sustainable fodder production. Energy, 261, 125274. https://doi.org/10.1016/J.ENERGY.2022.125274

Bradley, P., & Marulanda, C. (2001). Simplified hydroponics to reduce global hunger. Acta Horticulturae. https://doi.org/10.17660/ActaHortic.2001.554.31

Godde, C. M., Mason-D'Croz, D., Mayberry, D. E., Thornton, P. K., & Herrero, M. (2021). Impacts of climate change on the livestock food supply chain; a review of the evidence. Global Food Security, 28(October 2020), 100488. https://doi.org/10.1016/j.gfs.2020.100488

10 Testing of a Lab-Scale Electrically Charged Hybrid Latent-Sensible Thermal Storage System for High Temperature Applications

Rhys Jacob, Shane Sheoran, Ross Flewell-Smith, and Frank Bruno

10.1 INTRODUCTION

As we move into a carbon-constrained world, we require alternative forms of energy duration. Currently, the majority of energy for thermal loads is generated by fossil fuels (REN21, 2022), and mostly by gas. This is especially true for higher temperature applications (>250°C), which are seen as one of the most difficult sectors to decarbonise (Papadis and Tsatsaronis, 2020; Gross, 2021). The carbon footprint of gas coupled with its high cost make switching to renewable sources of energy more desirable. However, the current sources of low-cost renewable energy (i.e., solar photovoltaics and wind) are variable and non-dispatchable by nature, something that does not suit most energy users. Therefore, storage of the generated electricity is required.

A technology that is able to do this without geographic constraints or critical materials is thermal energy storage. In thermal storage, energy can be stored as heat in low-cost media such as rocks (Tiskatine et al., 2017; El Alami et al., 2020), molten salts (Caraballo et al., 2021; Bauer et al., 2021), phase change materials (Kenisarin, 2010; Jacob et al., 2019b, Liu et al., 2021b), or waste materials (Gutierrez et al., 2016; Jacob et al., 2022), where the energy can then be later dispatched as heat or converted to electricity using traditional power blocks. Furthermore, the generated heat can be produced from renewable sources such as solar photovoltaics (PV) or wind using resistive heaters (Stack et al., 2019; Okazaki, 2020) or turbomachinery (Trebilcock et al., 2020; Liang et al., 2022). In doing so, an electrically charged thermal energy storage (ECTES) system is both renewable and dispatchable.

DOI: 10.1201/9781003345558-12

Thermal storage has been well deployed for a number of applications such as concentrated solar power (Liu et al., 2016; Pelay et al., 2017; SolarPACES, 2022) or building heating and cooling (Heier et al., 2015, Navarro et al., 2016a, 2016b); however, only recently this technology has been coupled with electrically generated heat (Benato and Stoppato, 2018; Dumont et al., 2020).

For example, Zanganeh et al. (2015) built and tested a 42 kWh$_t$ lab-scale prototype for thermal storage above 575°C. In this system, a packed bed of rocks was combined with a layer of encapsulated aluminium alloy (AlSi$_{12}$). Testing of the system indicated that the AlSi$_{12}$ layer helped to stabilise the temperature outlet while thermal losses were less than 3.5%. Unfortunately, the thermal inertia and small tank-to-particle diameter ratio made it difficult to numerically verify the system. Stack et al. (2019) investigated the use of resistance heating and ceramic firebricks to store and deliver high-temperature heat as air (1000–1700°C). Using the firebricks as a sensible store, the system was simulated and shown to cycle daily with minimal issues and typically with a depth of discharge between 70–90%. Even with traditional insulation, the system was found to have a heat loss of less than 3% per day. Economically, the system was estimated to have a storage cost of $10/kWh while its use in the Northwestern Iowa market showed that it would achieve payback within two years. Knobloch et al. (2022) tested a 1 MWh$_t$ packed bed with a storage temperature of 675°C. Storage was in a packed bed of rock with air used as the heat transfer fluid (HTF). In keeping the storage partially underground, the system was well insulated and mechanically stable with a round-trip efficiency of 80.7% reported. As a first-of-its-kind design, the storage cost was relatively high (€189/kWh$_t$); however, it was mentioned this value should be reduced with scaling.

Therefore, to continue to build upon the promise of ECTES systems and provide further real-world operating results and experience with these systems, a small-scale system was designed, commissioned, and built at the University of South Australia (UniSA).

10.2 METHODOLOGY

Previous research has indicated that systems utilising both latent and sensible components (i.e., a hybrid system) can be more efficient and less costly (Jacob et al., 2019b; Liu et al., 2021a). Therefore, the system designed and built was to utilise this configuration.

10.2.1 MATERIALS

The system is designed to mainly store heat through two media, namely, quartzite rock and encapsulated aluminium (steel encapsulation), with the key properties of the materials given in Table 10.1.

10.2.1.1 Suppliers

The quartzite rock was sourced from Garden Grove with an average diameter of 20 mm. Pre-sorting of the rock ensured that smaller rocks were not included, while the media was also washed and dried before filling to reduce dust. The aluminium

TABLE 10.1
Selected Properties of the Storage System

Material	Parameter	Value	Material	Parameter	Value
HTF-Air	Density	0.3576 kg/m³	HTF-Air	Viscosity	4.3E-5 Pa.s
	Specific Heat Capacity	1137 J/kg·K		Thermal Conductivity	0.0665 W/m·K
Filler-Quartzite Rocks	Particle Diameter	20 mm	Filler-Quartzite Rocks	Density	2500 kg/m³
	Thermal Conductivity	5.69 W/m·K		Specific Heat Capacity	0.83 J/kg·K
PCM-Aluminium	Density-solid	2700 kg/m³	PCM-Aluminium	Density-liquid	2375 kg/m³
	Thermal Conductivity	205 W/m·K		Specific Heat Capacity	900 J/kg·K
	Latent Heat	321 kJ/kg			
Containment-Stainless Steel 316	Density	7900 kg/m³	Containment-Stainless Steel 316	Thermal Conductivity	15 W/m·K
	Specific Heat Capacity	0.5 J/kg·K			

was sourced from Bell Bay Aluminium with a purity > 99%. The packed bed tank, sheaths, and aluminium encapsulation material were all stainless steel 316. To minimise heat loss, the system was wrapped in rock-wool insulation.

10.2.1.2 Quartzite Rock Weight
Based on the density of the quartzite rocks and a filler fraction of 0.658, the estimated filler mass was 360 kg.

10.2.1.3 Encapsulated Aluminium

10.2.1.3.1 Tube Design
For the proposed design, stainless steel 316 tubes with an inner diameter of 26.64 mm and wall thickness of 1.68 mm were used. Each tube was 680 mm in length and inserted through two brackets mounted to the roof of the tank.

10.2.1.3.2 Internal Tube Pressure
The expansion of the aluminium from solid to liquid during cycling is likely to cause additional stress on the tube if allowances are not made. Using the solid and liquid density of aluminium, it is expected that the aluminium will increase in volume by 13%; therefore, a ≈20% void allowance will be left in each tube. The internal pressure after melting can be estimated by Equation 1:

$$P_2 = \frac{T_2 V_2}{T_1 V_1} P_1 \qquad [1]$$

Therefore, it is estimated that the pressure will rise 10 times to 1,013 kPa, well below the yield point of stainless steel at the experimental temperatures.

10.2.1.3.3 Aluminium and Steel Encapsulation Weight

Based on the aforementioned stainless-steel dimensions and tube void fraction, the estimated weight of aluminium was 10 kg while the weight of the stainless-steel tubes was 53 kg.

10.3 STORAGE DESIGN

As designed, the system was to store heat from 300°C to 700°C, while the maximum air flow rate during discharge was 0.1 kg/s. Of the internal volume, approximately 0.01 m³ was to be filled with the PCM, while the rest was filled with the inlet/outlet manifolds and quartzite rock.

10.3.1 TANK DESIGN

10.3.1.1 Dimensions

The tank dimensions were 700 mm ID and 700 mm length. The thickness of the storage tank was 6 mm and was constructed from stainless-steel 316.

10.3.1.2 Manifold Design

Similar to the tank and encapsulation materials, the manifold was fabricated from stainless-steel 316. To promote good flow distribution, a tubular design was chosen. Additionally, a stainless-steel mesh (> 70% open area) was placed around the inlet manifold and held up by four stainless-steel braces. In having such a design, the inlet manifold was protected from blockage from the quartzite rocks. A schematic of the manifold can be found in Figure 10.1.

The manifold information and dimensions can be found in Table 10.2.

10.3.1.3 Radiative Heater Design

To supply heat to the system, radiative elements were utilised. To estimate the radiative heat transfer, Equations 2–3 were used, while the element parameters are shown in Table 10.3.

$$h_{12} = \frac{\tilde{A}\left(T_2^2 - T_1^2\right)\left(T_1 + T_2\right)}{\frac{1-\mu_1}{\mu_1} + 1} \tag{2}$$

$$Q_{12} = h_{12} A_1 \left(T_1 - T_2\right) \tag{3}$$

From this information, the estimated maximum and minimum radiative heat transfer from the elements is 9.2 kWt and 5.1 kWt, respectively. However, due to the maximum energy output of the elements, the maximum heat transfer from the elements is 8.25 kW, while the actual radiative heat transfer will be based on the packed bed temperature at the element location. Each element was protected by a stainless-steel sheath (20 mm ID). The electrical elements chosen to achieve this heating were Watlow® LA straight stainless-steel braid (610 mm × ¾" 240V/2750W).

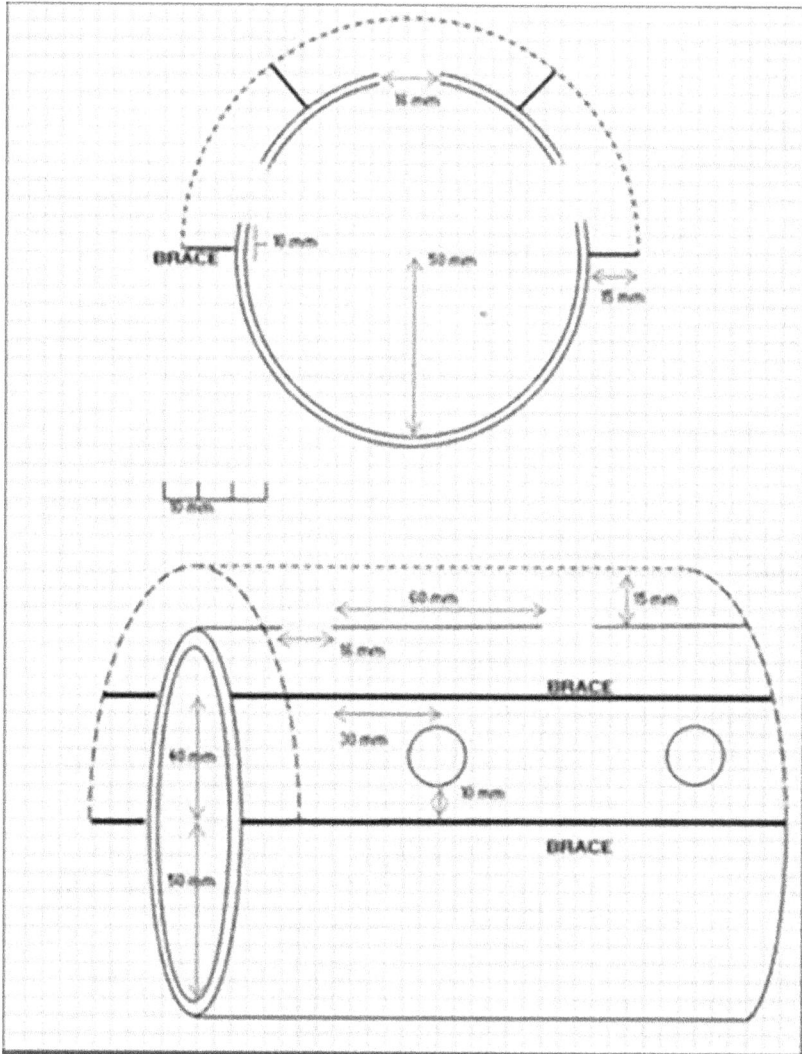

FIGURE 10.1 Manifold design.

TABLE 10.2
Manifold Information and Dimensions

Manifold		Holes	
Diameter	100 mm	Diameter	16 mm
Length	700 mm	Spacing	60 mm
Velocity	34.6 m/s	Number/space	3
Pressure Drop	70.9 Pa	Velocity	38.6 m/s
		Pressure Drop	329.3 Pa

TABLE 10.3
Radiative Heater Design Parameters

Parameter	Value	Parameter	Value
Element Diameter	20 mm	Emissivity	0.8
Element Length	600 mm	Element Temperature	900°C
No. of Elements	3	Storage Temperature	300–700°C

10.3.1.4 Storage Capacity

The storage capacity of the system can be calculated using Equation 4:

$$Q = \left(cp_f \times \Delta T \times m_f\right) + \left(cp_t \times \Delta T \times m_t\right) + m_{PCM}\left(cp_{PCM} + \Delta T + \Delta h\right) \qquad [4]$$

where Q, cp, ΔT, m, and Δh are the storage capacity, specific heat capacity, temperature difference, mass, and latent heat, respectively. The subscripts f, t, and PCM correspond to the filler (quartzite rock), stainless-steel tubes (for encapsulation), and PCM (aluminium), respectively.

Therefore, based on the storage material weights given in section 10.2.1, the storage capacity of the quartzite rock, aluminium, and steel is 38.06 kW$_t$.

10.3.1.5 Thermocouple Locations

To measure the temperature at various locations, thermocouples were placed axially and radially in three locations. This ensured an accurate temperature profile was recorded. Each thermocouple was protected by a stainless-steel sheath inserted into the packed bed at a depth of 350 mm. Each of these sheaths contained three thermocouples placed at 0 mm, 175 mm, and 350 mm. Additional 'temperature-measurement studs' were mounted around the outside of the tank to also measure temperature.

10.3.1.6 Pressure Drop

10.3.1.6.1 Packed Bed

In an effort to better predict the pressure drop for the irregularly shaped filler, Equation 5 has been suggested. In Equation 5, A, B, and ψ have values of 217, 1.83, and 0.6, respectively. These values have been experimentally verified for similar systems.

$$\Delta P = \frac{LG^2}{\rho fd}\left(A\frac{(1-\varepsilon)^2}{\varepsilon^3\psi^2}\frac{\mu}{Gd} + B\frac{1-\varepsilon}{\varepsilon^3\psi}\right) + \rho gL\frac{\Delta T}{T} \qquad [5]$$

In the prior equation, the last term is the pressure drop due to temperature; however, due to the low density of air, this term has been ignored.

As the tank is horizontal, the flow direction will be through a rectangular duct rather than a circular duct. To account for this, a correction factor (Equation 6) is used:

$$D_e = 1.3 \times \frac{(d \times l)^{0.625}}{(d + l)^{0.625}}$$ [6]

In this equation d and l are the width and length of the equivalent rectangular duct area.

The estimated pressure drop through the packed bed at the specified design conditions is therefore 331 Pa.

10.3.1.6.2 Tubes

The pressure drop through the tubes has been estimated using the Kerns Method (Equation 7):

$$\Delta P = 8 j_f N_{cv} \frac{\rho u_s^2}{2}$$ [7]

where j_f, N_{cv}, ρ, and u_s, are the friction factor, ratio of tank diameter to equivalent tube diameter, density of the HTF, and the superficial velocity, respectively. Using this methodology, the estimated pressure drop across the tubes is 10 Pa.

10.3.1.7 Tank Heat Loss

The tank heat loss was estimated as 1155 W based on an insulation R value of 1 K·m^2/W (300 mm thick, 0.3 W/mK). Based on previous testing with the high-temperature test facility at UniSA, the expected heat loss is likely to be approximately double the estimated heat loss.

10.3.1.8 System Summary and Storage Layout

A summary of the storage design and a simple schematic of the system is shown in Table 10.4 and Figure 10.2.

TABLE 10.4

Storage System Summary

Parameter	Value	Parameter	Value
Tank Height	0.7 m	Tank Diameter	0.7 m
PCM Volume	0.01 m³	Filler Diameter	20 mm
Filler Mass	360 kg [Quartzite] 10 kg [Aluminium] 53 kg [Stainless-steel]	Pressure Drop	741 Pa [Total] 331 Pa [Packed Bed] 10 Pa [Tubes] 400 Pa [Manifold]
Storage Capacity	38.06 kWh$_t$	Average Charge/ Discharge Rate	22.7 kW$_t$
Element Heating	5.1–8.25 kW$_t$		

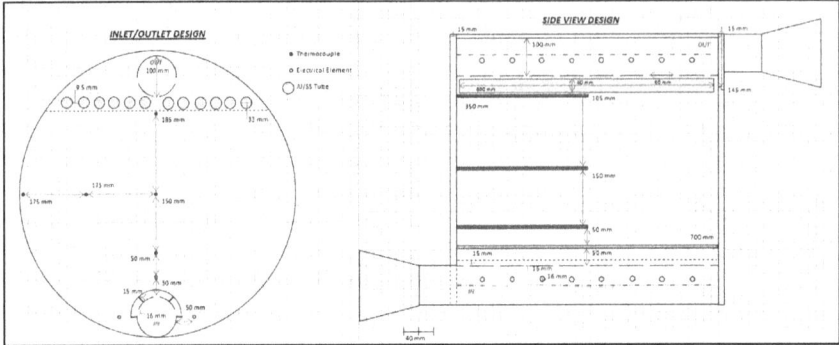

FIGURE 10.2   System layout.

10.4 RESULTS AND DISCUSSION

The following section documents the building, commissioning, and preliminary testing of the packed bed hybrid thermal storage system.

10.4.1 INITIALLY CONSTRUCTED PROTOTYPE

The initially built as-designed prototype is shown in Figure 10.3.

In this system, the three elements are clearly visible at the bottom of the system, while the port for the inlet and outlet manifolds can also be seen. Discharging was by way of an electrical fan using outside air.

FIGURE 10.3   Packed bed hybrid prototype [initial].

10.4.1.1 Initial Charge Test

To test if the system would operate as expected, an initial charge test was undertaken using electric elements located at the bottom of the tank. During charging, the inlet and outlet manifolds were sealed with insulation to prevent heat loss. It was found that under this configuration, the maximum bed temperature was 580°C (Figure 10.4).

In looking at the storage temperature by location (Figure 10.5), it can be seen that the area around the elements rises very fast in line with the element temperature and spreads throughout the bed. The centre of the bed heats up quicker due to losses through the wall or manifold. Unfortunately, the lower power of the elements was

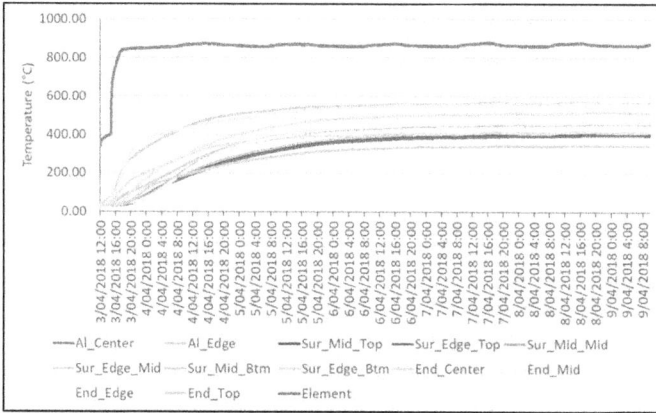

FIGURE 10.4   Initial packed bed charge test.

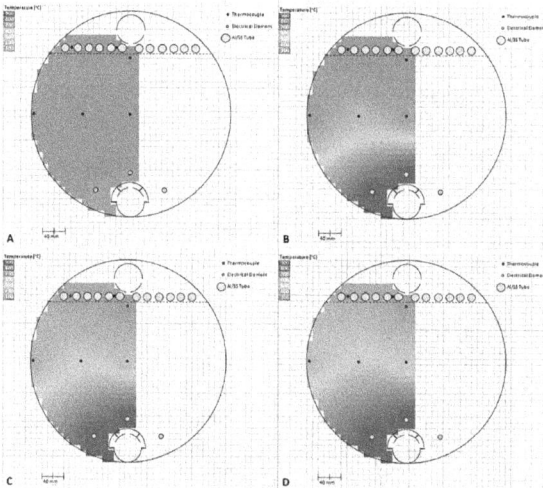

A) 0 min   B) 1800 min   C) 3600 mins   D) 8506 mins

FIGURE 10.5   Temperature results from initial charge test by location.

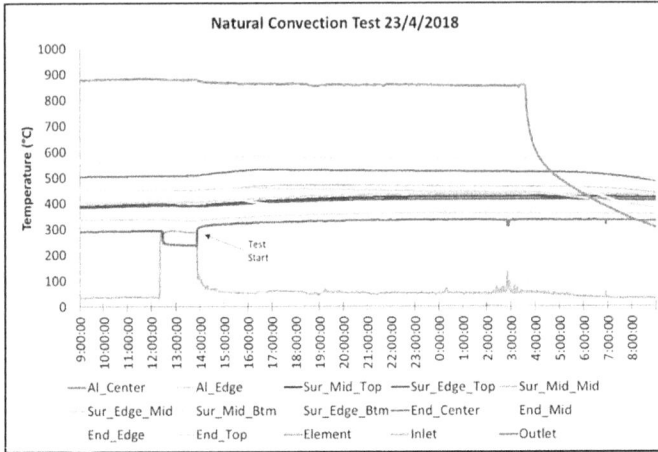

FIGURE 10.6   Second charge test with natural convection.

not able to heat up the entire bed, with a relatively stable temperature reached after approximately 3000 minutes.

A second charge test was also undertaken in which the inlet and outlet were left open in an effort to increase the natural convection in the system and therefore increase the packed bed temperature. Unfortunately, several of the elements failed during testing, resulting in the test being cut short. Nevertheless, during charging the bed temperature was as high as 600°C (Figure 10.6). Despite the higher temperature, the increase in power required to maintain the bed temperature meant that charging should be undertaken in a closed system.

10.4.1.2 Initial Discharge Test

The system was then discharged using ambient air at a rate of 0.04 kg/s. Under these conditions the system was able to maintain an outlet temperature of above 150°C for nearly four hours (Figure 10.7).

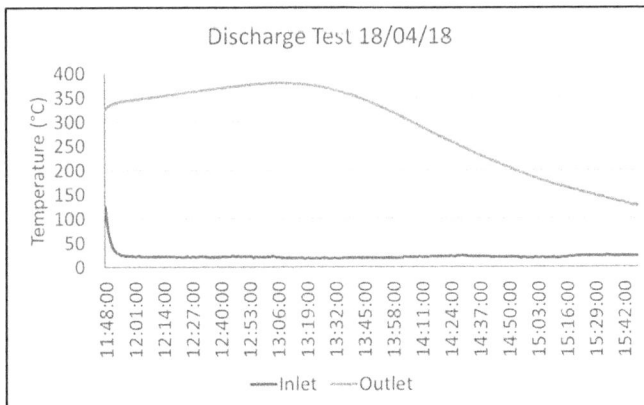

FIGURE 10.7   Packed bed discharge test.

10.5 IMPROVED DESIGN

As previously mentioned, the initial design of the prototype was unable to deliver the temperatures required to adequately charge and discharge. Therefore, three new elements were added into the system just below the encapsulated aluminium layer (Figure 10.8). However, one of the elements was unable to be replaced, so only five elements could be used.

Additionally, due to the low heat transfer from the original design, the resistive element sheath diameter was increased from 20 mm to 50 mm to increase radiative transfer.

10.5.1 SUBSEQUENT CHARGE TESTS

With the new elements installed, a charge test was conducted, with the melting of aluminium clearly visible (Figure 10.9).

FIGURE 10.8   Packed bed hybrid prototype [final].

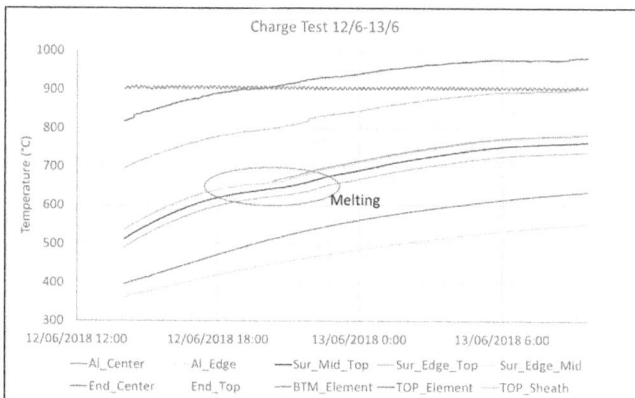

FIGURE 10.9   Packed bed charge test [12/6–13/6].

10.5.2 SUBSEQUENT DISCHARGE TESTS

With the aluminium melted, a discharge test using ambient air at 0.02 kg/s was initiated. The results can be seen in Figure 10.10. Similar to before, it is possible to see
the influence of the aluminium in stabilising the temperature. It is predicted that this
influence will be greater with an increase in PCM.

With the system able to charge (albeit slowly) and discharge, several more discharge tests using the same conditions at various charge capacities were undertaken
(Figure 10.11).

From these tests, the influence of aluminium is clear, with the outlet temperature
being stabilised for between 1–2 hours. Additionally, in all tests the temperature of
the outlet air is well over 200°C for more than four hours.

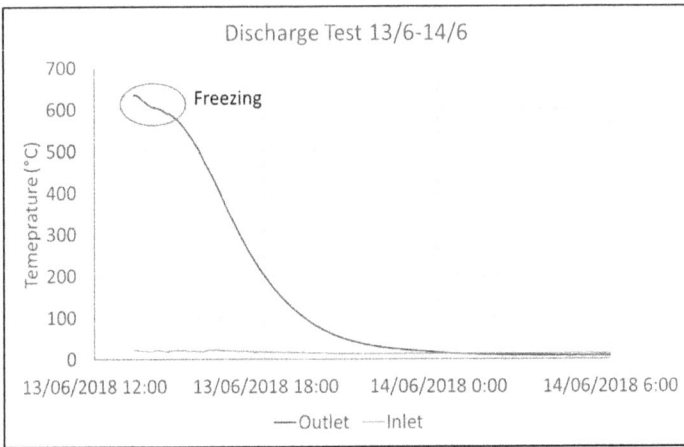

FIGURE 10.10   Discharge test [13/6–14/6].

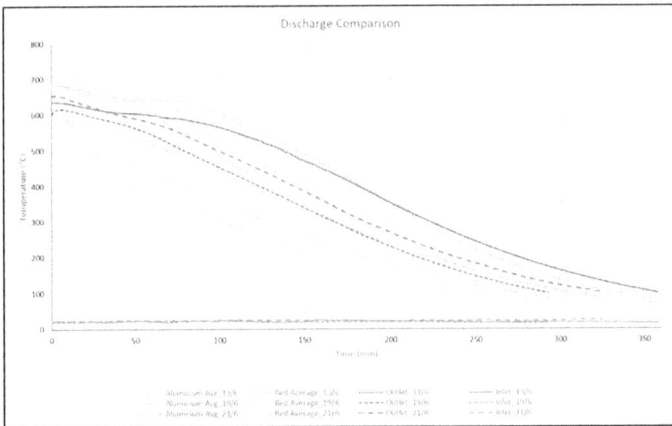

FIGURE 10.11   Comparison of various discharge tests.

Additionally, by extrapolating the results from the discharge test, it was estimated that a sensible-only system would result in an approximate 10% loss of discharge capacity (Figure 10.12).

10.5.3 System Power

The power requirements for the charge test performed on 19–21 June can be seen in Figure 10.13.

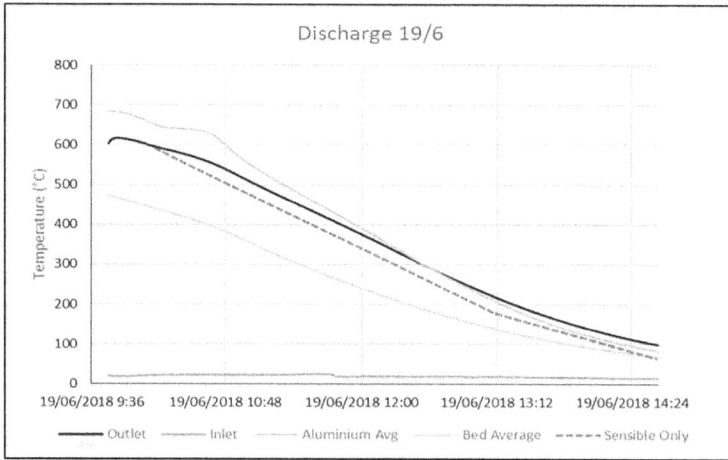

FIGURE 10.12   Hybrid vs. sensible-only discharge.

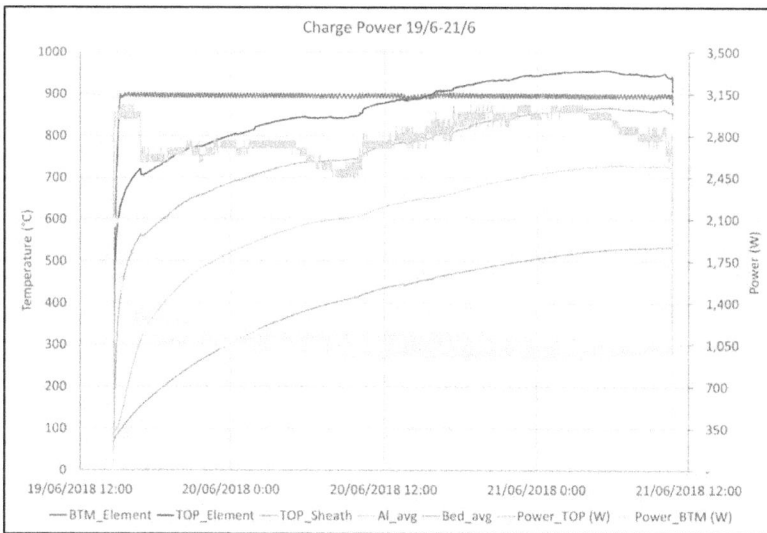

FIGURE 10.13   Power requirements for charging [19/6–21/6].

From this it can be seen that the bottom elements quickly reach the maximum temperature (900°C) and only require approximately 50% capacity to maintain its temperature. This results in a much slower charging for the bottom half of the tank. Conversely, the top elements rise much slower in temperature (albeit with deliberately throttled power) but continue to operate at near full power even when the element temperature approaches the maximum. Therefore, it can be concluded that the larger sheath diameter allows a significantly higher heat transfer to be achieved. Furthermore, the sheath temperature is consistently 100°C lower than the element temperature, further reducing the charge rate but minimising the storage material exposure temperature.

10.6 CONCLUSIONS AND IMPLICATIONS

While the system was unable to charge and discharge in a timely manner, it was able to reach the required temperatures and subsequently discharge at a range of flow rates. From the initial and subsequent tests on this system, many lessons have been learnt that can be applied in future systems. The main conclusions from these tests are as follows:

- Natural convection in the bed is negligible compared to radiation; therefore, element placement should be optimised for this issue.
- Element temperature should be 250°C higher than the desired temperature to maintain good heat transfer.
- Sheath diameter should be maximised to maximise heat transfer. Additionally, the usage of fins should be explored.
- Usage of PCM resulted in a 10% increase in discharge capacity compared to a theoretical sensible-only system.
- While forced convection was able to increase heat transfer, the loss of energy through the outlet was larger than the increase in heat transfer but led to a more uniform bed temperature. Circulation of the outlet air would likely require a high-temperature fan, which has many issues of its own.

10.6.1 ACKNOWLEDGEMENTS

This research was performed as part of the Australian Solar Thermal Research Institute (ASTRI), a project supported by the Australian Government, through the Australian Renewable Energy Agency (ARENA). The team gratefully acknowledges the work of Pat Miller and Ivan Zanatta for their help with the construction and testing of the system. Rhys Jacob would gratefully like to acknowledge the Alexander von Humboldt Foundation for providing funding to support this work.

REFERENCES

Bauer T, Odenthal C, Bonk A. *Molten salt storage for power generation*. Chemie Igenieur Technik (2021); **93**(4): 534–546.
Benato A, Stoppato A. *Pumped thermal electricity storage: A technology overview*. Thermal Science and Engineering Progress (2018); **6**: 301–315.

Caraballo A, Galán-Casado S, Caballero Á, Serena S. *Molten salts for sensible thermal energy storage: A review and an energy performance analysis*. Energies (2021); **14**(4): 1197. https://doi.org/10.3390/en14041197

Dumont O, Frate GF, Pillai A, Lecompte S, De paepe M, Lemort V. *Carnot battery technology: A state-of-the-art review*. Journal of Energy Storage (2020); **32**: 101756.

El Alami K, Asbik M, Agalit H. *Identification of natural rocks as storage materials in thermal energy storage (TES) system of concentrated solar power (CSP) plants – A review*. Solar Energy Materials and Solar Cells (2020); **217**: 110599.

Gross S (2021). *The Challenge of Decarbonizing Heavy Industry*. [Online]. Available at: www. brookings.edu/wp-content/uploads/2021/06/FP_20210623_industrial_gross_v2.pdf [Last Accessed: 18/10/2022].

Gutierrez A, Miró L, Gil A, Rodríguez-Aseguinolaza J, Barreneche C, Calvet N, Py X, Inés Fernández A, Grágeda M, Uschak S, Cabeza LF. *Advances in the valorization of waste and by-product materials as thermal energy storage (TES) materials*. Renewable and Sustainable Energy Reviews (2016); **59**: 763–783.

Heier J, Bales C, Martin V. *Combining thermal energy storage with buildings – a review*. Renewable and Sustainable Energy Reviews (2015); **42**: 1305–1325.

Jacob R, Liu M, Sun Y, Belusko M, Bruno F. *Characterisation of promising phase change materials for high temperature thermal energy storage*. Journal of Energy Storage (2019b); **24**: 100801.

Jacob R, Sergeev D, Müller M. *Valorisation of waste materials for high temperature thermal storage: A review*. Journal of Energy Storage (2022); **47**: 103645.

Kenisarin MM. *High-temperature phase change materials for thermal energy storage*. Renewable and Sustainable Energy Reviews (2010); **14**(3): 955–970.

Knobloch K, Muhammad Y, Solar Costa M, Matya Moscoso F, Bahl C, Alm O, Engelbrecht K. *A partially underground rock bed thermal energy storage with a novel air flow configuration*. Applied Energy (2022); **315**: 118931.

Liang T, Vecchi A, Knobloch K, SciAcovelli A, Engelbrecht K, Li Y, Ding Y. *Key components for carnot battery: Technology review, technical barriers and selection criteria*. Renewable and Sustainable Energy Reviews (2022); **163**: 112478.

Liu M, Jacob R, Belusko M, Riahi S, Bruno F. *Techno-economic analysis on the design of sensible and latent heat thermal energy storage systems for concentrated solar power plants*. Renewable Energy (2021a); **178**: 443–455.

Liu M, Omaraa ES, Qi J, Haseli P, Ibrahim J, Sergeev D, Müller M, Bruno F, Majewski P. *Review and characterisation of high-temperature phase change material candidates between 500 C and 700 °C*. Renewable and Sustainable Energy Reviews (2021b); **150**: 111528.

Liu M, Tay NHS, Bell S, Belusko M, Jacob R, Will G, Saman W, Bruno F. *Review on concentrating solar power plants and new developments in high temperature thermal energy storage technologies*. Renewable and Sustainable Reviews (2016); **53**: 1411–1432.

Navarro L, de Gracia A, Colclough S, Browne M, McCormack SJ, Griffiths P, Cabeza LF. *Thermal energy storage in building integrated thermal systems: A review. Part 1. Active storage systems*. Renewable Energy (2016a); **88**: 526–547.

Navarro L, de Gracia A, Niall D, Castell A, Browne M, McCormack SJ, Griffiths P, Cabeza LF. *Thermal energy storage in building integrated thermal systems: A review. Part 2. Integration as passive system*. Renewable Energy (2016b); **85**: 1334–1356.

Okazaki T. *Electric thermal energy storage and advantage of rotating heater having synchronous inertia*. Renewable Energy (2020); **151**: 563–574.

Papadis E, Tsatsaronis G. *Challenges in the decarbonization of the energy sector*. Energy (2020); **205**: 118025.

Pelay U, Luo L, Fan Y, Stitou D, Rood M. *Thermal energy storage systems for concentrated solar power plants*. Renewable and Sustainable Reviews (2017); **79**: 82–100.

REN21 (2022). *Renewables 2022 Global Status Report*. [Online]. Available at: www.ren21.net/wp-content/uploads/2019/05/GSR2022_Full_Report.pdf [Last Accessed: 18/10/2022].

SolarPACES (2022). *Concentrating Solar Power Projects*. [Online]. Available at: https://solarpaces.nrel.gov/by-project-name [Last Accessed: 18/10/2022].

Stack DC, Curtis D, Forsberg C. *Performance of firebrick resistance-heated energy storage for industrial heat applications and round-trip electricity storage*. Applied Energy (2019); **242**: 782–796.

Tiskatine R, Oaddi R, Ait El Cadi R, Bazgaou A, Bouirden L, Aharoune A, Ihlal A. *Suitability and characteristics of rocks for sensible heat storage in CSP plants*. Solar Energy Materials and Solar Cells (2017); **169**: 245–257.

Trebilcock F, Ramirez M, PasCual C, Weller T, Lecompte S, Hassan AA (2020). *Development of a Compressed Heat Energy Storage System Prototype*. IIR International Rankine 2020 Conference-Heating, Cooling and Power Generation. Available at: https://elib.dlr.de/136589/1/R2020%20Paper%20No1178.pdf

Zanganeh G, Khanna R, Walser C, Pedretti A, Haselbacher A, Steinfeld A. *Experimental and numerical investigation of combined sensible–latent heat for thermal energy storage at 575 °C and above*. Solar Energy (2015); **114**: 77–90.

Part III

Thermal Energy
Global Scenario and Future Directions

11 Overview of Hydrogen Production Methods from Solar Energy

R. Kumar, Anil Kumar, and Atul Sharma

11.1 INTRODUCTION

Today, most energy production uses fossil fuels, but these fuels generate pollution and make a harmful environment by growing toxic byproducts that affect climate change and environmental degradation [1]. The primary energy is provided from fossil fuels to more than 80% of the world, of which 32% receives energy from oil, which is still the largest part of fuel for the transport sector [2].

The global primary consumption of fuel depends on conventional energy sources, such as natural gas known as methane, coal, and oil. The use of conventional energy develops economic and environmental issues such as ozone depletion, acid rain, global warming, and local pollution that lead to economic and political crises [3]. The use of renewable energy sources becomes more important than in earlier contexts, but fossil-based fuels are still more important in various sectors such as industry fields. For example, there are an estimated 225 million cars and other light vehicles, which are traveling over seven million miles in a day and the consumption of eight million barrels of fuel a day in the U.S. Due to the increase of vehicles, imported fuel (oil) is expected to increase by 68% by 2025 in the U.S., even though it is third place in oil produced in the world [4].

After the era of fossil-based fuels, renewable energy (especially solar energy) has become a powerful driving force to maintain energy availability through the use of hydrogen as a fuel [5]. Hydrogen is considered an alternative energy carrier for the next decades as a result of its having more energy density based on mass, less concern for the environment, its abundance in various forms in the universe, and its potential to be transformed into electricity and valuable chemicals. It is a light element in the universe that is lacking odor, color, and taste, and non-toxic, and it also has more heating value than other fuels such as coal (4 times), gasoline (2.8 times), and methane (2.4 times) [6].

Hydrogen energy is the main chain between H_2-consuming industries such as ethanol and ammonia production plants and some other sectors such as the gas grid, electricity grid, agriculture, residential, transportation, and energy storage system, as shown in Figure 11.1 [2]. Hydrogen energy has an incorporation role between these areas, which increases the performance of the electricity grid [7]. It can be generated

DOI: 10.1201/9781003345558-14

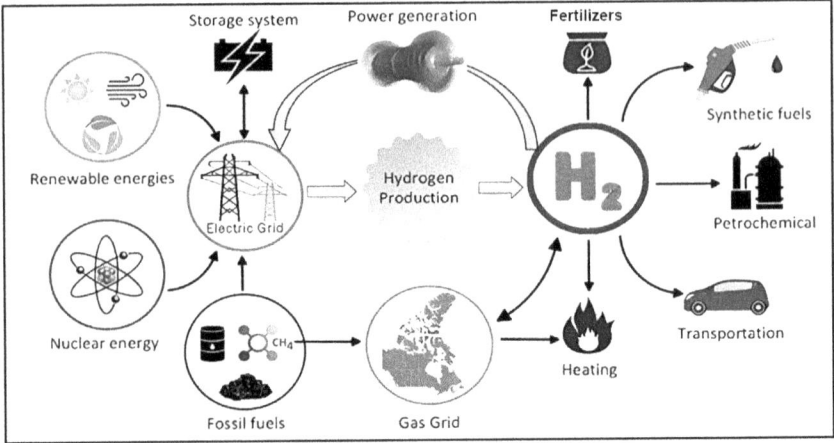

FIGURE 11.1   Hydrogen participation in energy sectors [2].

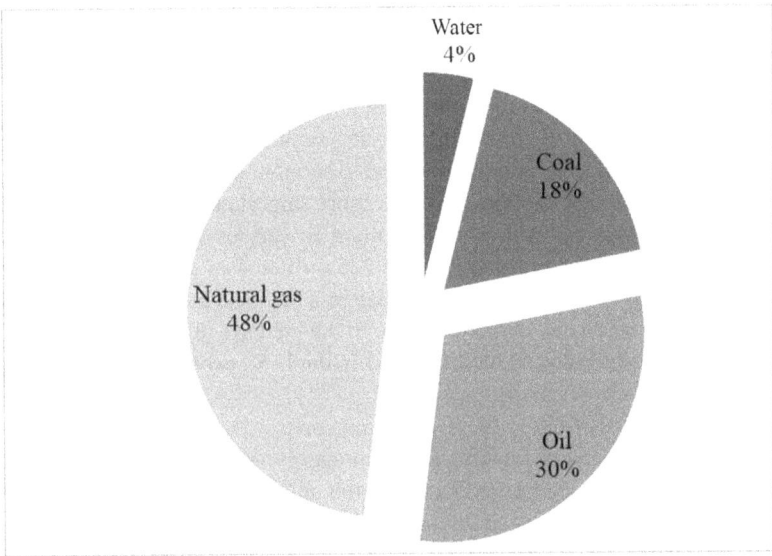

FIGURE 11.2   Modes of global hydrogen production [5, 10].

from renewable sources such as water and biomass, and most other important sources, and it is environmentally friendly at all routes using hydrogen [8]. A study by the Hydrogen Council reported on the assessment of hydrogen potential and the roadmap to its deployment. The study presumes that 18% of energy demand (nearly 78 EJ) worldwide can be fulfilled from hydrogen up to 2050. The current demand has been increased for pure hydrogen by nearly 70 million tons and for syngas by nearly 45 million tons in 2019 [9].

The maximum production of hydrogen is generated from fossil-based fuels through advanced technology, as shown in Figure 11.2. It is realized that 48% use natural gas through a steam reforming process, for hydrogen production and others such as coal (18%) and oil (30%) obtained through the partial oxidation process. There is minimum water utilization at 4% for hydrogen production through the electrolysis process [5, 10].

Plentiful renewable energy (solar energy) is accessible to us, and consideration of solar-based hydrogen production is one of the alternate solutions. Numerous types of research have elaborated on the various hydrogen production methods from solar energy based on energy and exergy analysis.

The literature covers studies on hydrogen production through renewable energy sources such as solar energy. Wang et al. [11] have studied solar hydrogen production in which thermochemical processes consume heat to split water molecules, photochemical methods develop photon-activated electrons to fragmented water molecules, and electrolysis employs electric potential to riven molecules of water. Joshi et al. [12] investigated exergy analysis for solar hydrogen production and discussed various types of hydrogen production from solar energy on the bases of energy input, numerous chemical reactants, and the involvement of various techniques such as gasification, reforming, electrolysis, and cracking. They have also analyzed hydrogen production based on solar PV from the exergy efficiency and sustainability index. Pregger et al. [13] studied hydrogen production from solar thermal energy and discuss the potential of economic and technological advancement for their future hydrogen supply. The main processes are the involvement of water electrolysis with high temperature, thermochemical, methane cracking, and steam methane reforming from solar energy. Liu et al. [19] examined hydrogen production through methanol steam reforming using a 5,000-watt solar reactor, maintaining a temperature range from 150–300°C and achieving thermochemical efficiency from 30–50% for chemical energy conversion by solar thermal energy.

Among the several methods of hydrogen production from solar energy identified, some of them have touched a phase of industrial and commercial maturity level, whereas others need more research. Generally, there are three main technologies for solar hydrogen production, such as thermochemical, photochemical, and electrochemical, which comprise auspicious alternatives to the solar energy storage system. In current years, several examiners have carried out studies regarding solar-based hydrogen production methods as well as those comprehensive studies cited earlier. Although a few studies have considered a comparative approach, the main aim of the current book chapter is to discuss and compare some hydrogen production methods from solar energy.

11.2 LITERATURE REVIEW

11.2.1 Hydrogen as an Energy Carrier

Hydrogen is an environmentally friendly, renewable energy source carrier that has the potential to switch to fossil-based fuels as global fuel energy [10, 14, 15]. Hydrogen is the lightest and most plentiful component in the universe and constitutes 75%

availability of all matters based on mass. Although hydrogen occurs in abundant amounts all over the Earth, the Earth's atmosphere comprises only around 1 ppm of H_2 by volume. It is available as one of the two elements of water, which constitute 75% of the Earth's surface and also exists in several chemical components based on carbon such as petrochemicals and organic matter [16]. Due to its potential and versatility, hydrogen can be easily converted into electrical, mechanical, and thermal energy, making it an excellent energy carrier to use in fuel cells, internal combustion engines, and steam engines [17, 18]. It can also be utilized in existing internal combustion engines with low-cost modifications [19].

Hydrogen has more advantages using vehicles such as clean burning fuel, low-producing emissions, low viscosity than other fluids, and burns more efficiently than fossil-based fuels [20]. It also seems well suited for industrial and domestic purposes as heat applications [21]. Hydrogen has demonstrated its capability to provide a marvelous source of energy in the aerospace industry since the ignition of hydrogen and oxygen liberates the largest quantity of energy per unit mass of fuel [22]. For example, during combustion, hydrogen releases more heat at 142.26 KJ/g compared to petroleum at 35.15–43.10 KJ/g and wood at 17.57 KJ/g [22]. It is noted that lower ignition energy than methane and gasoline due to the presence of lower and higher ignition limits of hydrogen provides low explosion energy [17]. Additionally, it is being graded as a safer fuel due to its low flame emissivity and high ignition temperature, and it generates low toxic emissions when combustion occurs compared to methane and gasoline fuels [17]. Another advantage of H_2 as an energy carrier is more efficient arrangements of hydrogen storage than batteries. It seems no negative effect occurs as the deep discharge of metal hydrides or hydrogen gas cylinders, whereas the deep discharge of battery systems is well-known to unfavorably affect the capacity of batteries [23, 24].

11.2.2 HYDROGEN APPLICATIONS

Although hydrogen offers a tremendous possibility as an energy carrier, applications of H_2 are based on its ability to chemically react with other molecules. Most of the studies [25, 26] have investigated hydrogen applications in all parts of daily life as well as for industrial and domestic purposes. Generally, H_2 is primarily used in the production of ammonia, refining of petroleum, and refining of metals such as lead, uranium, zinc, copper, molybdenum, tungsten, and nickel [27, 28]. In the coming decade, H_2 as a fuel will be used in replace of fossil fuels where all applications exist by fossil fuel. Hydrogen would propose instant profits in terms of decreased pollution and a clean environment [26].

Mostly, H_2 is consumed for ammonia production, petro-chemistry, and other chemicals. Hydrogen is utilized out of 500 billion cubic meters (Bm^3) as 250 Bm^3 in ammonia production, 65 Bm^3 in other chemical products, and 185 Bm^3 in petro-chemistry and level of accounting a50%, 13%, and 37%, respectively [29]. A recent study exploring countless hydrogen applications confirms their implications for electricity generation, jet planes, hydrogen industries, cooking food, and fuel for automobiles [15]. Figure 11.3 represents the integration of several production approaches and energy sources in numerous fuel cell applications.

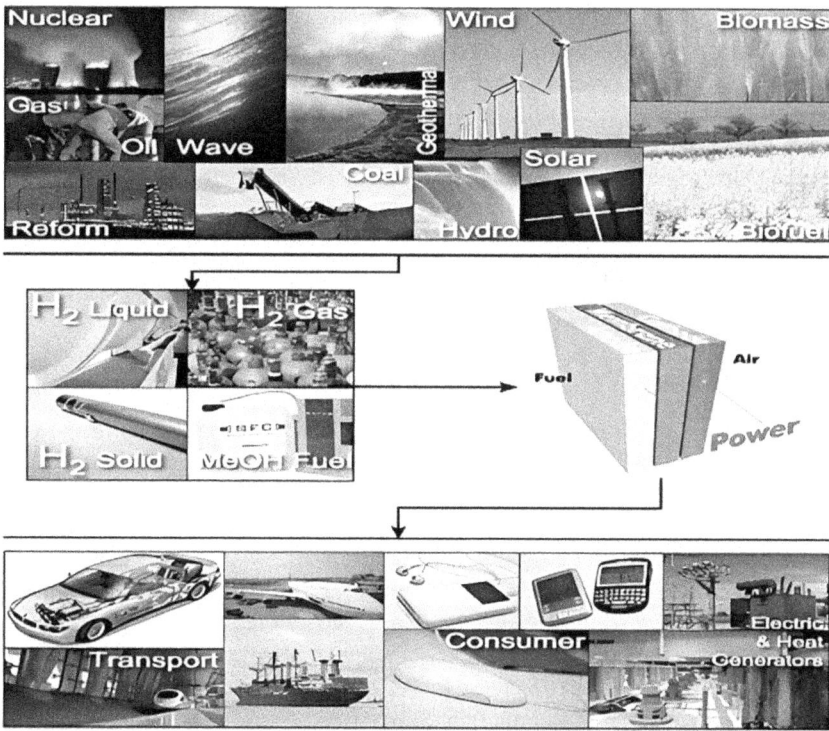

FIGURE 11.3   Hydrogen as an energy carrier involving multiple production approaches and sources to several fuel cell applications [30, 31].

11.2.3 RISK AND SECURITY ISSUES

Nejat et al. [32] presented that H_2 has a higher coefficient of diffusion and lower density that is much safer than other known fuels. In addition, the high specific heat of H_2 efficiently reduces the increasing temperature to a certain heat input. Normally, there are low energies, higher ignition temperatures, and wide ignition limits from the fuels that are considered less safe, as they raise the level of fire initialization. A less safe fuel is measured due to its high temperature of the flame, flame emissivity, and explosion energy as its fire could be more detrimental. A comparison of hydrogen (H_2) along with methane (CH_4) and gasoline fuel exposed that hydrogen was found the less harmful fuel with a factor of safety close to 1. However, the factor of safety of gasoline and methane fuels is declared to be 0.54 and 0.81, respectively [32].

Gaseous hydrogen having very light density (6.9% of air density) is more dispersive compared to air. Hence, an H_2 gas leak quickly diffuses as it develops from its source and lowers the risk of fire or blast. The leak of H_2 is not responsible for damaging climate conditions due to its non-toxicity [33]. H_2 can create an ignition or blast in a confined space by mixing with the air. Any type of fire could start and burn out quickly as the hydrogen gas becomes dissipated. It is very tedious to prepare a mixture of hydrogen and air for detonation. Hydrogen power-driven motor vehicles detected minor risks compared with petrol-driven motor vehicles in that leakage of petrol usually creates a longer cloud of flammable gas in confined spaces.

An additional security benefit of H_2 is connected attributed to a clear flame that cannot scorch skin from a distance. The main cause is a flame that emits a small thermal radiation in the absence of soot material. Moreover, it has fewer chances to leak from the source due to the storage of hydrogen containers being much stronger than petrol containers [34]. In addition, H_2 can also be safely carried in pipelines. However, hydrogen burns at low concentrations by mixing with air, and this can also be a reason for safety purposes [35].

11.2.4 H₂ Storage

The possibilities of hydrogen storage consist of physical and chemical methods that include compressed gas, cryogenically freezing or liquefying, metal hydrides, chemical hydrides, and sorbents respectively, as shown in Figure 11.4. There is a much higher energy requirement of hydrogen for freezing or liquefying processes due to its having a low boiling point ($-252.9°C$) and also low melting point ($-259.2°C$) [22]. For example, hydrogen storage requires more energy consumption for a liquefied or solid state of hydrogen, difficult-to-attain insulation requirements for such type of storage, and a 30% potential energy requirement of hydrogen stored for the solid or liquid state. However, a compressed form of hydrogen needs more storage space and less energy compared to liquefying or freezing hydrogen. Currently, hydrogen in compressed form has become a mature and commonly used method in storage technology, and it can be stored in pressurized gas from 35–70 MPa [22].

A metal hydride system stores hydrogen in the solid phase, which bonds the hydrogen to the metal on a molecular level as a result of adopting a safer method than others such as compressed liquid and gaseous [36]. The key research of metal

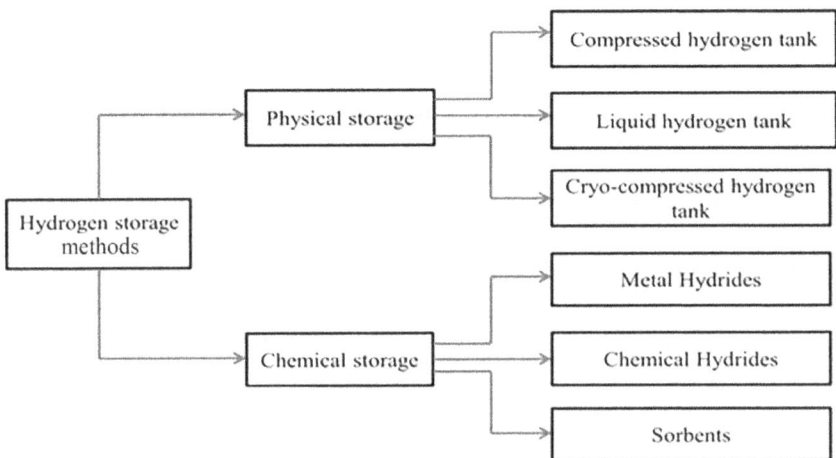

FIGURE 11.4 Outline of various H_2 storage methods, modified from Ref. [1].

hydrides is concentrated on cost-reducing systems to proprieties of adsorption and desorption, and magnesium hydrides are offering the best solution for their storage in the solid state as a result of their favorable characteristics [36]. However, the system of magnesium hydride needs to eject gas above 27°C in the barrier of the transportation field [22].

The most serious issues of hydrogen storage must be resolved before technically setting up and arranging economically feasible hydrogen fuel [37]. There are essentially two routes to running a vehicle with hydrogen fuel. First, H_2 is burnt rapidly in an engine cylinder by mixing air. Second, hydrogen fuel is burnt in a fuel cell electrochemically with oxygen from the air, which generates heat and electricity and an electric engine run by it [38]. The main problem is that using H_2 in transportation is connected with its heavy storage tank due to the lower density of H_2 [25]. Power-driven vehicles must be lightweight, compact, inexpensive, and safe for energy storage.

Generally, a commercial car is not optimized for prestige even if it is for mobility. For a vehicle arrangement in the 400-kilometer range, a standard vehicle consumes around 24 kg of gasoline during combustion in an engine cylinder, while 4 kg H_2 are required in a fuel cell for an electric engine and 8 kg H_2 for combustion in an engine to travel off the same range of vehicle [38]. Hydrogen is the simplest, most plentiful, and lightest element in nature. It can be stored physically through the changing state conditions along with pressure and temperature and also physio-chemically in several solid and liquid compounds, such as carbon nanostructure, metal hydrides, borohydrides, methanol, alanates, methane, and light HCs [39]. Both parameters such as gravimeter and volumetric density of H_2 about storage material are crucial in stationary and mobile applications. It is a possibility to store by various routes: (1) highly pressurized gas in cylinders around 800 bar, (2) at a very low temperature of about 21K in cryogenic tanks as hydrogen in liquid form, (3) adsorbed hydrogen through the materials of surface area at a temperature less than 100K, (4) absorbed host metal by interstitial sites with temperature and pressure, (5) ionic compounds chemically bonded by covalent with normal pressure, and (6) the reactive metals by oxidation, e.g., Al, Mg, Li, and Zn with H_2O [40].

11.3 SOLAR H_2 PRODUCTION METHODS

Maximum hydrogen production is generated by nearly 99% based on natural gas and fossil fuels. The rate of hydrogen production is nearly 50 million tons globally, which denotes only 2% energy demand worldwide [41]. The cost of hydrogen production from various production methods is still high. As such, there is a cost of $2 USD per kg H_2 as coal gasification process, $7 USD per kg H_2 as solar hydrogen through the photovoltaic cell, and water electrolysis. According to the U.S. Department of Energy, the expected price will reduce by $1.40 USD per kg H_2 and $4 USD per kg H_2, respectively [4]. The utility of fossil fuels is expected to decline over the next few years, as renewable sources are completely developed and accompanied by the advancement of other production methods that can attain sustainability.

Figure 11.5 represents the hydrogen production methods from solar energy [5]. Some methods are established for industrial purposes and others are in the laboratory

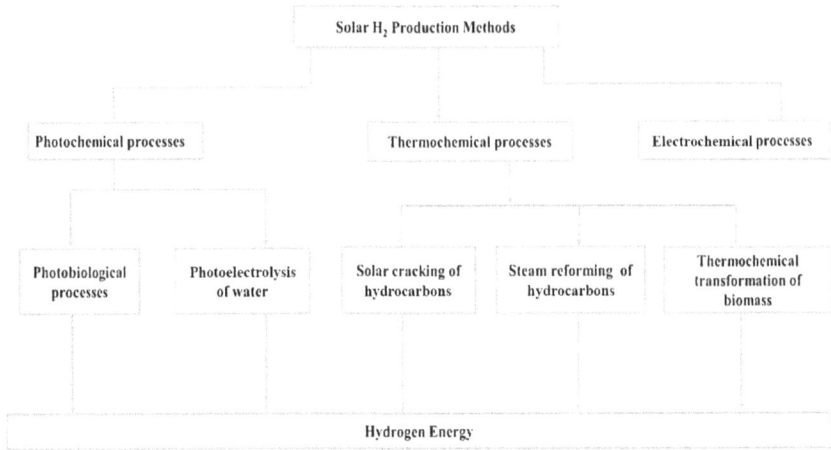

FIGURE 11.5   Classification of hydrogen production from solar, adopted by [5].

phase, which is required for more research and advancement. The techniques of solar energy permit the conversion of solar radiation that transforms into heat in the range between 200–2000°C. This primary heat is transformed into hydrogen energy. Several different kinds of projects have been adopted to promote solar energy for hydrogen production as the primary source of energy [5].

11.3.1 PHOTOCHEMICAL PROCESSES

Photochemical routes use solar energy to develop hydrolysis of water, which consists of breaking the water molecules into one or more chemical bonds through a chemical reaction. Currently, two techniques are recognized for this method: the photobiological and the photoelectrochemical.

11.3.1.1 Photobiological Processes

Hydrogen production from the photobiological process is considered a huge potential renewable source, and fuel is free from pollution, so it is also suitable for the environment as a result of using solar energy and not producing CO_2 during combustion from this process [4]. Photobiological processes are established on the ability of certain organisms, for example, photosynthetic bacteria, green algae, and cyanobacteria, to behave as biological catalysts in hydrogen production through water and several enzymes such as hydrogenase and nitrogenase [42]. Bio-photolysis and photobiology processes are the water decomposition into other molecules as H_2 and O_2, during sunlight under anaerobic situations through the cyanobacteria and microalgae [41]. In biophotolysis, the microalgae discharge solar energy to produce electrons that are facilitated by the electron, and ferredoxin is established from hydrogenase enzyme to yield hydrogen [43].

In the photobiological process, it is the photofermentation of organic compounds via photosynthetic bacteria. It is a grouping of dark fermentation and photofermentation.

Hydrogen is formed via anaerobic bacteria that decompose the organic compounds deprived of light in dark fermentation, whereas photofermentation is possible by light [44].

11.3.1.2 Photoelectrolysis of Water

The cycle started in the 1970s with hydrogen production through water electrolysis from photovoltaic (PV) cells using electricity [45]. Solar heat is first converted into electrical energy from a PV cell, and electrolysis of water occurs through the current developed by the PV cell. The current efficiency of the photo converter and electrolyzer is 20% and 80%, respectively [46]. However, the overall efficiency of solar radiant converted to chemical hydrogen energy is close to 16% [47]. This technology is more expensive, and PV cells are also costly. Therefore, more extensive research is required in this field. Water electrolysis technology consists of dissociating water molecules into hydrogen and oxygen. The main benefit of this method can achieve 99.99 Vol% of pure hydrogen [48]. An illustration of a PV unit to hydrogen production is referred to in Figure 11.6. Additional challenges for the wide-ranging uses of water electrolysis are the drop in energy consumption, cost and system maintenance, and upturn energy efficiency, durability, safety, and reliability [49].

11.3.2 THERMOCHEMICAL PROCESSES

Thermochemical processes include methods for solar hydrogen production such as thermochemical cycles, direct thermolysis of water and solar cracking, gasification, and reforming of hydrocarbons. These methods depend on concentrated solar radiation and produce high temperatures through the solar device to perform the endothermic reaction. Various kinds of devices can be used to achieve higher ratios of solar

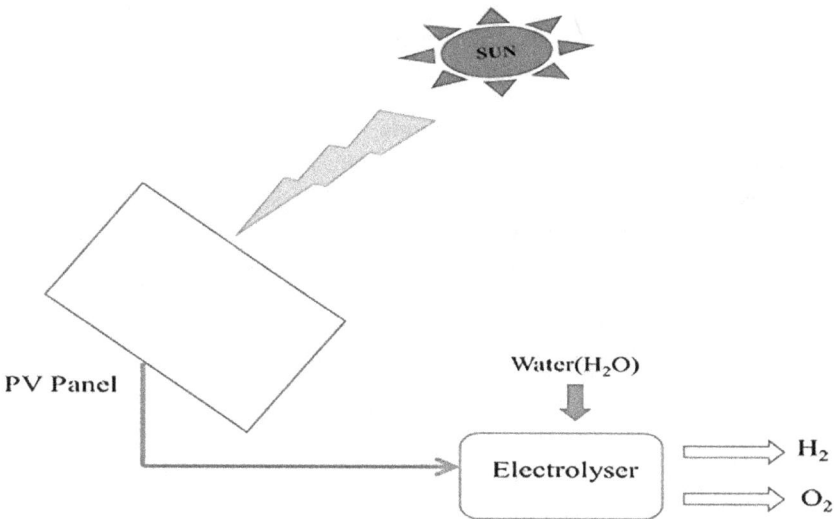

FIGURE 11.6 A systematic process for hydrogen production from a PV unit.

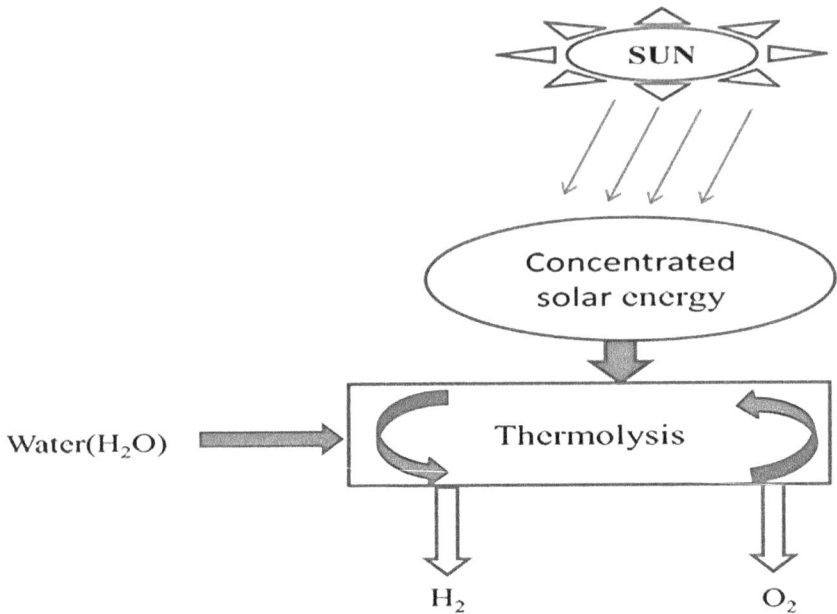

FIGURE 11.7   Process of solar thermal hydrogen production.

concentration, such as tower systems, parabolic disks, and solar furnaces [5]. Figure 11.7 shows a decomposition of water for hydrogen production during the operation of a concentrated solar power (CSP) system. Despite the different kinds of CSP systems, they all depend on concentrated solar light using hundreds of mirrors and converting them into solar heat or electricity. The required heat energy is absorbed into H_2 production from various CSP systems such as solar towers, a parabolic dish, and others known as heliostats [50].

11.3.2.1 Solar Cracking of Hydrocarbons

The solar thermal cracking of hydrocarbons (HCs) contains the recognition of the co-synthesis of H_2 and black carbon. This process consists of the thermal decarbonization of hydrocarbons as natural gas, as shown in Figure 11.8.

Hydrogen production through methane decomposition was started in the 1970s [51]. The separation of methane was permitted at a temperature of 2100K from the developed reactor, and this method was again encouraged in 1993 [52, 53]. The general equation of this method is given as follows in Equation 1 [5]:

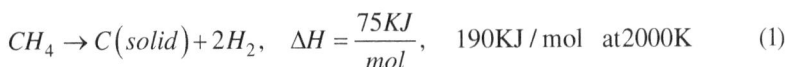

$$CH_4 \rightarrow C(solid) + 2H_2, \quad \Delta H = \frac{75KJ}{mol}, \quad 190KJ/mol \quad at 2000K \quad (1)$$

Dahl et al. [54] achieved 90% methane conversion at 2133K through an Ar/CH_4 mixture inside the reactor, and Hirsch and Steinfeld [55] experimented with a very low temperature on cracking of methane in a reactor directly irradiated with a power of 5000 W under solar flux intensity of 2.8 MW/m^2. Currently, solar hydrogen production

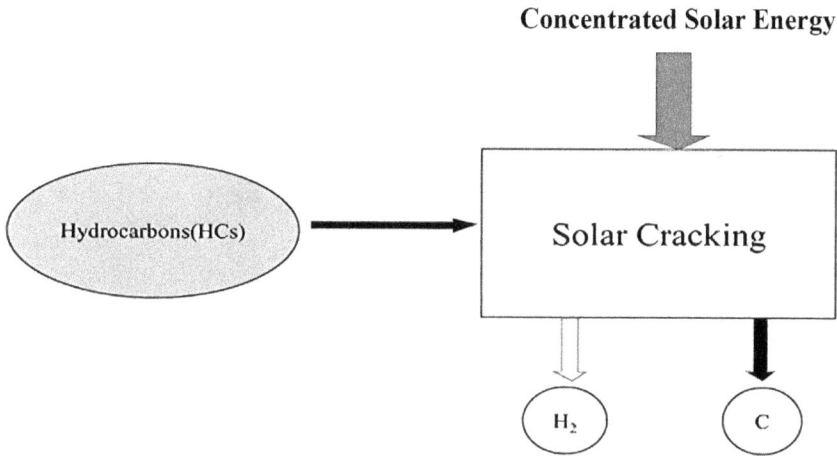

FIGURE 11.8   Graphical presentation of the solar cracking process [5].

from methane cracking in coming decades is disclosed to an exploited arrangement, i.e., consume fossil-based fuels and releases waste byproducts as pollution.

11.3.2.2 Steam Reforming of Hydrocarbons

Steam reforming technology is developed for hydrogen production using various feedstocks such as propane, butane, methanol, jet fuel, and diesel naphtha [56]. The process is presented in Figure 11.9 and Figure 11.10.

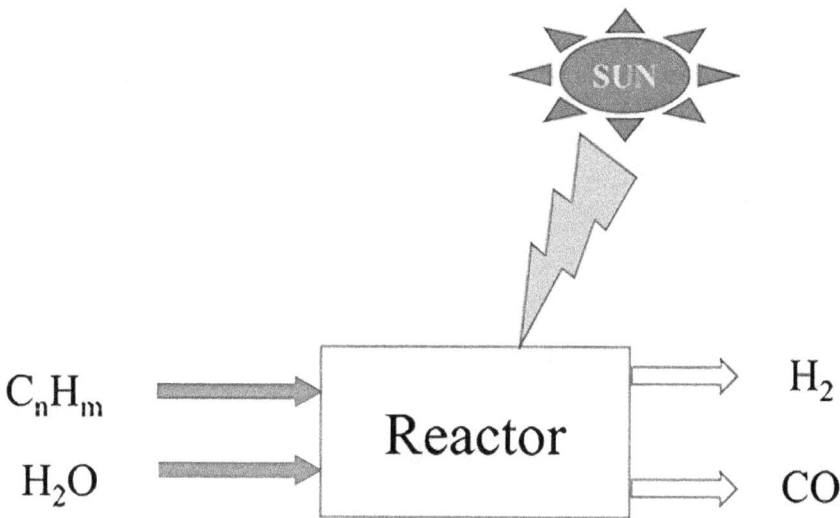

FIGURE 11.9   Principle diagram for solar reformation, adopted and modified [5].

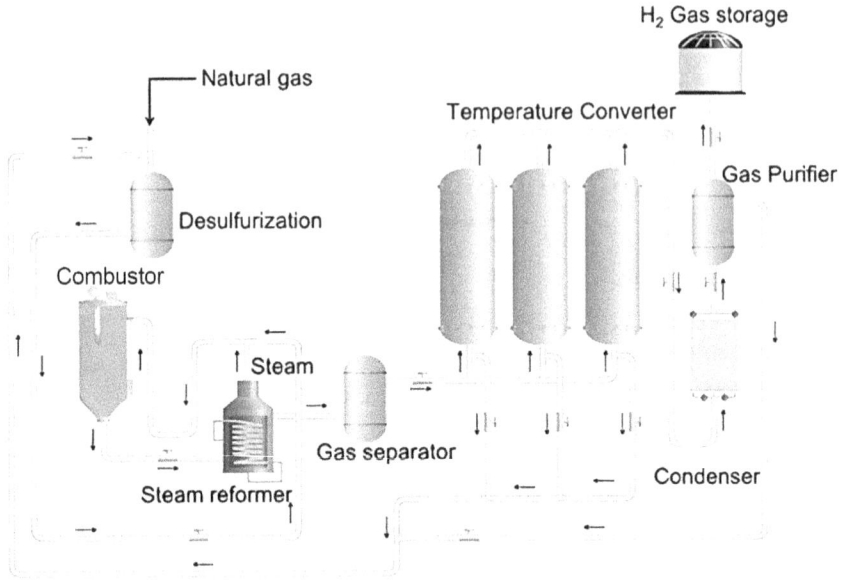

FIGURE 11.10   The flow process of steam reforming of natural gas (methane) [56].

It is the process of the conversion of syngas from various hydrocarbons using steam with the presence of a catalyst such as nickel [10]. The reaction takes place at a pressure range from 20–30 bar with a high temperature of 840–950°C. It is a complex production chain and endothermic reaction that includes various operations to yield pure hydrogen. Generally, the desulphurization process is done before entering into the unit for all hydrocarbons, mainly methane [5]. In this method, feedstock as natural gas (methane) contains different mixtures of gases such as CH_4 (95%), N_2 (1%), CO_2 (0.5%), and small traces of sulfur elements. A higher H/C ratio initiates reduced emission of CO_2, while a lower H/C ratio refers to a negative impact by producing more CO_2 as 7.05 kg CO_2/kg H_2 during hydrogen production to insufficient feedstock characteristics [56]. The basic equations for this process are as follows (Eqs. 2 and 3) [57]:

Reformer Equation:

$$CH_4 + H_2O_{(Steam)} \rightarrow CO + 3H_2, \Delta H_{298K} = +206.2\left(\frac{kJ}{mol}\right) \tag{2}$$

Water Gas Shift Reaction (WGSR):

$$CO + H_2O_{(Steam)} \rightarrow CO_2 + H_2, \Delta H_{298K} = -41.2\left(\frac{kJ}{mol}\right) \tag{3}$$

The reaction (Equation 2) is referring to the endothermic reaction, steam reforming phase, and close to 3 for the H_2/CO_2 stoichiometric ratio. CO enclosed in gas being hazardous is removed by the next reaction (Equation 3), which allows the restoration of hydrogen. Reaction (Equation 3) refers to exothermic reaction and CO_2 conversion from CO known as WGSR. In this stage, the involvement of high- and low-temperature shift reactions at 400°C and 200°C obtained gas product as H_2, CO_2 (nearly 16–20% in Vol.), CH_4, and small traces of CO [58].

Marty [59] performed a study on propane steam reforming with a non-catalytic at a higher temperature using a solar reactor. Tamme et al. [60] also performed a steam methane reforming operation through a solar reactor and attained a methane conversion of approximately 87% with an H_2 molar fraction of 49.3%.

11.3.2.3 Thermochemical Transformation of Biomass

Biomass comprises all types of plants growing on the surface of the earth. It can derive from the photosynthesis of CO_2, H_2O, and sunlight; it produces molecules of the same composition as $C_6H_9O_4$ such as cellulose, lignocellulose, and lignin. The stored energy can then be recovered as fuel through sufficiently efficient energy conversion and economic points [61]. This occurs through alcoholic fermentation, methanation, combustion, and thermochemical transformation [62]. The scheme is mainly suitable for the valorization of lignocellulosic products such as wood or straw. This area leads to the gasification of carbon-based elements and constitutes a consecutive chain of processes that require the simultaneous transfer of physical amounts, proportion control, and reaction time contacts at a given prompt [63]. Figure 11.11

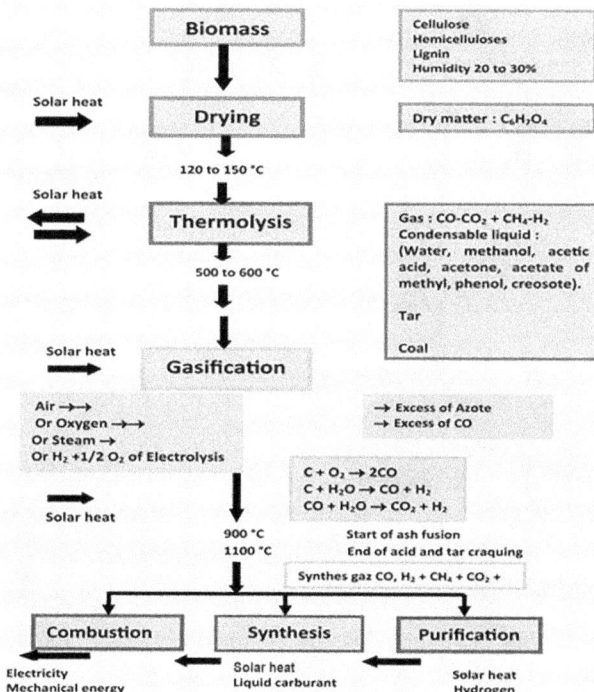

FIGURE 11.11   Processes involved in the thermochemical transformation of biomass [5].

26

etails the various operation steps. Generally, H_2 can be produced from the following reaction (Eq. 4):

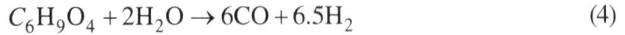

$$C_6H_9O_4 + 2H_2O \rightarrow 6CO + 6.5H_2 \tag{4}$$

In that case, supplementary hydrogen can be produced from a gas shift reaction for their favorable condition, the reaction is given in Eq. 5:

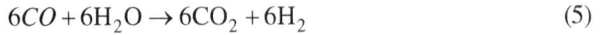

$$6CO + 6H_2O \rightarrow 6CO_2 + 6H_2 \tag{5}$$

In the end, among 12.5 hydrogen molecules obtained, 4.5 were obtained from biomass and 8 molecules come from water that was used as a reactive agent. In addition, 6 molecules of CO_2 emitted can be considered a non-effective greenhouse because it is the same molecule that is captured and recycled by photosynthesis in the atmosphere for plant growth [5].

11.3.3 Electrochemical Processes

Apart from the reformation of the gas system, water electrolysis is the most developed and applicable method for the industrial production of hydrogen. Water electrolysis is an electrochemical process in which water can be split into its constituent components hydrogen and oxygen using electrical energy as a result of two chemical reactions occurring separately from the anode and cathode [64]. The principle of the electrochemical process is cited in Figure 11.12. The general equation of electrolysis reaction (Eq. 6) is as follows. For anode and cathode reactions, refer to Equations 7 and 8, respectively [5]:

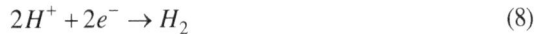

$$H_2O + \text{electricity} \rightarrow H_2 + 0.5O_2 \tag{6}$$

$$H_2O + \text{electricity} \rightarrow 2H^+ + 0.5O_2 + 2e^- \tag{7}$$

$$2H^+ + 2e^- \rightarrow H_2 \tag{8}$$

The electrodes are detached through an electrolytic conductor composed of ions, which can easily transport ionic particles between the electrodes. The electrical energy entering the system is converted into chemical energy in the form of hydrogen. Although water electrolysis is well known, it is not economically viable because electricity must be obtained from renewable sources. Some important technology of water electrolysis is polymer membranes, alkaline electrolytes, electrolysis of steam, and oxide electrolytes of ceramic that are obtained from the electrolysis process in the range from 65–85% [64].

Grigoriev et al. [65] reviewed electrolyzers PEM (Proton Exchange Membrane) and showed the various technologies from water electrolysis and found that one of the electrolyzers PEM technology is practical for pure hydrogen. Agbli et al. [66]

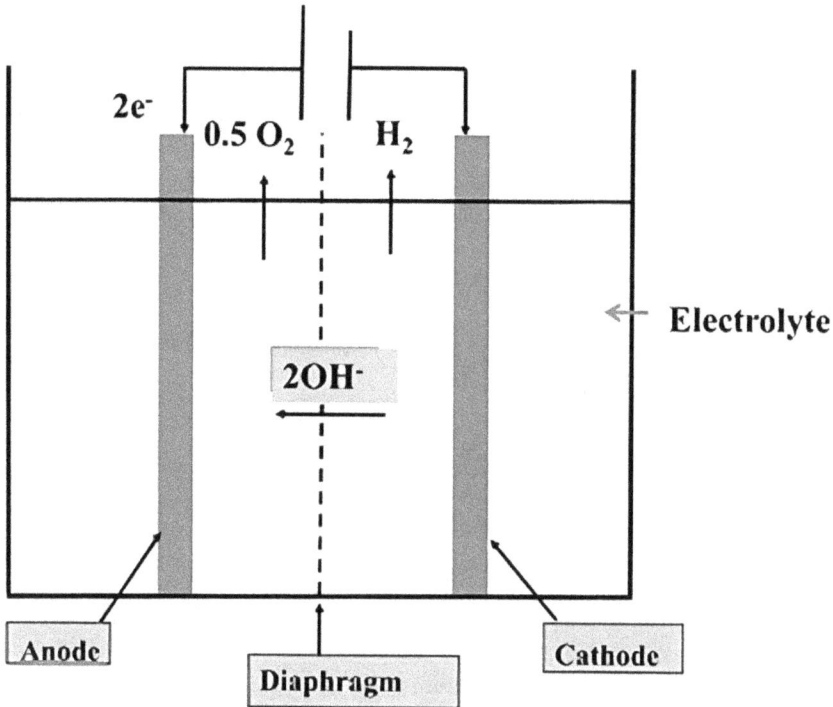

FIGURE 11.12   Principle of alkaline water electrolysis process [5].

have established a graphical model of the electrolyzer as a macroscopic energetic representation. Much of the work has been done on the solar-based electrical electrolyzers model [67]. The main commercial advantages of electrolytic hydrogen production are scalability and zero-emission hydrogen production (when produced from renewable energy sources). It is difficult to establish the main applications of electrolyzers for hydrogen production, especially the high capital and electricity costs. For example, a solar power PV cell costs about $30/kWh, ten times the cost of competitive electrolysis, and the cost of the electrolyzer must be significantly reduced to allow for large-scale application [5].

11.4 SUMMARY OF SOLAR H₂ PRODUCTION METHODS

Table 11.1 offers the relevant important studies that are considered for this study. The various ways of using solar energy for hydrogen production are deliberated, and evaluation of several parameters in terms of energy and exergy performances, efficiencies, and hydrogen production rates are carried out, as referred to in Table 11.2. The efficiency of converting solar energy into electricity is a major barrier to improving the overall efficiency of hydrogen production. By comparison, solar electrolysis, photo electrochemical and photochemical technologies for hydrogen filling stations

TABLE 11.1
Several Kinds of Literature on Solar-Based H$_2$ Production

Method	Study Type	Source	Authors	Country	Reference
Hybrid thermochemical	Review	Concentrated solar	Ngoh and Njomo	Cameroon	[5]
PV/fuel cell hybrid	Review and case study	Concentrated solar	Yilanci et al.	Turkey	[47]
Thermochemical cycle	Review	Solar energy	Xiao et al.	China	[68]
Photochemical, electrochemical, and thermochemical	Review	Solar energy	González et al.	Spain	[69]
Thermochemical	Review	Concentrated solar radiation (high temperature)	Steinfeld A	Switzerland	[50]
Biological, electrochemical, and thermochemical	Review	Renewable	Chaubey et al.	India	[70]
Thermochemical cycles	Review	Concentrated solar power	Agrafiotis et al.	Germany	[44]

TABLE 11.2
Several Studies on Solar-Based Hydrogen Production Systems

Source	Process	Performance Analysis	Efficiency	Hydrogen production	Year	Authors	Ref.
Solar energy	Thermochemical (Mg–Cl)	Energy and Exergy	18.8	—	2014	Ozcan and Dincer	[71]
Solar photovoltaic/thermal energy	Electrolysis	Exergy	—	0.064 (kg/s)	2014	Ngoh et al.	[72]
Solar energy	Thermochemical	Energy	20.4	—	1999	Steinfeld et al.	[73]
Solar and wind	Electrolysis	—		26–19 (kWh/ m^2 yr)	2011	Dagdougu et al.	[74]
Solar energy	Thermochemical (Cu–Cl), photocatalytic	Energy and Exergy	46.60– 56.41	366.74–328.1 (kg/day)	2014	Ratlamwala and Dincer	[75]
Solar energy	Thermochemical (H$_2$SO$_4$)	Energy	34	—	2012	Liberatore et al.	[76]
Solar energy	Photovoltaic, photoelectrolysis, biophotolysis	Exergy	—		2010	Joshi et al.	[12]

TABLE 11.2 *(Continued)*
Several Studies on Solar-Based Hydrogen Production Systems

Source	Process	Performance Analysis	Efficiency	Hydrogen production	Year	Authors	Ref.
Solar energy	Photocatalytic	Energy and Exergy	0.085	0.1–0.21 (mmol/h)	2014	Shamim et al.	[77]
Solar energy	Solar process, hybrid process, Claus process	Energy	29–36	81–138 (million-kW h per year)	2010	Villasmil and Steinfeld	[78]
Solar energy	Thermochemical Electrolysis	Energy	79	—	2014	Mallapragada and Agrawal	[79]
Solar energy	Electrolyte membrane (PEM) electrolyzer	Energy	80.81	—	2014	Scamman et al.	[80]
Solar energy	Thermochemical	Energy	20–28	—	2012	Hong et al.	[81]
Solar energy	Photocatalytic (H_2S)	Energy and Exergy	39	—	2014	Shamim et al.	[82]
Solar energy	Power cycles/ polymer exchange membrane	Energy and exergy	52.9	—	2013	Ozturk and Dincer	[83]
Solar energy	PV-Electrolyzer	Exergy		17 kg/h	2012	Bozoglan et al.	[84]
Solar energy	PV/PV/proton exchange membrane (PEM)	—		20–29 (m^3/ year)	2013	Ghribi et al.	[85]
Solar energy	Thermochemical (Cu–Cl)	Energy	49	—	2010	Ghandehariun et al.	[86]

can be cost-effective because they require fewer processes and avoid external energy sources and additional systems for distributing hydrogen.

High-temperature methods require additional material development, focusing on high-temperature membranes and heat exchangers for solar processes. Thus, the global solar hydrogen system mainly consists of solar hydrogen systems for transport and stationary applications. The environmentally friendly production of hydrogen using solar energy is very important since no greenhouse gases are generated during operation. The low exergy efficiency of photovoltaic generators, and hence the overall exergy efficiency of solar-hydrogen systems, is currently a challenge for researchers and scientists. Numerous renewable energy sources have their own social or environmental issues that need to be addressed. However, all of these obstacles are being actively addressed by organizations and industries, and it is worth considering whether the benefits justify further investment now.

11.5 CONCLUSION

The current study commenced a review of the literature concerning solar hydrogen production approaches. The following conclusions are offered through this research study:

- Among the various hydrogen production processes studied and associated, solar hydrogen production is one of the most important sources of renewable hydrogen production, with zero or low greenhouse gas emissions in the production process.
- According to the latest technologies, the production of hydrogen based on solar energy requires further research and development in terms of the system implementation process, a more "cost-effective" system, and more efficiency for their efficiencies.
- Although the production of solar hydrogen in a thermochemical cycle is in the development phase, it offers countless benefits. In particular, it offers eco-friendly and sustainable opportunities.
- Existing research shows that solar hydrogen production from numerous methods can play an important role in decreasing greenhouse gas emissions and provide opportunities for sustainable development and environmental improvement.

11.5.1 ACKNOWLEDGMENT

The authors are highly grateful to the Department of Mechanical Engineering and the Centre for Energy and Environment, Delhi Technological University, Delhi (India), for providing basic facilities to compile this work.

REFERENCES

[1]. Burton, N. A., Padilla, R. V., Rose, A., & Habibullah, H. (2021). Increasing the efficiency of hydrogen production from solar powered water electrolysis. *Renewable and Sustainable Energy Reviews*, *135*, 110255.
[2]. Safari, F., & Dincer, I. (2020). A review and comparative evaluation of thermochemical water splitting cycles for hydrogen production. *Energy Conversion and Management*, *205*, 112182.
[3]. Dincer, I. (1998). Energy and environmental impacts: Present and future perspectives. *Energy Sources*, *20*(4–5), 427–453.
[4]. Satyapal, S., Petrovic, J., Read, C., Thomas, G., & Ordaz, G. (2007). The US Department of Energy's National Hydrogen Storage Project: Progress towards meeting hydrogen-powered vehicle requirements. *Catalysis Today*, *120*(3–4), 246–256.
[5]. Ngoh, S. K., & Njomo, D. (2012). An overview of hydrogen gas production from solar energy. *Renewable and Sustainable Energy Reviews*, *16*(9), 6782–6792.
[6]. Pagliaro, M., & Konstandopoulos, A. G. (2012). *Solar hydrogen: Fuel of the future*. Royal Society of Chemistry, Cambridge.
[7]. Staffell, I., Scamman, D., Abad, A. V., Balcombe, P., Dodds, P. E., Ekins, P., et al. (2019). The role of hydrogen and fuel cells in the global energy system. *Energy & Environmental Science*, *12*(2), 463–491.

[8]. Abdalla, A. M., Hossain, S., Nisfindy, O. B., Azad, A. T., Dawood, M., & Azad, A. K. (2018). Hydrogen production, storage, transportation and key challenges with applications: A review. *Energy Conversion and Management, 165,* 602–627.

[9]. Council, H. (2017). *Hydrogen scaling up: A sustainable pathway for the global energy transition.* Hydrogen Knowledge Centre, IGEM. Retrieved October 31, 2023, From https://www.h2knowledgecentre.com/content/policypaper1201

[10]. Kumar, R., Kumar, A., & Pal, A. (2021). An overview of conventional and non-conventional hydrogen production methods. *Materials Today: Proceedings, 46,* 5353–5359.

[11]. Wang, Z., Roberts, R. R., Naterer, G. F., & Gabriel, K. S. (2012). Comparison of thermochemical, electrolytic, photoelectrolytic and photochemical solar-to-hydrogen production technologies. *International Journal of Hydrogen Energy, 37*(21), 16287–16301.

[12]. Joshi, A. S., Dincer, I., & Reddy, B. V. (2010). Exergetic assessment of solar hydrogen production methods. *International Journal of Hydrogen Energy, 35*(10), 4901–4908.

[13]. Pregger, T., Graf, D., Krewitt, W., Sattler, C., Roeb, M., & Möller, S. (2009). Prospects of solar thermal hydrogen production processes. *International Journal of Hydrogen Energy, 34*(10), 4256–4267.

[14]. Mazloomi, K., & Gomes, C. (2012). Hydrogen as an energy carrier: Prospects and challenges. *Renewable and Sustainable Energy Reviews, 16*(5), 3024–3033.

[15]. Jain, I. P. (2009). Hydrogen the fuel for 21st century. *International Journal of Hydrogen Energy, 34*(17), 7368–7378.

[16]. Mariolakos, I., Kranioti, A., Markatselis, E., & Papageorgiou, M. (2007). Water, mythology and environmental education. *Desalination, 213*(1–3), 141–146.

[17]. Singh, S., Jain, S., Venkateswaran, P. S., Tiwari, A. K., Nouni, M. R., Pandey, J. K., & Goel, S. (2015). Hydrogen: A sustainable fuel for future of the transport sector. *Renewable and Sustainable Energy Reviews, 51,* 623–633.

[18]. Ball, M., & Wietschel, M. (2009). The future of hydrogen–opportunities and challenges. *International Journal of Hydrogen Energy, 34*(2), 615–627.

[19]. Sopena, C., Diéguez, P. M., Sainz, D., Urroz, J. C., Guelbenzu, E., & Gandía, L. M. (2010). Conversion of a commercial spark ignition engine to run on hydrogen: Performance comparison using hydrogen and gasoline. *International Journal of Hydrogen Energy, 35*(3), 1420–1429.

[20]. Verhelst, S. (2014). Recent progress in the use of hydrogen as a fuel for internal combustion engines. *International Journal of Hydrogen Energy, 39*(2), 1071–1085.

[21]. Dodds, P. E., Staffell, I., Hawkes, A. D., Li, F., Grünewald, P., McDowall, W., & Ekins, P. (2015). Hydrogen and fuel cell technologies for heating: A review. *International Journal of Hydrogen Energy, 40*(5), 2065–2083.

[22]. Zhang, Y. H., Jia, Z. C., Yuan, Z. M., Yang, T., Qi, Y., & Zhao, D. L. (2015). Development and application of hydrogen storage. *Journal of Iron and Steel Research International, 22*(9), 757–770.

[23]. Krieger, E. M., & Arnold, C. B. (2012). Effects of undercharge and internal loss on the rate dependence of battery charge storage efficiency. *Journal of Power Sources, 210,* 286–291.

[24]. Scrosati, B., & Garche, J. (2010). Lithium batteries: Status, prospects and future. *Journal of Power Sources, 195*(9), 2419–2430.

[25]. Balat, M. (2008). Potential importance of hydrogen as a future solution to environmental and transportation problems. *International Journal of Hydrogen Energy, 33*(15), 4013–4029.

[26]. Barbir, F. (2009). Transition to renewable energy systems with hydrogen as an energy carrier. *Energy, 34*(3), 308–312.

[27]. Eliezer, D., Eliaz, N., Senkov, O. N., & Froes, F. H. (2000). Positive effects of hydrogen in metals. *Materials Science and Engineering: A, 280*(1), 220–224.

[28]. Eliaz, N., Eliezer, D., & Olson, D. L. (2000). Hydrogen-assisted processing of materials. *Materials Science and Engineering: A, 289*(1–2), 41–53.

[29]. Dupont, V. (2007). Steam reforming of sunflower oil for hydrogen gas production/oxidación catalítica del aceite de girasol en la producción del gas hidrógeno/reformage à la vapeur de l'huile de tournesol dans la production de gaz hydrogène. *Helia, 30*(46), 103–132.

[30]. Edwards, P. P., Kuznetsov, V. L., David, W. I., & Brandon, N. P. (2008). Hydrogen and fuel cells: Towards a sustainable energy future. *Energy Policy, 36*(12), 4356–4362.

[31]. Sharma, S., & Ghoshal, S. K. (2015). Hydrogen the future transportation fuel: From production to applications. *Renewable and Sustainable Energy Reviews, 43*, 1151–1158.

[32]. Veziroğlu, T. N., & Şahi, S. (2008). 21st century's energy: Hydrogen energy system. *Energy Conversion and Management, 49*(7), 1820–1831.

[33]. Sharma, S., & Ghoshal, S. K. (2015). Hydrogen the future transportation fuel: From production to applications. *Renewable and Sustainable Energy Reviews, 43*, 1151–1158.

[34]. Lattin, W. C., & Utgikar, V. P. (2007). Transition to hydrogen economy in the United States: A 2006 status report. *International Journal of Hydrogen Energy, 32*(15), 3230–3237.

[35]. Midilli, A., Ay, M., Dincer, I., & Rosen, M. A. (2005). On hydrogen and hydrogen energy strategies: I: Current status and needs. *Renewable and Sustainable Energy Reviews, 9*(3), 255–271.

[36]. Barthélémy, H., Weber, M., & Barbier, F. (2017). Hydrogen storage: Recent improvements and industrial perspectives. *International Journal of Hydrogen Energy, 42*(11), 7254–7262.

[37]. Ball, M., & Wietschel, M. (2009). The future of hydrogen–opportunities and challenges. *International Journal of Hydrogen Energy, 34*(2), 615–627.

[38]. Schlapbach, L., & Züttel, A. (2011). Hydrogen-storage materials for mobile applications. In *Materials for sustainable energy: A collection of peer-reviewed research and review articles from nature publishing group* (pp. 265–270). Nature Publishing Group, London.

[39]. Shakya, B. D., Aye, L., & Musgrave, P. (2005). Technical feasibility and financial analysis of hybrid wind–photovoltaic system with hydrogen storage for Cooma. *International Journal of Hydrogen Energy, 30*(1), 9–20.

[40]. Richards, M., & Shenoy, A. (2007). H_2-MHR pre-conceptual design summary for hydrogen production. *Nuclear Engineering and Technology, 39*(1), 1.

[41]. Midilli, A., & Dincer, I. (2007). Key strategies of hydrogen energy systems for sustainability. *International Journal of Hydrogen Energy, 32*(5), 511–524.

[42]. Hallenbeck, P. C., & Benemann, J. R. (2002). Biological hydrogen production; fundamentals and limiting processes. *International Journal of Hydrogen Energy, 27*(11–12), 1185–1193.

[43]. Guo, L. J., Zhao, L., Jing, D. W., Lu, Y. J., Yang, H. H., Bai, B. F., et al. (2009). Solar hydrogen production and its development in China. *Energy, 34*, 1079–1090.

[44]. Androga, D. D., Capsulatus, R., & Eroğlu, I. (2009). *Biological hydrogen production on acetate in continues panel photo-bioreactors using Rhodobacter capsulatus.* M.Sc. Thesis in Chemical Engineering. Department, Middle East Technical University, Turkey.

[45]. Khan, S. U., Al-Shahry, M., & Ingler, W. B., Jr. (2002). Efficient photochemical water splitting by a chemically modified n-TiO_2. *Science, 297*(5590), 2243–2245.

[46]. Rzayeva, M. P., Salamov, O. M., & Kerimov, M. K. (2001). Modeling to get hydrogen and oxygen by solar water electrolysis. *International Journal of Hydrogen Energy, 26*(3), 195–201.

[47]. Yilanci, A., Dincer, I., & Ozturk, H. K. (2009). A review on solar-hydrogen/fuel cell hybrid energy systems for stationary applications. *Progress in Energy and Combustion Science*, *35*(3), 231–244.

[48]. Abdin, Z., Webb, C. J., & Gray, E. M. (2015). Solar hydrogen hybrid energy systems for off-grid electricity supply: A critical review. *Renewable and Sustainable Energy Reviews*, *52*(C), 1791–1808.

[49]. AlZaharani, A., Dincer, I., & Naterer, G. F. (2013). Performance evaluation of a geo-thermal based integrated system for power, hydrogen and heat generation. *International Journal of Hydrogen Energy*, *38*, 14505–14511.

[50]. Steinfeld, A. (2005). Solar thermochemical production of hydrogen—a review. *Solar Energy*, *78*(5), 603–615.

[51]. Matovich, E. (1978). *U.S. Patent No. 4,095,974*. Washington, DC: U.S. Patent and Trademark Office.

[52]. Steinberg, M. (1986). The direct use of natural gas for conversion of carbonaceous raw materials to fuels and chemical feedstocks. *International Journal of Hydrogen Energy*, *11*(11), 715–720.

[53]. Muradov, N. Z. (1993). How to produce hydrogen from fossil fuels without CO_2 emission. *International Journal of Hydrogen Energy*, *18*(3), 211–215.

[54]. Rodat, S., Abanades, S., Coulié, J., & Flamant, G. (2009). Kinetic modelling of methane decomposition in a tubular solar reactor. *Chemical Engineering Journal*, *146*(1), 120–127.

[55]. Hirsch, D., & Steinfeld, A. (2004). Solar hydrogen production by thermal decomposition of natural gas using a vortex-flow reactor. *International Journal of Hydrogen Energy*, *29*(1), 47–55.

[56]. Kannah, R. Y., Kavitha, S., Karthikeyan, O. P., Kumar, G., Dai-Viet, N. V., & Banu, J. R. (2021). Techno-economic assessment of various hydrogen production methods–A review. *Bioresource Technology*, *319*, 124175.

[57]. Goyal, H. B., Seal, D., & Saxena, R. C. (2008). Bio-fuels from thermochemical conversion of renewable resources: A review. *Renewable and Sustainable Energy Reviews*, *12*(2), 504–517.

[58]. Kumar, R., Kumar, A., & Pal, A. (2022). Overview of hydrogen production from bio-gas reforming: Technological advancement. *International Journal of Hydrogen Energy*, *47*(82), 34831–34855.

[59]. Pen, M. A., Gomez, J. P., & Fierro, J. G. (1996). New catalytic routes for syngas and hydrogen production. *Applied Catalysis A: General*, *144*(1–2), 7–57.

[60]. Marty, P. (2002). *Procédé et dispositif de génération d'hydrogène par conversion à haute température avec vapeur d'eau* (Doctoral dissertation, Perpignan).

[61]. Tamme, R., Buck, R., Epstein, M., Fisher, U., & Sugarmen, C. (2001). Solar upgrading of fuels for generation of electricity. *Journal of Solar Energy Engineering*, *123*(2), 160–163.

[62]. Boutin, O., Ferrer, M., & Lédé, J. (2002). Flash pyrolysis of cellulose pellets submitted to a concentrated radiation: Experiments and modelling. *Chemical Engineering Science*, *57*(1), 15–25.

[63]. Albertazzi, S., Basile, F., Brandin, J., Einvall, J., Hulteberg, C., Fornasari, G., et al. (2005). The technical feasibility of biomass gasification for hydrogen production. *Catalysis Today*, *106*(1–4), 297–300.

[64]. Padin, J., Veziroglu, T. N., & Shahin, A. (2000). Hybrid solar high-temperature hydrogen production system. *International Journal of Hydrogen Energy*, *25*(4), 295–317.

[65]. Grigoriev, S. A., Porembsky, V. I., & Fateev, V. N. (2006). Pure hydrogen production by PEM electrolysis for hydrogen energy. *International Journal of Hydrogen Energy*, *31*(2), 171–175.

[66]. Agbli, K. S., Péra, M. C., Hissel, D., Rallières, O., Turpin, C., & Doumbia, I. (2011). Multiphysics simulation of a PEM electrolyser: Energetic macroscopic representation approach. *International Journal of Hydrogen Energy*, *36*(2), 1382–1398.

[67]. Atlam, O. (2009). An experimental and modeling study of a photovoltaic/protonex-change membrane electrolyzer system. *International Journal of Hydrogen Energy*, *34*, 6589–6595.

[68]. Xiao, L., Wu, S. Y., & Li, Y. R. (2012). Advances in solar hydrogen production via two-step water-splitting thermochemical cycles based on metal redox reactions. *Renewable Energy*, *41*, 1–12.

[69]. Suárez-González, M. A., Blanco-Marigorta, A. M., & Peña Quintana, J. A. (2011). Review on hydrogen production technologies from solar energy. *Renewable Energy and Power Quality Journal*, *1*(9), 347–351.

[70]. Chaubey, R., Sahu, S., James, O. O., & Maity, S. (2013). A review on development of industrial processes and emerging techniques for production of hydrogen from renewable and sustainable sources. *Renewable and Sustainable Energy Reviews*, *23*, 443–462.

[71]. Ozcan, H., & Dincer, I. (2014). Energy and exergy analyses of a solar driven Mg–Cl hybrid thermochemical cycle for co-production of power and hydrogen. *International Journal of Hydrogen Energy*, *39*(28), 15330–15341.

[72]. Ngoh, S. K., Ohandja, L. A., Kemajou, A., & Monkam, L. (2014). Design and simulation of hybrid solar high-temperature hydrogen production system using both solar photovoltaic and thermal energy. *Sustainable Energy Technologies and Assessments*, *7*, 279–293.

[73]. Steinfeld, A., Sanders, S., & Palumbo, R. (1999). Design aspects of solar thermochemical engineering – a case study: Two-step water-splitting cycle using the Fe_3O_4/FeO redox system. *Solar Energy*, *65*(1), 43–53.

[74]. Dagdougui, H., Ouammi, A., & Sacile, R. (2011). A regional decision support system for onsite renewable hydrogen production from solar and wind energy sources. *International Journal of Hydrogen Energy*, *36*(22), 14324–14334.

[75]. Ratlamwala, T. A. H., & Dincer, I. (2015). Comparative energy and exergy analyses of two solar-based integrated hydrogen production systems. *International Journal of Hydrogen Energy*, *40*(24), 7568–7578.

[76]. Liberatore, R., Lanchi, M., Giaconia, A., & Tarquini, P. (2012). Energy and economic assessment of an industrial plant for the hydrogen production by water-splitting through the sulfur-iodine thermochemical cycle powered by concentrated solar energy. *International Journal of Hydrogen Energy*, *37*(12), 9550–9565.

[77]. Shamim, R. O., Dincer, I., Naterer, G. F., & Zamfirescu, C. (2014). Experimental investigation of a solar tower based photocatalytic hydrogen production system. *International Journal of Hydrogen Energy*, *39*(11), 5546–5556.

[78]. Villasmil, W., & Steinfeld, A. (2010). Hydrogen production by hydrogen sulfide splitting using concentrated solar energy–thermodynamics and economic evaluation. *Energy Conversion and Management*, *51*(11), 2353–2361.

[79]. Mallapragada, D. S., & Agrawal, R. (2014). Limiting and achievable efficiencies for solar thermal hydrogen production. *International Journal of Hydrogen Energy*, *39*(1), 62–75.

[80]. Scamman, D., Bustamante, H., Hallett, S., & Newborough, M. (2014). Off-grid solar–hydrogen generation by passive electrolysis. *International Journal of Hydrogen Energy*, *39*(19), 19855–19868.

[81]. Hong, H., Liu, Q., & Jin, H. (2012). Operational performance of the development of a 15 kW parabolic trough mid-temperature solar receiver/reactor for hydrogen production. *Applied Energy*, *90*(1), 137–141.

[82]. Shamim, R. O., Dincer, I., & Naterer, G. (2014). Thermodynamic analysis of solar-based photocatalytic hydrogen sulphide dissociation for hydrogen production. *International Journal of Hydrogen Energy, 39*, 15342–15351.

[83]. Ozturk, M., & Dincer, I. (2013). Thermodynamic analysis of a solar-based multigeneration system with hydrogen production. *Applied Thermal Engineering, 51*(1–2), 1235–1244.

[84]. Bozoglan, E., Midilli, A., & Hepbasli, A. (2012). Sustainable assessment of solar hydrogen production techniques. *Energy, 46*(1), 85–93.

[85]. Ghribi, D., Khelifa, A., Diaf, S., & Belhamel, M. (2013). Study of hydrogen production system by using PV solar energy and PEM electrolyser in Algeria. *International Journal of Hydrogen Energy, 38*(20), 8480–8490.

[86]. Ghandehariun, S., Naterer, G. F., Dincer, I., & Rosen, M. A. (2010). Solar thermochemical plant analysis for hydrogen production with the copper–chlorine cycle. *International Journal of Hydrogen Energy, 35*(16), 8511–8520.

12 Efficient Hydrogen Production Using Solar Thermal Energy for a Sustainable Future
Challenges and Perspectives

Shubham Raina, Krishma Kumari,
Deepak Pathania, and Richa Kothari

12.1 INTRODUCTION

Global energy demands have been considerably impacted by population growth and urbanisation. The use of conventional resources such as fossil fuels in energy infrastructure results in air pollution, acid rain, and greenhouse gas (GHG) emissions that include different gases such as methane (CH_4), carbon dioxide (CO_2), nitrous oxide (N_2O), water vapour, and ozone (O_3). This results in serious environmental problems including global warming, environmental degradation, and dwindling energy resources that significantly contribute to climate change (Amin et al., 2022). The increase in GHGs is also a contributing factor to both non-infectious and infectious diseases, poor diet, shortage of water, and other societal problems. The global average temperature is gradually rising over the years, and the greenhouse effect's augmentation has caused the atmospheric CO_2 concentration to exceed 400 ppm (Mikhaylov et al., 2020). Climate change and air pollution are already causing severe harm to children's mental and physical health, because the foetus, newborns, and kids are especially susceptible to the adverse effects of these factors (Perera and Nadeau, 2022). Excessive fossil fuel consumption has also had a significant adverse effect on GHG emissions, which are the primary factor triggering global warming (Nayak et al., 2022). The ongoing exploitation of fossil fuels and issues with their carbon emissions have encouraged researchers to think more about sustainability, which creates a need for exploring alternate energy resources that are environmentally friendly, such as biomass, hydro, solar, wind, geothermal, and tidal energy (Callegari et al., 2020). Hydrogen (H_2) is a renewable energy carrier that has the capacity for reducing dependency on fossil fuels and lowering CO_2 emissions to counter the impacts of global warming (Hosseini and Wahid, 2020). H_2 in itself is a clean fuel with almost zero emissions during burning. H_2 can be synthesized using various energy forms

                                    DOI: 10.1201/9781003345558-15

such as nuclear and renewable energy sources including geothermal, solar, wind, and hydroelectric power to split water, as well as fossil fuels such as coal and natural gas (preferably with carbon capture, utilisation, and storage) (Kayfeci et al., 2019). H_2 is a diverse energy carrier that can assist many industrial operations, storage of energy, and power generation, as well as can be used as fuel for vehicles. Multiple energy-consuming sectors can use H_2 for to enhance energy flexibility, improve energy security, and decrease environmental emissions. As the H_2 production systems are diverse, they can help in supplying surplus energy to the power grid (Wang et al., 2018). H_2 has many positive properties as it is a clean fuel with zero carbon emission, reduces GHG emissions, is non-toxic, highly efficient, and has minimised carbon footprints if produced from renewable energy resources. H_2 production from renewable energy resources also supports Sustainable Development Goal 7 (SDG 7), which is affordable and clean energy for all (Ahmad et al., 2023). Although production is a challenging task and has major constraints with storage and supply too, that will be discussed in the upcoming sections.

12.2 HYDROGEN PRODUCTION

H_2 being a versatile fuel, it can be prepared from different methods. Further, these methods can be categorised on the basis of the source used for their production. The fossil fuel–based methods include steam methane reforming (SMR), partial oxidation and carbon dioxide reforming, and autothermal reforming (ATR). Biological methods include fermentation (dark and photofermentation), and biophotolysis. The thermochemical methods are gasification and pyrolysis. Figure 12.1 summarises different methods of H_2 production.

12.2.1 HYDROGEN PRODUCTION FROM FOSSIL FUEL

The current trends show the highest use of fossil fuels for H_2 production. Different energy sources used include coal, natural gas, and oil (Wang and Han, 2022). The different approaches to H_2 production from fossil fuels include steam methane

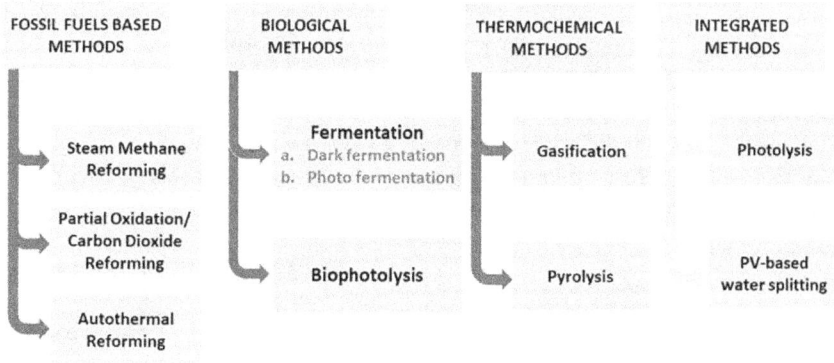

FIGURE 12.1 Hydrogen production methods.

reforming (SMR), partial oxidation and carbon dioxide reforming, and autothermal reforming (ATR).

12.2.2 STEAM METHANE REFORMING (SMR)

SMR is an endothermic process that uses methane or natural gas at elevated temperatures and pressure. In this process, steam is used along with catalysts to produce H_2 (Muritala et al., 2020). In SMR, methane and steam are combined with a catalyst to generate H_2, CO, and a negligibly minor quantity of CO_2 under a range of pressures from 3–25 bars (1 bar = 14.5 psi). As SMR is an endothermic process, constant external heat is required to run the process. At 450°C temperature and 300 kilopascals (kpa) of pressure, the natural gas reforming reaction achieves more than an 80% conversion rate of the higher hydrocarbons and more than 65% H_2 recovery, outperforming the steam methane reforming reaction (Anzelmo et al., 2018). In the context of the H_2 economy, the endothermic and equilibrium-limited reaction of SMR is crucial for the synthesis of syngas and the creation of H_2 (Chompupun et al., 2018). Palladium (Pb) is being used as an excellent material in industrial sectors for developing inorganic metallic membranes to convert methane and natural gas into H_2 (Jokar et al., 2022).

SMR is a commercial process for hydrogen production but is responsible for damages in the environment. According to the literature, every kilogram of hydrogen produced with SMR releases about 7 kilograms of CO_2.

12.2.3 PARTIAL OXIDATION AND CARBON DIOXIDE REFORMING

The partial oxidation of methane is another process for the production of syngas at lower temperatures (Goodman et al., 2018). Syngas can also be produced from natural gas through its partial oxidation and carbon dioxide reforming. The reforming of carbon dioxide involves the reaction of CO_2 with short-chain hydrocarbons such as methane in the presence of metal catalysts such as nickel or its alloys. It also results in the syngas production having H_2 and carbon monoxide (CO) as major products.

Direct conversion of methane (CH_4) and CO_2 into syngas is mainly used in industrially utilised methane partial oxidation reactors. As per Chen and Gan (2022), the ideal preheating temperature is 923K and input velocity is 200–300 m/s, respectively, and the ideal reactant ratio is 5:4:1 for CH_4, O_2, and CO_2. The percentage conversion value of CO_2 and CH_4 is about 47% and 99.88% respectively, and yields of CO and H_2 are about 91% and 65%. CH_4 is partially oxidised without catalysts in steam and H_2 at atmospheric pressure in a sorption-enhanced gasification (SEG) environment with a non-premixed burner (Kertthong et al., 2022).

12.2.4 AUTOTHERMAL REFORMING (ATR)

ATR involves the use of O_2 with CO_2 steam for reacting with CH_4 to produce syngas. Autothermal reforming is a viable technology that combines steam reforming with partial oxidation for cost-effective H_2 production. It has many advantages over steam reforming due to its easy operational conditions and ability to be operated in small systems. It also requires less input energy and has lower operational temperatures. Moreover, an air separation unit (ASU) is necessary to provide a supply of pure O_2, which prevents the use of ATR in industrial applications due to the expensive initial investment and operating costs (Kim et al., 2021). The two processes that produce H_2 at the lowest and highest costs, respectively, are ATR with carbon capture and storage (CCS) and SMR with CCS. The life cycle of GHG emissions of blue H_2 from autothermal reforming are the lowest at 3.91 $kgCO_2eq/kg\ H_2$ (Oni et al., 2022). A similar study by Zhang et al. (2022) used $Pd-Zn/Al_2O_3$ catalyst for the autothermal reforming process because of its excellent oxidation and reforming activities, and it increased H_2 selectivity.

12.3 BIOLOGICAL METHODS OF HYDROGEN PRODUCTION

Biological methods of H_2 production utilise living organisms such as microbes and algae. There are numerous methods for the synthesis of biological H_2, such as fermentation (including dark and photofermentation) and bio-photolysis.

12.3.1 FERMENTATION

Fermentation induces a metabolic change in organic matter by microbial activity. In fermentation, sugar molecules are digested enzymatically into simpler molecules and compounds. It can be divided into two types: dark fermentation and photofermentation (Kothari et al., 2012). Dark fermentation proceeds anaerobically while photofermentation operates under light conditions in the presence of specially abled photosynthetic microbes. Substrates including carbohydrates, lipids, and proteins are consumed to produce H_2, organic acids, and CO_2.

12.3.1.1 Dark Fermentation

Dark fermentation is a biological process for H_2 production that involves facultative and obligate anaerobic microorganisms in a light and oxygen-deficient environment. A wide range of substrates ranging from lignocellulosic biomass, carbohydrate-rich materials including industrial wastewater rich in sugars, and municipal wastewater (MWW) can be used in fermentation. In the case of complex lignocellulosic materials such as bagasse and crop residues, pretreatment and hydrolysis are carried out to break down complex biomass forms into simple organic molecules. These important steps lead to greater efficiency in the overall process. Further, microorganisms take up the electrons present in the organic substrate rich in H_2. The metabolic pathway results in the production of excess electrons that reduces protons (H^+) into biohydrogen (H_2).

12.3.1.2 Photofermentation

Photosynthetic bacteria are the catalysts in producing biohydrogen (H_2) using a light source. The source of light can be natural sunlight or artificial light. H_2 is majorly produced using purple non-sulfur bacteria (PNSB), and rarely, strains of green and purple bacteria are also used. Electrons are generated via oxidation of biomolecules. Mainly nitrogenase enzyme, and in some cases hydrogenase enzyme, catalyses the H_2 production.

12.3.2 BIOPHOTOLYSIS

Another process involving microorganisms for H_2 production is biophotolysis. It involves different microorganisms including cyanobacteria (blue-green algae) and green algae. Biophotolysis can be classified into direct and indirect types. In direct biophotolysis, green algae are the main driver for H_2 production involving hydrogenase. Species including *Chlamydomonas reinhardtii* and *Anabaena sp.* can be used for biohydrogen production (Goria et al., 2023). The latter, indirect biophotolysis, involves two stages. First, photosynthetic microbes produce a large amount of biomass, and second, the blue-green algae (BGA) produce H_2 using nitrogenase enzyme. The efficiency of direct biophotolysis is lesser as H_2 and O_2 are produced in separate stages in indirect biophotolysis (Melitos et al., 2021).

12.4 THERMOCHEMICAL METHODS

The thermochemical methods of H_2 production involve the use of heat for degrading biomass into simpler compounds. The two thermochemical methods mainly used for H_2 production are pyrolysis and gasification.

12.4.1 GASIFICATION

Gasification is a process of turning solid biomass or liquid hydrocarbons into gas (syngas) that can be utilised for a variety of purposes, including power generation and H_2 production. Syngas is the major product of gasification systems that are being employed mostly for heat and electricity generation (Singh et al., 2022). High temperature and low-pressure conditions are favourable for syngas production (Meramo-Hurtado et al., 2020). H_2 can be produced by gasifying carbon-based feedstock such as biomass, or municipal solid waste to produce syngas. The H_2 and other components of the syngas are subsequently extracted and refined. The gasification process is often divided into five stages: drying, pyrolysis, combustion, cracking, and reduction (Singh et al., 2022). The feedstock is dried in order to remove any moisture present. The feedstock is pyrolyzed by heating it without oxygen, resulting in the separation of its constituent solid, liquid, and gaseous phases. The gaseous components undergo additional heating and chemical transformation into syngas during the gasification phase. Syngas is burned at the combustion stage to produce heat and electricity. Many benefits of gasification make it preferable to other techniques of producing H_2. It is flexible enough to handle a variety of feedstocks, including waste materials, and to work in tandem with other processes such as CCS (Meramo-Hurtado et al., 2020). Also, gasification can be more efficient than other processes, such as SMR, for producing H_2. However, there are several difficulties with gasification for H_2 production. The setup and running costs of the process can be high, and the syngas produced may need to be purified before it can be utilised to make H_2 (Yadav et al., 2015).

12.4.2 Pyrolysis

Pyrolysis is the chemical process of heating organic material in the absence of or limited O_2 conditions. Typically, organic matter is plant biomass. In the absence of oxygen, the thermochemical pyrolysis of biomass or other organic materials produces H_2 gas. This method of producing H_2 is considered renewable and sustainable because it employs organic materials that can be grown again or recycled. The end result of pyrolysis is liquid fuel (bio-oil), volatile gases, solid char, and tar. Usually, pyrolysis is used at the industrial scale for converting biomass into energy. The pyrolysis process can be separated into three stages on the basis of the temperature range. At 120–200°C, a slight weight reduction of biomass is observed due to bond breaking, rearrangement, and dehydration. In the second stage, there is significant weight reduction due to thermal decomposition. The third stage (end stage) is marked by the breaking of carbon-hydrogen bonds (C-H bonds) and carbon-oxygen bonds (C-O bonds) (Singh et al., 2022). Slow pyrolysis favours higher production of char while fast pyrolysis yields more bio-oil.

In order to produce H_2 through pyrolysis, biomass must be heated at higher temperature ranges. Aiming H_2 production through pyrolysis can be achieved using two different methods. In one method, raw biomass can be subjected to suitable catalysts in the pyrolysis reactors. Another method of H_2 production is a multi-step approach that starts with producing bio-oil via pyrolysis and then subjecting it to catalytic steam reforming (García et al., 2015). When biomass is heated to the point where the molecular bonds holding it together are disrupted, H_2, CO, and other gases including CH_4 and CO_2 are released. This process produces an H_2 gas with an exceptionally high degree of purity as a byproduct.

The production of H_2 through pyrolysis has the benefit of being adaptable to a wide range of feedstocks, including food residues, wood scraps, and energy crops. This serves to reduce reliance on finite fossil fuels and diversify sources of H_2 fuel.

Biochar and bio-oil, for instance, have multiple applications in agriculture, carbon sequestration, and as feedstocks for chemicals and biofuels, and can be generated as a byproduct of this process. Pyrolysis reduces landfill waste and carbon dioxide emissions by producing new products from otherwise useless organic resources.

Nonetheless, the production of H_2 via pyrolysis is not without its challenges. The process requires high temperatures and specialised apparatus, which can be expensive and energy-intensive. The nature of the feedstock and pyrolysis conditions influence both the yield and quality of H_2.

Thermochemical and biological methods of hydrogen production are mostly based on biomass materials, which convert to more complex and non-degradable derivatives after final hydrogen production. So, the approach of selection of biomass is needed to reduce the production cost and full utilization of materials.

H_2 is a flexible source of energy that can be used to power a wide range of things, from fuel cell cars to industrial processes. These methods have their own merits and demerits. While determining which strategy to utilise, it is essential to take into consideration both the benefits and drawbacks of each process. Table 12.1 summarises the advantages and disadvantages of the methods adopted by researchers.

TABLE 12.1

Advantages and Disadvantages of Hydrogen Production Methods

S. No.	Methods	Advantages	Disadvantages	References
1.	Steam Methane Reforming (SMR)	Steam availability enhances the efficiency; the operating temperature is less than ATR; extensively used and commercially available; no need for O_2; H_2 is produced in high concentration.	Energy balance is not a favouring factor as in ATR; excessive air emissions; CH_4 concentrations are high; coke mostly forms.	Aouad et al., 2018; Wang, 2012
2.	Partial oxidation (POX)	No need for supplying external heat; commercially accessible; CH_4 concentrations are low; quick startup and response time.	Low H_2 to CO ratios; formation of soot; quick phase sintering; pure O_2 is required.	Aouad et al., 2018; Wang, 2012
3.	Autothermal Reforming (ATR)	Low operational temperature; low CH_4 concentrations; no need for supplying external heat.	Limited commercial availability; air or O_2 is required; risk of process deactivation.	Wang, 2012
4.	Dry reforming	CO_2 can be used.	A constant supply of pure CO_2 is required; catalyst coking is a common phenomenon; high input of energy.	Aouad et al., 2018
5.	Direct Biophotolysis	H_2 is directly produced from H_2O and sunlight.	It decreased in photochemical efficiency and light conversion.	Germscheidt et al., 2021
6.	Indirect Biophotolysis	Blue-Green Algae (BGA) can be used for H_2 production from water.	Removal of hydrogenates uptake.	Kothari et al., 2017
7.	Dark Fermentation	Different substrates can be used; light is not required.	Low yield and inhibitory effect of O_2 towards hydrogenase enzyme.	Yadav et al., 2015
8.	Light Fermentation	Any wastewater can be decomposed; light is required.	Low photochemical efficiency and hydrogen yield.	Chaurasia and Mondal, 2021

TABLE 12.1 (*Continued*)
Advantages and Disadvantages of Hydrogen Production Methods

S. No.	Methods	Advantages	Disadvantages	References
9.	Gasification	Low emissions; high H_2 production; biochar also produced along with H_2.	High operating and installation costs; purification of syngas adds to the product cost; high energy input	Yadav et al., 2015
10.	Pyrolysis	Emission less, simple and reduced-step process.	High energy input; carbon-rich biochar is the major product.	(Nikolaidis and Poullikkas, 2017)

12.5 INTEGRATION OF SOLAR ENERGY FOR H_2 PRODUCTION

About 99% of H_2 is still being produced through the direct use of fossil fuels or using electricity originating from fossil fuels. Also, 96% of H_2 put into commercial use directly comes from fossil fuels. An estimate suggests the emission of 2 to 5 tonnes of CO_2 per tonne of H_2 produced (Burton et al., 2021). All of these concerns are valid for thinking of novel approaches for H_2 production. Integrating solar energy with modern H_2-producing systems is expected to address environmental and sustainability concerns.

Technological innovations in solar concentrating systems have influenced researchers to explore H_2 production via solar thermal energy. Solar thermochemical processes are among the most promising methods currently being researched for H_2 production. The solar-based thermochemical method allows utilisation of the entire solar spectrum. Since no precious metal catalysts are needed, it provides a favourable thermodynamic route for the production of solar fuels with the potential for high sun-to-fuel efficiencies. Thermochemical processes utilise the highly concentrated solar spectrum supplied by concentrating devices to carry out endothermic chemical reactions at elevated temperatures. The following describes the basic structure of thermochemical processes. Initially, solar energy is passed through optical concentrating devices that allow for the generation of high temperatures. These systems include highly reflecting structures that track the sun's path and concentrate its light on a fixed spot. Concentrating structures such as parabolic dishes, power towers, and solar furnaces are ideal for utilising the solar spectrum. The concentrated solar energy is further absorbed in the solar reactor to raise its temperature. Further thermolysis of H_2O splits water to produce H_2 and O_2. However, thermolysis of water takes place above 2500 K (Villafán-Vidales et al., 2019). The desired temperature can be reached by concentrating structures but faces tough constraints due to the limited availability of materials that can withstand such temperature ranges.

Another approach to H_2 production is water splitting using photovoltaic (PC) cells as a source of electricity. Photovoltaic solar electrolysis offers H_2 with higher energy content in comparison to the conventional SMR method (Pathak et al., 2020). Using an electrical current generated from PV cells, H_2O molecules are separated into H_2

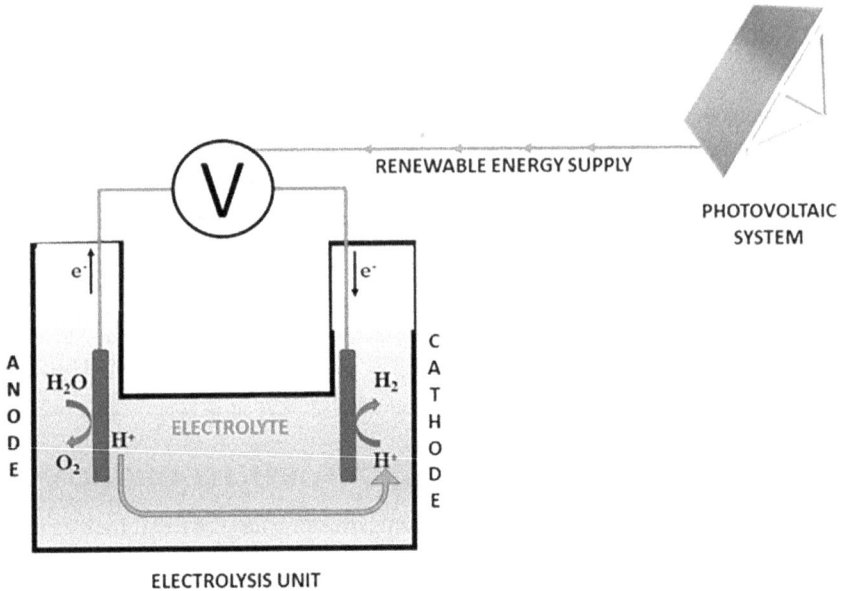

FIGURE 12.2   Diagram representing water splitting using solar PV cells for H_2 production.

and O_2 in this method. This approach offers various advantages over conventional methods of H_2 production, such as avoiding the use of biomass and fossil fuels. The electricity from PV cells is supplied to the electrolysis unit. The O_2 and H^+ ions are liberated at the anode, and further H^+ ions move towards the cathode to form H_2 gas. The setup can be visualized in Figure 12.2.

12.6 HYDROGEN CONSUMPTION IN DIFFERENT SECTORS

H_2 being a versatile fuel has the potential to play an important part in the transition to a low-carbon economy and can be utilised in a variety of industries. Its demand is rising day by day. Different industries need H_2 for a variety of purposes. For instance, in the food processing industry, H_2 is necessary for the conversion of vegetable oil into fats by hydrogenation; in the petroleum sector for sulphurous compounds removal during the refining process; in the power sector as a coolant in power plants for electricity generation; in aviation and automobile sectors, as fuel; and in welding industries as atomic hydrogen torches and oxyhydrogen torches for cutting and welding processes (Goria et al., 2022). It is also needed for aviation fuels and laboratory gas for running different instruments (Agyekum et al., 2022; Osman et al., 2022). According to a World Health Assembly WHA, International Inc (2020) report in 2020, around the world 55% of hydrogen is used in the production of ammonia, 25% of hydrogen is used for refining, 10% is used to make methanol, and only 10% of other applications such as food processing, metalworking, welding, flat glass production, electronic manufacturing (displays, LEDs, etc.), and medical, etc. Figure 12.3 illustrates the various industrial applications of H_2.

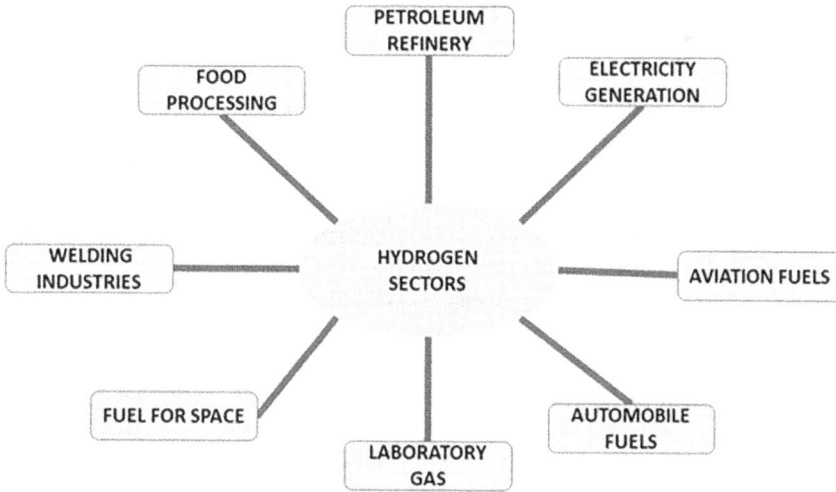

FIGURE 12.3   Different sectors using hydrogen.

12.7 INDIAN SCENARIO

Currently, almost all of the H_2 used in India originates from fossil sources using natural gas. The H_2 is mostly used for fertilizer industries and refineries. The cost of low-carbon H_2 (green and blue H_2) at present is expensive in comparison to grey hydrogen. Although it is expected that by 2050, approximately 80% of India's H_2 will be produced through renewable electricity mainly involving electrolysis. In India, H_2 was previously produced on a modest scale through the electrolysis of water, but this practice is decreased as the demand for electricity has increased alongside the availability of natural gas. A limited amount of hydrogen is still produced by splitting for on-site applications or as a byproduct of the chlor-alkali industry.

The cost of electrolysers will decrease in the upcoming years. For instance, it is anticipated that the cost of alkaline electrolysers will decrease from approximately Rs. 6.3 Cr/MW presently to approximately Rs. 2.8 Cr/MW by the end of 2030. A virtuous circle amid lowering costs and tightening policy to support H_2 will help in reducing electrolyser costs, which will be partially fuelled by large-scale distribution in India and around the world. The cost of green H_2 would be reduced to less than Rs. 150/kg ($2/kg) by 2030 from today's Rs. 300–440/kg ($4–6/kg) by boosting solar plant load factors and electrolyser efficiency (Hall et al., 2020).

In 2030, home use will have the largest electrolysis (green H_2 generation) capacity in the world, at over 60 GW/5 million tonnes. This will assist India in achieving its goal of 500 GW of renewable energy. Also, the cost of natural gas in India is higher in comparison to electricity, which will also favour shifting towards green H_2. The production of green steel will reach its peak globally by 2030, announcing between 15 and 20 million tonnes, in a ground-breaking endeavour to make green steel more widely used. By 2028, India and the rest of the globe will have access to the biggest annual production capacity of 25 GW of electrolysers. By 2030, India will have the greatest green ammonia output in the world, which will aid in the decarbonisation of India's partners.

Some pioneer institutes of India working in the electrolysis sector are Bhabha Atomic Research Centre (BARC), Mumbai; CSIR-CECRI, Karakudi; and ONGC Energy Centre. To understand the magnitude of the potential demand for H_2 in the fertiliser industry, it is necessary to evaluate past and anticipated fertiliser production and consumption. Fertiliser consumption is projected to increase from approximately 45 kg per capita today to 75 kg per capita by 2050. But the demand for fertilisers saturates between 85 and 135 kg per capita. The Indian fertiliser industry uses around 3 metric tonnes of H_2 at present, and this demand is expected to rise up to 7.5 Mt by the end of 2050. Green H_2 may be needed in amounts of up to 100 GW. An investment of $1 billion in H_2 research and development could enable game-changing innovations for the world at scale and at the required speed (Niti Aoyog, 2022). The use of H_2 in the transport sector is very limited; a few pilot projects are started but are in the developmental stage. Overall, the consumption pattern of H_2 in India shows dominance by the industrial sector, with H_2 consumption in transportation in the early stages. Further H_2 production and consumption in residential areas is also expected by the development of electrolysers. The government is also planning to blend 15% of H_2 with natural gas by the year 2070.

12.8 CHALLENGES

H_2 as a fuel source has a large number of challenges that limit its viability. The environment is being harmed by rising populations and growing urbanisation, while most developing nations face significant issues from waste generation and energy crises. One of the most promising energy sources, H_2 is able to supply the world's energy needs, replace fossil fuels, and simultaneously minimise hazardous emissions (Chaurasia and Mondal, 2021). The high cost of manufacturing and the large storage needs are the major obstacles in the path of hydrogen energy commercialisation. H_2 is low energy density and difficult to transport and store for long periods of time.

The biggest environmental challenge of H_2 is its production from fossil-based fuels, including technologies such as SMR. A low H_2 yield and a high manufacturing cost are the two main obstacles to bio-H_2 synthesis (Pal et al., 2021). The sensitivity of this method to O_2, the key inhibitor of the hydrogenase enzyme responsible for producing H_2, is one of the largest hindrances to the production of H_2 via biophotolysis (Javed et al., 2022). Economic challenges are also vast, such as the low cost-competitiveness of the H_2 economy is the fundamental barrier to its growth. The green H_2 technology (Maroušek, 2022) is in the development phase so efficiency is quite low. Still, in order to shift from fossil fuels, the facilitation of green H_2 initiatives is required in India with a steady approach.

12.9 FUTURE PERSPECTIVES

H_2 is considered a future energy source as of the development of H_2 production technologies and renewable feedstock. Due to poor infrastructure and high costs, there are mainly problems with H_2 production (Amin et al., 2022). More than 95% of H_2 is produced by reforming fossil fuels, which often results in excessive CO_2 emissions (Finke at al., 2021). About 90% of the H_2 that is produced worldwide is of the grey type that is made by reforming fossil fuels. The grey H_2 is a less expensive H_2 production approach, but the CO_2 emissions are high (Qureshi et al., 2022). As far as sustainability is concerned, green H_2 is the future of energy with long-term decarbonising goals (Xu et al., 2022). Green H_2 does not involve fossil fuel burning and can be produced from different methods such as electrolysis, fermentation, photolysis, etc. It is expected that the cost of renewable H_2 will decrease in future along with curtailment in its production cost (Aziz et al., 2021; Rasul et al., 2022). All of the developed and developing countries are showing great interest in green energy projects. The feasibility of the H_2 energy network must be assessed, together with the network's tools, production processes, storage, fuel transport, dispensing, and consumption, before H_2 fuel cell vehicles may be widely adopted in the road transportation sector (Habib and Arefin, 2022).

In the future, the market for green H_2 is expected to rise as the energy markets shift towards low-carbon fuels to combat climate change. As green H_2 has low carbon emissions, it is an attractive alternative for nations working to reduce their carbon footprint. Green H_2 also has the potential to be used as a transportation fuel for commercial and private vehicles. Rapid advancement in technologies will also favour its high production, which can be used further as an energy source for producing electricity during peak hours.

12.10 CONCLUSION

H_2 production holds a prominent space at the present time. Different methods are available for its production. At present, commercial H_2 is being produced by fossil fuels and renewable methods are in the developmental phase. Present viable technologies are more oriented towards SMR for H_2 production The current dependency of H_2 production through fossil fuels must be curtailed in the upcoming time. The demand for green H_2 is increasing day by day and has solutions to numerous environmental problems such as gaseous emissions and climate change. The demand and acceptance of H_2 is rising, which can be supported by renewable approaches. Many renewable H_2 production technologies are in the pilot stage. Electrolysis for H_2O splitting is gaining momentum, but supply of renewable energy is a constraint. Many developed and developing countries including India have launched special policies that will support and endeavour current efforts in research and development (R&D). Another viable solution to green H_2 production can be an integration of solar energy using photovoltaic systems for water splitting and H_2 production. At present, stringent R&D is being done to increase efficiency of the green H_2 production.

REFERENCES

Agyekum, E. B., Nutakor, C., Agwa, A. M., & Kamel, S. (2022). A critical review of renewable hydrogen production methods: Factors affecting their scale-up and its role in future energy generation. *Membranes*, *12*(2), 173. https://doi.org/10.3390/membranes12020173

Ahmad, S., Bharti, A., Haq, M. I., & Kothari, R. (2023). Bioeconomy: Current status and challenges. In *Sustainable butanol biofuels* (pp. 57–75). CRC Press, Boca Raton. https://doi.org/10.1201/9781003165408

Amin, M., Shah, H. H., Fareed, A. G., Khan, W. U., Chung, E., Zia, A., & Lee, C. (2022). Hydrogen production through renewable and non-renewable energy processes and their impact on climate change. *International Journal of Hydrogen Energy*. https://doi.org/10.1016/j.ijhydene.2022.07.172

Anzelmo, B., Wilcox, J., & Liguori, S. (2018). Hydrogen production via natural gas steam reforming in a Pd-Au membrane reactor. Comparison between methane and natural gas steam reforming reactions. *Journal of Membrane Science, 568*, 113–120. https://doi.org/10.1016/j.memsci.2018.09.054

Aouad, S., Labaki, M., Ojala, S., Seelam, P., Turpeinen, E., Gennequin, C., & Abi Aad, E. (2018). A review on the dry reforming processes for hydrogen production: Catalytic materials and technologies. In *Catalytic materials for hydrogen production and electrooxidation reactions: 2 (frontiers in ceramic science)* (pp. 60–128). https://doi.org/10.2174/9781681087580118020007

Aziz, M., Darmawan, A., & Juangsa, F. B. (2021). Hydrogen production from biomasses and wastes: A technological review. *International Journal of Hydrogen Energy, 46*(68), 33756–33781. https://doi.org/10.1016/j.ijhydene.2021.07.189

Burton, N. A., Padilla, R. V., Rose, A., & Habibullah, H. (2021). Increasing the efficiency of hydrogen production from solar powered water electrolysis. *Renewable and Sustainable Energy Reviews, 135*, 110255. https://doi.org/10.1016/j.rser.2020.110255

Callegari, A., Bolognesi, S., Cecconet, D., & Capodaglio, A. G. (2020). Production technologies, current role, and future prospects of biofuels feedstocks: A state-of-the-art review. *Critical Reviews in Environmental Science and Technology, 50*(4), 384–436. https://doi.org/10.1080/10643389.2019.1629801

Chaurasia, A. K., & Mondal, P. (2021). Hydrogen production from waste and renewable resources. In *Hydrogen fuel cell technology for stationary applications* (pp. 22–46). IGI Global. https://doi.org/10.4018/978-1-7998-4945-2.ch002

Chompupun, T., Limtrakul, S., Vatanatham, T., Kanhari, C., & Ramachandran, P. A. (2018). Experiments, modeling and scaling-up of membrane reactors for hydrogen production via steam methane reforming. *Chemical Engineering and Processing-Process Intensification, 134*, 124–140. https://doi.org/10.1016/j.cep.2018.10.007

Finke, C. E., Leandri, H. F., Karumb, E. T., Zheng, D., Hoffmann, M. R., & Fromer, N. A. (2021). Economically advantageous pathways for reducing greenhouse gas emissions from industrial hydrogen under common, current economic conditions. *Energy & Environmental Science, 14*(3), 1517–1529. https://doi.org/10.1039/D0EE03768K

García, L., Abrego, J., Bimbela, F., & Sánchez, J. L. (2015). Hydrogen production from catalytic biomass pyrolysis. In *Production of hydrogen from renewable resources* (pp. 119–147). https://doi.org/10.1007/978-94-017-7330-0_5

Germscheidt, R. L., Moreira, D. E., Yoshimura, R. G., Gasbarro, N. P., Datti, E., dos Santos, P. L., & Bonacin, J. A. (2021). Hydrogen environmental benefits depend on the way of production: An overview of the main processes production and challenges by 2050. *Advanced Energy and Sustainability Research, 2*(10), 2100093. https://doi.org/10.1002/aesr.202100093

Goodman, E. D., Latimer, A. A., Yang, A. C., Wu, L., Tahsini, N., Abild-Pedersen, F., & Cargnello, M. (2018). Low-temperature methane partial oxidation to syngas with modular nanocrystal catalysts. *ACS Applied Nano Materials, 1*, 5258–5267. https://doi.org/10.1021/acsanm.8b01256

Goria, K., Kothari, R., Singh, H. M., Singh, A., & Tyagi, V. V. (2022). Biohydrogen: Potential applications, approaches, and hurdles to overcome. In *Handbook of biofuels* (pp. 399–418). Academic Press. https://doi.org/10.1016/B978-0-12-822810-4.00020-8

Goria, K., Singh, H. M., Singh, A., Kothari, R., & Tyagi, V. V. (2023). Insights into biohydrogen production from algal biomass: Challenges, recent advancements and future directions. *International Journal of Hydrogen Energy*. https://doi.org/10.1016/j.ijhydene.2023.03.174

Habib, M. S., & Arefin, P. (2022). Adoption of hydrogen fuel cell vehicles and its prospects for the future (a review). *Oriental Journal of Chemistry*, *38*, 621–631. http://dx.doi.org/10.13005/ojc/380311

Hall, W. I. L. L., Spencer, T. H. O. M. A. S., Renjith, G., & Dayal, S. H. R. U. T. I. (2020). The potential role of hydrogen in India. In *A pathway for scaling-up low carbon hydrogen across the economy*. New Delhi: The Energy and Resources Institute (TERI).

Hosseini, S. E., & Wahid, M. A. (2020). Hydrogen from solar energy, a clean energy carrier from a sustainable source of energy. *International Journal of Energy Research*, *44*(6), 4110–4131. https://doi.org/10.1002/er.4930

Javed, M. A., Zafar, A. M., Hassan, A. A., Zaidi, A. A., Farooq, M., El Badawy, A., & Al-Zuhair, S. (2022). The role of oxygen regulation and algal growth parameters in hydrogen production via biophotolysis. *Journal of Environmental Chemical Engineering*, *10*(1), 107003. https://doi.org/10.1016/j.jece.2021.107003

Jokar, S. M., Farokhnia, A., Tavakolian, M., Pejman, M., Parvasi, P., Javanmardi, J., . . . Basile, A. (2022). The recent areas of applicability of palladium based membrane technologies for hydrogen production from methane and natural gas: A review. *International Journal of Hydrogen Energy*. https://doi.org/10.1016/j.ijhydene.2022.05.296

Kayfeci, M., Keçebaş, A., & Bayat, M. (2019). Hydrogen production. In *Solar hydrogen production* (pp. 45–83). Academic Press. https://doi.org/10.1016/B978-0-12-814853-2.00003-5

Kertthong, T., Schmid, M., & Scheffknecht, G. (2022). Non-catalytic partial oxidation of methane in biomass-derived syngas with high steam and hydrogen content optimal for subsequent synthesis process. *Journal of the Energy Institute*, *105*, 251–261. https://doi.org/10.1016/j.joei.2022.09.007

Kim, J., Park, J., Qi, M., Lee, I., & Moon, I. (2021). Process integration of an autothermal reforming hydrogen production system with cryogenic air separation and carbon dioxide capture using liquefied natural gas cold energy. *Industrial & Engineering Chemistry Research*, *60*(19), 7257–7274. https://doi.org/10.1021/acs.iecr.0c06265

Kothari, R., Kumar, V., Pathak, V. V., Ahmad, S., Aoyi, O., & Tyagi, V. V. (2017). A critical review on factors influencing fermentative hydrogen production. *Frontiers in Bioscience*, *22*(8), 1195–1220. https://doi.org/10.2741/4542

Kothari, R., Singh, D. P., Tyagi, V. V., & Tyagi, S. K. (2012). Fermentative hydrogen production–An alternative clean energy source. *Renewable and Sustainable Energy Reviews*, *16*(4), 2337–2346. https://doi.org/10.1016/j.rser.2012.01.002

Maroušek, J. (2022). Nanoparticles can change (bio) hydrogen competitiveness. *Fuel*, *328*, 125318. https://doi.org/10.1016/j.fuel.2022.125318

Melitos, G., Voulkopoulos, X., & Zabaniotou, A. (2021). Waste to sustainable biohydrogen production via photo-fermentation and biophotolysis–A systematic review. *Renewable Energy and Environmental Sustainability*, *6*, 45. https://doi.org/10.1051/rees/2021047

Meramo-Hurtado, S. I., Puello, P., & Cabarcas, A. (2020). Process analysis of hydrogen production via biomass gasification under computer-aided safety and environmental assessments. *ACS Omega*, *5*(31), 19667–19681. https://doi.org/10.1021/acsomega.0c02344

Mikhaylov, A., Moiseev, N., Aleshin, K., & Burkhardt, T. (2020). Global climate change and greenhouse effect. *Entrepreneurship and Sustainability Issues*, *7*(4), 2897. http://doi.org/10.9770/jesi.2020.7.4(21)

Muritala, I. K., Guban, D., Roeb, M., & Sattler, C. (2020). High temperature production of hydrogen: Assessment of non-renewable resources technologies and emerging trends. *International Journal of Hydrogen Energy*, *45*(49), 26022–26035. https://doi.org/10.1016/j.ijhydene.2019.08.154

Nayak, S., Goveas, L. C., Selvaraj, R., Vinayagam, R., & Manickam, S. (2022). Advances in the utilisation of carbon-neutral technologies for a sustainable tomorrow: A critical review and the path forward. *Bioresource Technology*, 128073. https://doi.org/10.1016/j.biortech.2022.128073

Nikolaidis, P., & Poullikkas, A. (2017). A comparative overview of hydrogen production processes. *Renewable and Sustainable Energy Reviews*, *67*, 597–611.

NITI Aayog. (2022, June). *Harnessing green hydrogen: Opportunities for deep decarbonisation in India (report)*. (Accessed on 24 November 2022). www.niti.gov.in/sites/default/files/2022-06/Harnessing_Green_Hydrogen_V21_DIGITAL_29062022.pdf

Oni, A. O., Anaya, K., Giwa, T., Di Lullo, G., & Kumar, A. (2022). Comparative assessment of blue hydrogen from steam methane reforming, autothermal reforming, and natural gas decomposition technologies for natural gas-producing regions. *Energy Conversion and Management*, *254*, 115245. https://doi.org/10.1016/j.enconman.2022.115245

Osman, A. I., Mehta, N., Elgarahy, A. M., Hefny, M., Al-Hinai, A., Al-Muhtaseb, A. A. H., & Rooney, D. W. (2022). Hydrogen production, storage, utilisation and environmental impacts: A review. *Environmental Chemistry Letters*, 1–36. https://doi.org/10.1007/s10311-022-01432-x

Pal, D. B., Singh, A., & Bhatnagar, A. (2021). A review on biomass-based hydrogen production technologies. *International Journal of Hydrogen Energy*. https://doi.org/10.1016/j.ijhydene.2021.10.124

Pathak, A. K., Kothari, R., Tyagi, V. V., & Anand, S. (2020). Integrated approach for textile industry wastewater for efficient hydrogen production and treatment through solar PV electrolysis. *International Journal of Hydrogen Energy*, *45*(48), 25768–25782. https://doi.org/10.1016/j.ijhydene.2020.03.079

Perera, F., & Nadeau, K. (2022). Climate change, fossil-fuel pollution, and children's health. *New England Journal of Medicine*, *386*(24), 2303–2314. http://doi.org/10.1056/NEJMra2117706

Qureshi, F., Yusuf, M., Kamyab, H., Vo, D. V. N., Chelliapan, S., Joo, S. W., & Vasseghian, Y. (2022). Latest eco-friendly avenues on hydrogen production towards a circular bio-economy: Currents challenges, innovative insights, and future perspectives. *Renewable and Sustainable Energy Reviews*, *168*, 112916. https://doi.org/10.1016/j.rser.2022.112916

Rasul, M. G., Hazrat, M. A., Sattar, M. A., Jahirul, M. I., & Shearer, M. J. (2022). The future of hydrogen: Challenges on production, storage and applications. *Energy Conversion and Management*, *272*, 116326. https://doi.org/10.1016/j.enconman.2022.116326

Singh, H. M., Raina, S., Pathak, A. K., Goria, K., Kothari, R., Singh, A., . . . Tyagi, V. V. (2022). Bioenergy: Technologies and policy trends. In *Biomass, bioenergy & bioeconomy* (pp. 209–231). Singapore: Springer Nature Singapore. https://doi.org/10.1007/978-981-19-2912-0_11

Villafán-Vidales, H. I., Arancibia-Bulnes, C. A., Valades-Pelayo, P. J., Romero-Paredes, H., Cuentas-Gallegos, A. K., & Arreola-Ramos, C. E. (2019). Hydrogen from solar thermal energy. In *Solar hydrogen production* (pp. 319–363). Academic Press. https://doi.org/10.1016/B978-0-12-814853-2.00010-2

Wang, D., Muratori, M., Eichman, J., Wei, M., Saxena, S., & Zhang, C. (2018). Quantifying the flexibility of hydrogen production systems to support large-scale renewable energy integration. *Journal of Power Sources*, *399*, 383–391. https://doi.org/10.1016/j.jpowsour.2018.07.101

Wang, F. (2012). *Hydrogen production from steam reforming of ethanol over an Ir/ceria-based catalyst: catalyst ageing analysis and performance improvement upon ceria doping*. Doctoral dissertation, Université Claude Bernard-Lyon I.

Wang, Q., & Han, L. (2022). Hydrogen production. In *Handbook of climate change mitigation and adaptation* (pp. 1855–1900). Cham: Springer International Publishing. https://doi. org/10.1007/978-3-030-72579-2_29

WHA International, Inc. (2020, September 29). *Hydrogen applications in industry.* Author. (Accessed on 15 March, 2023). https://wha-international.com/hydrogen-in-industry

Xu, X., Zhou, Q., & Yu, D. (2022). The future of hydrogen energy: Bio-hydrogen production technology. *International Journal of Hydrogen Energy.* https://doi.org/10.1016/j. ijhydene.2022.07.261

Yadav, Y. K., Kumar, S., & Rao, R. (2015). *Production of biological hydrogen in India: A potential source for the future.* www.researchgate.net/publication/335724927

Zhang, T. Q., Malik, F. R., Jung, S., & Kim, Y. B. (2022). Hydrogen production and temperature control for DME autothermal reforming process. *Energy, 239*, 121980. https://doi. org/10.1016/j.energy.2021.121980

13 Solid Particle Thermal Energy Storage for High-Temperature CSP Applications

Karunesh Kant and R. Pitchumani

13.1 INTRODUCTION

The utilization of renewable energies, particularly solar energy, necessitates deploying a thermal or electrical storage system, depending upon the technology, to compensate for the mismatch that might develop between the supply of renewable energy and its usage. Solar energy may be stored in various methods, depending on the temperature range, the quantity of energy to be stored, the duration of storage (ranges from a few hours per day to many months), and the final usage of the energy. Concentrating solar power (CSP) efficiently converts solar energy into electricity with cost-effective energy storage for grid-scale, dispatchable renewable power generation. The prime advantage is the ability of the technology to store the sun's energy as heat, which can then be transferred to a power block for conversion to electricity. However, to compete with other power production technologies, CSP plants are required to be cost-effective. Using low-cost materials in the system and supporting a high-efficiency power conversion system are two strategies to decrease the cost of CSP (Ma *et al.*, 2015).

Conventional CSP systems, called Generation 1 or Gen1 CSP systems, are based on a field of large parabolic trough concentrators with a maximum operating temperature of around 400°C to reduce the thermal deterioration of oil used as the heat transfer fluid (HTF) medium (Stekli, Irwin and Pitchumani, 2013). A heliostat field that reflects solar beam radiation onto a central receiver is another type of CSP system, referred to as the Generation 2 or Gen2 CSP system (Nithyanandam and Pitchumani, 2014). This type of CSP plant employs molten nitrate salt eutectics as the HTF, which allows for a higher maximum operating temperature of up to 565°C (Rodríguez-Sánchez *et al.*, 2014). There is now a significant effort in advancing CSP technologies to higher operating temperatures in excess of 650°C toward improving plant efficiency and reducing the levelized cost of electricity. These Generation 3 (Gen3) CSP systems call for HTFs, energy storage materials, and technologies that allow for higher maximum working temperatures, up to around 1000°C (Ho, 2017; Ho *et al.*, 2017). The existing HTF has limitations, including a restricted operating

10.1201/9781003345558-16

temperature domain for nitrate salts (usually 240–565°C), extremely high pressure for steam, inadequate heat transfer capacity for air, corrosion of materials in contact with higher temperature molten carbonate and chloride salts (Kondaiah and Pitchumani, 2021, 2022a, 2022b), and high cost. To circumvent these disadvantages, the use of solid particles as HTF and thermal energy storage is being actively pursued (Flamant *et al.*, 2014). While there are several treatises on the liquid pathway for heat transfer and energy storage, a dedicated treatment of the emerging solid particle heat transfer and energy storage media is less available, which is sought to be filled by this chapter.

This chapter presents an overview of solid-particle-based thermal energy storage (TES) for CSP systems for economic, continuous, dispatchable, grid-scale electricity generation. Solid particles of bauxite, silicon sand, or silicon carbide can be utilized as HTF in CSP systems in direct or indirect heating receiver configurations. In the direct heating configuration, solid particles receive concentrated solar radiation directly, whereas a heat transfer wall absorbs solar radiation and transfers it to a heat transfer medium in the indirect heating configuration (Flamant *et al.*, 2014). Solid particle solar receivers used in conjunction with solar tower CSP systems provide exciting possibilities for high-temperature and high-efficiency power cycles, thermal energy storage (utilizing the same particles as storage media and HTF), and chemical applications of concentrated solar energy (water splitting for hydrogen production, cement processing, etc.).

Figure 13.1a depicts the integration of a solid particle CSP thermal system into a power tower solar field. An enclosed particle receiver (Figure 13.1b), a particle heat exchanger, low-cost, high-temperature solid particle containment comprising of a cold silo (integrated into the tower) and a hot silo for TES, and auxiliary equipment (a bucket lifter and particle distribution system) are some of the primary components. Inside a solid particle receiver, the particles are heated by focused solar radiation. The particles exit the receiver and fall into a hot silo, where they are used as a storage medium. High-temperature particles discharged from thermal storage flow via a heat exchanger, heating the power cycle's working fluid.

13.2 PARTICLE SELECTION FOR THERMAL ENERGY STORAGE

TES development aims to lower the cost of TES while enhancing its performance in terms of usable temperature and storage efficiency. Numerous TES technologies, including thermocline, phase-change, and chemical energy storage, have been proposed and are actively being developed. The particle-TES system may employ almost any stable particle with good fluidization capabilities. In practice, particle selection must consider particle stability, fluidization performance and flowability, cost, design knowledge of heat transfer, material handling, and equipment compatibility. Furthermore, the following considerations govern the selection of solid particles for the particle-TES system:

- Composition, softening temperature, density, heat capacity, particle size, and void fraction are the properties affecting the CSP thermal system's overall performance, stability, and energy density.

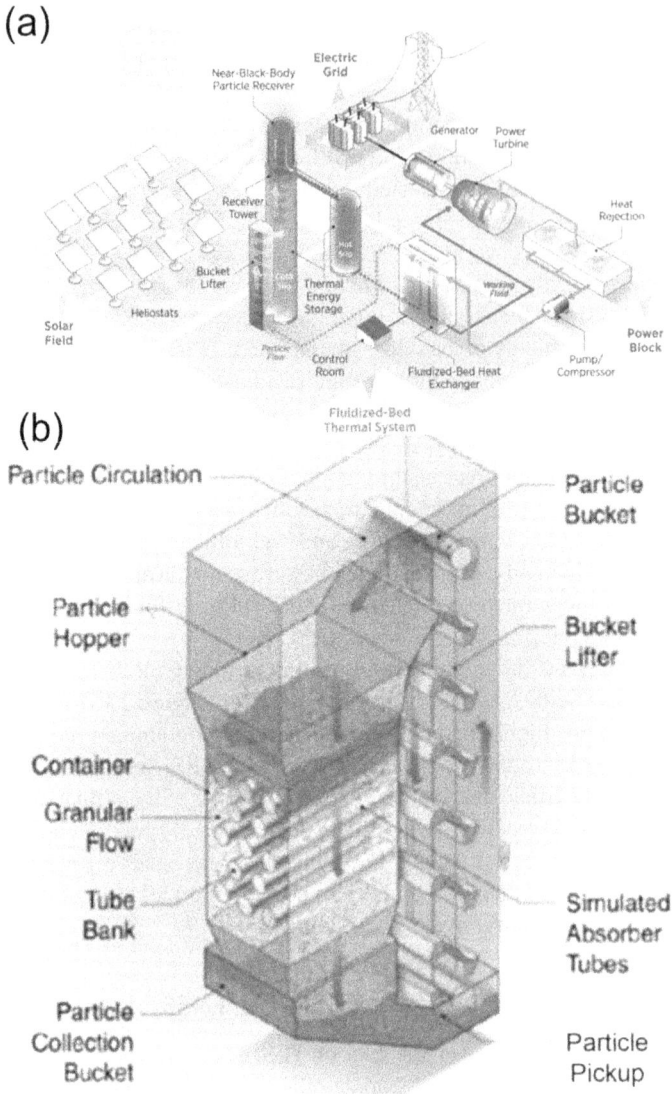

FIGURE 13.1 Schematic: (a) fluidized-bed, solid particle thermal energy storage CSP system; (b) a near-blackbody enclosed particle receiver.

Source: Reproduced with permission from Elsevier (Ma et al., 2015).

- Particle sizes need to be chosen appropriately. Although particle storage is not sensitive to the size of the particles, particle size is crucial for thermal transport, heat exchanger, and particle receiver performance. The heat transfer coefficient is a function of the particle size, with higher values being associated with the smaller particle size.

TABLE 13.1

Comparison of Solid-Particle Thermal Energy Storage Materials

Material	Composition	Specific Heat (J/kg°C)	Density (kg/m³)	Cost	Benefits	Concern
Silica sand	SiO_2	710	2610	Low	Stable	Low thermal conductivity
Graphite pebble	C	710	2250	High	High heat capacity	Oxidization, easy attrition
Silicon carbide	SiC	670	3210	High	High thermal conductivity	Hard for surfaces
Fly ash	$SiO_2+Al_2O_3$	720	2100	Low	Moderate	Containing corrosive minerals
Alumina powder	Al_2O_3	880	3960	High	Stable	High cost
Quartz sand	SiO_2	755	2650	Medium	Purity	Quartz inversion
CARBO proppants	75% Al_2O_3, 11% SiO_2, 9% Fe_2O_3, 3% TiO_2	1,150	3,300	High	Highly stable	—
Calcined flint clay	SiO_2, Al_2O_3, TiO_2, Fe_2O_3	1,050	2,600	—	—	Mined abundant

For details on the optical and thermal properties, refer to (Calderón *et al.*, 2019).

Sand and ash are suitable alternatives among regularly available particle materials in Table 13.1. Ash may outperform silicon sand due to its Al_2O_3 composition, which contributes more substantial heat conductivity and stability. Ash and sand are abundant, inexpensive, and stable.

13.3 THERMAL ENERGY STORAGE SYSTEM BASED ON SOLID PARTICLE

Particle storage CSP systems are being developed all over the world (Ma, Glatzmaier and Mehos, 2014; Wu *et al.*, 2015; Ho, 2016). Particle-based TES for CSP application provides numerous advantages over typical molten-salt-based TES systems, including eliminating issues of low-temperature salt freezing, high-temperature salt stability, and corrosion of metal parts. Solid particle receivers offer a way to directly irradiate a curtain of falling ceramic particles without using pipes. Particle-TES may handle a variety of power cycles, ranging from the traditional steam-Rankine (Ma, Zhang and Sawaged, 2017) to the more efficient Brayton combined power cycle (BCPC) adopted from gas turbine combined cycle (GTCC) or a developing supercritical carbon dioxide (sCO_2) Brayton cycle (Turchi *et al.*, 2013). When combined with highly efficient power cycles, TES becomes economically viable for electric grid storage.

The operation of a particle-TES system entails heating particles to high temperatures to store surplus energy that can subsequently be discharged to power a power cycle. The particle-TES system has a robust thermal performance and can store hot particles well over 900°C. The high particle temperatures enhance the TES capacity because they widen the temperature gap between hot and cold particles. The particle-based CSP system further supports the supercritical carbon dioxide Brayton power cycle, which achieves >50% thermal-electric conversion efficiency. Figure 13.2 shows how these key characteristics can be included in a particle-TES design framework that considers the planning and technology screening processes. The framework explains how storage needs from a CSP plant, nuclear power plant, or grid energy storage application constrain TES operating parameters and particle selection to serve a specific power system, such as a sCO_2 Brayton cycle, steam-Rankine cycle, or BCPC derived from a GTCC system. The choice of thermal insulation, containment silo structural design, and particle-handling technology is influenced by operating conditions and particle selection.

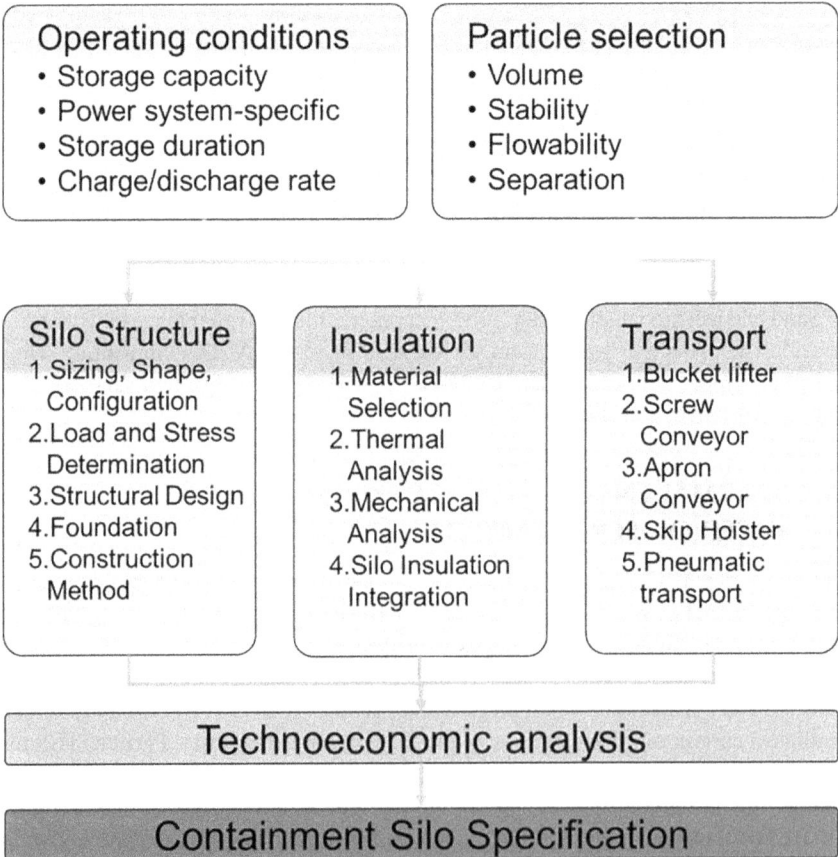

Operating conditions
- Storage capacity
- Power system-specific
- Storage duration
- Charge/discharge rate

Particle selection
- Volume
- Stability
- Flowability
- Separation

Silo Structure
1. Sizing, Shape, Configuration
2. Load and Stress Determination
3. Structural Design
4. Foundation
5. Construction Method

Insulation
1. Material Selection
2. Thermal Analysis
3. Mechanical Analysis
4. Silo Insulation Integration

Transport
1. Bucket lifter
2. Screw Conveyor
3. Apron Conveyor
4. Skip Hoister
5. Pneumatic transport

Technoeconomic analysis

Containment Silo Specification

FIGURE 13.2 Design framework of the particle-TES system for CSP/grid energy storage requirements (Ma, Davenport and Zhang, 2020).

The TES system capacity can be calculated by multiplying the power-cycle design point thermal load by the desired storage period (Δt_{TES}), then dividing the power-cycle efficiency (η_p) and TES exergetic efficiency ($\eta_{TES,II}$), as shown in Eq. (1):

$$\Delta E = \frac{\dot{W}_p \Delta t_{TES}}{\eta_p \cdot \eta_{TES,II}} \tag{1}$$

The power-system operating parameters and particle type determine the possible stored energy density. Because the TES system includes sensible storage, the net stored energy can be represented as the total of sensible heating and cooling, as shown in Eq. (2):

$$\Delta E = \sum_i \left(m_i \int_{T_c}^{T_h} C_{p,i} dT \right) \tag{2}$$

where m and C_p denote the mass of the storage medium and specific heat, respectively, and T_c and T_h are the temperature range in which the TES operates (hot and cold). The lowest desired mass of the TES medium is calculated by taking into account the operating conditions and incorporating the system TES capacity from Equation (1) into Equation (2). It should be noted that the storage medium used must be stable and suitable with the containment materials above the charging temperature.

Ma et al. (Ma, Zhang and Sawaged, 2017) developed a particle-based TES for a CSP system to indicate that particle-based TES can realize high thermal, storage, and energetic efficiency. Particle-TES design emphasizes foundation, containment silos, and material choices. The hot solid particles discharged to the heat exchanger drive the power cycles. Gas fluidizes solid particles within the fluidized bed when they come into contact with the boiler's heat exchanger. The gas/solid flow enters a unit that separates the solid particles from the carrying gas after the working fluids have been heated. The cooled particles are collected in a cold particle silo. When solar radiation is available, the cold solid particles are raised to a particle receiver, which absorbs solar radiation and heats the particles, which subsequently drop into the hot silo and are transported to the heat exchanger.

Reyes-Belmonte (Reyes-Belmonte et al., 2018) presented a particle-based TES system for CSP and applied it to different CSP plant layout scenarios. The system's main component is the fluidized-bed heat exchanger (DPS-HX), which connects the particle-based storage system to the solar loop and the power block. It is conceivable to include no temperature constraints, no freezing or temperature deterioration, simplicity of handling, and lack of toxicity as advantages of particle-based thermal energy storage. The efficiency of the particle heat exchanger was determined to be high (between 91–99% in most situations), and the amount of power needed for fluidization is little compared to the thermal power transmitted to the work transfer fluid. Instead of transferring complete particles and work transfer fluid via a single heat exchanger component, it is preferable for large power plants to distribute the particles across several heat exchangers connected in parallel.

Diago et al. (2018) characterized desert sand to be used as a high-temperature TES medium in particle solar receiver technology. The thermophysical and mechanical parameters of many samples taken from different desert regions in the United Arab Emirates (UAE) are analyzed. Additionally, the optical characteristics of desert sand are examined to assess how well it acts as a direct solar absorber. Following an initial mass loss during the first heating cycle, thermogravimetric measurements demonstrate that the samples appear to be thermally stable between about 650–1000°C. The conversion of calcium carbonate to calcium oxide at higher temperatures during the initial heating phase reduces the sand's solar absorption. Furthermore, the high calcium concentration causes sand agglomeration, which considerably impacts receiver design and operation. As a result, finding sand collection locations with low carbonate concentrations is crucial.

Trevisan et al. (Trevisan, Wang and Laumert, 2022) conducted a high-temperature thermal stability and optical property study of inorganic coatings on ceramic particles for potential TES applications. The thermal stability of the six inorganic coatings presented in Table 13.2 on ceramic particles was examined at 1000°C. Figure 13.3

TABLE 13.2

Coatings for Thermal Stability and Thermal Cyclic Test (Trevisan, Wang and Laumert, 2022)

Sample Name	Main Material (Adhesive)	Reference
Base	Denstone® 2000	(NorPro, 2000)
Coating 1	HIE-Coat 840MX	(Aremco Products Incorporated, 2018)
Coating 2	Pyromark 2500	(Ho et al., 2014)
Coating 3	Pyro-Paint 634-ZO	(Aremco, n.d.)
Coating 4	MgO (Aremco 571)	(Aremco, 2015)
Coating 5	TiO_2 (Aremco 643-2)	(Aremco, 2000)
Coating 6	SS 304 (Durabond 954)	(Strength, n.d.)

FIGURE 13.3 Photographs of the cured coated samples (MP 3). The uncoated Denstone 2000® is seen in the base sample. The average hydraulic diameter of a particle is 6 mm.

shows photographs of the samples after curing. The findings demonstrate that HIE-coated 840MX and Pyropaint 634 ZO have outstanding thermal stability at high temperatures (1000°C), as well as thermal cycle temperatures ranging from 400–800°C. Titanium oxide (TiO_2) coatings may be a feasible alternative if the powder has been pre-treated to avoid polymorph transition throughout the operation. Because the adhesive suppresses oxidation and preserves the coating from corrosion, stainless-steel 304 powder-based coating may also be a suitable alternative. In contrast, Pyromark 2500 and MgO-based coatings exhibited various degradation issues, limiting their use in high-temperature applications subjected to thermal cycles. The thermal emissivity of the coatings studied ranged from 0.6–0.9, with steady or decreasing trends with temperature. This allowed for a 20% modification in the packing structure's effective thermal conductivity.

Chung et al. (Rovense *et al.*, 2019) measured the thermal conductivity of CARBOBEAD HSP 40/70 and CARBOBEAD CP 40/100 (with average particle sizes of ~405 μm and ~275 μm, respectively, and thermal conductivities ranging from ~0.25 W m^{-1} K^{-1} to ~0.50 W m^{-1} K^{-1}) in N_2 gas and air environment under different gaseous pressures. A custom-made high-temperature transient hot wire (THW) setup was developed to operate within a temperature range of room temperature to 700°C, the target operating temperature of the next-generation CSP systems. The Zehner, Bauer, and Schlünder (ZBS) model calculations agreed well with the measurements. According to the model, the effective thermal conductivity of packed particle beds is dominated by gas conduction, with solid conduction and radiation accounting for roughly 20% of the effective thermal conductivity at high temperatures.

In order to simulate prolonged operation at a high temperature in the air between 700–1000°C, Siegel *et al.* (2014) measured the radiative properties, solar-weighted absorptance, and thermal emittance for several commercially available particle candidates both in the as-received state and after thermal exposure. Within 24 hours of exposure to air at 1000°C, heating the particles was demonstrated to dramatically diminish the solar-weighted absorptance of as-received particles, although heating at 700°C in the air had virtually little impact. Solar-weighted absorptance in the as-received condition can reach up to 93% before dropping to 84% after 192 hours at 1000°C. At 700 °C, particle stability is greater, and after 192 hours of exposure, the solar absorptance was still over 92%. In the sintered bauxite particle materials, which comprise oxides of aluminum, silicon, titanium, and iron, analysis utilizing X-ray diffraction (XRD) reveals evidence of several chemical modifications after heating in the air. Table 13.3 displays the physical properties of the three proppant formulations examined. The data in Table 13.3 do not represent a comprehensive set of properties for a thermal storage media. Hruby et al. (Hruby, 1986) provide an in-depth examination of these features.

Wünsch *et al.* (2020) conducted numerical investigations and bulk heat conductivity estimation of fluidized-bed particle-TES (FP-TES). This paper focuses on geometrical and fluidic design through numerical studies using computational particle fluid dynamics (CPFD). During the process, a controlled transient simulation approach known as co-simulation of FP-TES is established, which serves as the foundation for test bench design and future co-simulation. This procedure involves developing an enhanced design of rotating symmetric hoppers with extra baffles in the heat exchanger (HEX) and internal pipes to stabilize the particle mass flow.

TABLE 13.3

Summary of the Physical Properties of Commercially Available Proppants Investigated

Composition/Physical Properties	CARBOHSP®	CARBOACCUCAST®	CARBOPROP®
Al_2O_3	83	75	72
SiO_2	5	11	13
TiO_2	3.5	3	4
Fe_2O_3	7	9	10
Other components	1.5	2	1
Median diameter [μm]	697	300	443
Heat capacity [$J·kg^{-3}·K^{-1}$]	1275 (700°C)	1175 (700°C)	1175 (700°C)
Bulk density [$g·cm^{-3}$]	2	2	1.88
Solar-weighted absorptance as received	93%	91%	89%

Furthermore, a contribution of bulk heat conductivity was shown to indicate reduced thermal losses and minimum thermal insulation requirements by accounting for the thermal insulation of the hopper's outer layer. Thermal insulation analysis suggested that the outer layer of the hopper should be considered. When quartz sand is used as a storage medium, the thermal conductivity of the bed is double that of a typical insulating material. As a result, quartz sand is an excellent choice for long-term storage, but alternative materials with higher heat capacity and conductivity are preferable for short-term storage. A second part (Sulzgruber, Wünsch, Walter, *et al.*, 2020) of the foregoing study confirmed the numerical results using a cold test rig. Future computations should be undertaken to fine-tune the FP-TES design and analyze the temperature profiles in the hopper. Furthermore, one of the most challenging issues is constructing a reliable mass flow measurement to provide the groundwork for creating a control algorithm and demonstrating the FP-TES operation in a real-world industrial setting.

Sment *et al.* (2022) performed a cost analysis of commercial-scale particle-based TES systems. The heat exchanger temperature difference was shown to have the greatest influence on the overall TES cost, followed by particle cost. Narrow towers (<20 m) can minimize tower materials; however, they may not be a choice in seismically active locations or for towers with significant capacity. It is expensive and technically challenging to install enough refractory insulation to keep concrete surface temperatures at 100°C. With compressive strengths of more than >65 MPa (9000 psi), calcium aluminate-based, heat-resistant concrete may withstand cycle temperatures of up to 1100°C. Thermal models without insulation demonstrate that particles lost 3.4% of their heat after 14 hours of storage, whereas insulation lost <1%.

Rovense *et al.* (2019) performed thermodynamics modeling of a closed Brayton cycle coupled with particle-based TES and power block concept for flexible electricity dispatch in a CSP plant. To manage the turbine inlet temperature (TIT), the closed

Brayton cycle employs a mass flow regulation system based on pressure regulation (auxiliary compressor and bleed valve). As a result, the system can modulate turbine electricity production in response to fluctuations in solar resources and changes in power energy demand. It was found that the suggested power block can fully cover the electricity demand curve on days with abundant solar resources. When a particle-based high-temperature TES system is integrated, the power block may extend its output until the next day by following the electricity curve demand throughout the summertime. Because of the lower solar resource and more significant electric curve demand load during the winter, the power plant can extend its output for a few hours.

Sment *et al.* (2020) report ongoing development at the National Solar Thermal Test Facility (NSTTF) of Sandia National Laboratories on a Generation 3 Particle Pilot Plant (G3P3) that utilizes falling sand-like particles as the heat transfer medium. The system will comprise a 6 MWht TES bin that will need 120,000 kg of circulating particles. Testing and modeling were carried out to provide a validated modeling tool for understanding the temporal and spatial temperature variations within the storage bin as it charges and discharges. Particle discharge temperatures obtained from thermocouples installed throughout a small steel container were used to validate model findings. The model was then utilized to forecast heat loss during the G3P3 scale's operational modes of charging, storing, and discharging. The model captures many of the key features of the transient particle outlet temperature over time, according to comparative results from modeling and testing of the small bin.

Sulzgruber et al. (Sulzgruber, Wünsch, Haider, *et al.*, 2020) performed a numerical investigation on the flow behavior of a novel fluidization-based particle thermal energy storage (FP-TES). The most significant benefit of the FP-TES over conventional particle-based storage systems is the substitution of mechanical transport mechanisms with superior fluidization technology. Numerical simulations were used to prove and develop this particle transport concept. As a result, an optimum shape for a cold test rig using 800 kg of quartz sand is constructed, and its behavior, particle mass flow, and pressure drops are predicted. In addition, the outcomes of experimental studies conducted using the test rig are compared to numerical calculations. The modeling and the experiment reveal that the proposed advanced fluidization technique can achieve regulated stable particle mass flow. Finally, the energetic efficiency of an exemplary application was assessed, and a basic layout of the application was built.

Trevisan et al. (Corona *et al.*, 2022) used optical techniques to characterize particle-based thermal storage performance for usage in next-generation CSP reactors. An optical-based thermal analysis approach was used to determine particle beds' heat exchanger efficiency by measuring their near-wall thermal conductivity. Photothermal modulation radiometry was used to monitor the dynamic temperature, which allows the extraction of the most important thermal parameters, such as thermal conductivity, specific heat, and effusivity. The system employs a modulated laser source to generate a damped periodic heat flow, resulting in a frequency and thermal property-dependent surface temperature detected via radiometry. Lock-in procedures (Breitenstein, Warta and Langenkamp, 2010) are used to extrapolate the signal's amplitude. Plotting the amplitude against the root angular frequency may be used to measure the effusivity as a ratio to a known sample. The thermal conductivity of HSP16/30 was measured by modulated photothermal radiometry (MPTR), and thermal plan source (TPS) techniques and results show good agreement. The error between the techniques is around 0.1%. The simplicity of MPTR to probe through the depth of the bed is ideal for use in CSP for dynamic thermal performance monitoring. However, TPS are often applied to other bulk materials but can be difficult to apply to particle beds due to size differences between probes and particles, resulting in a large source of measurement error.

13.4 SUMMARY AND CONCLUSIONS

The utilization of solid particle materials has enormous potential for developing a new generation of solar power towers. The storage medium extends the operating temperature range and can help CSP thermal power plants lower electricity costs while also increasing flexibility to the electric grid, thereby enabling greater renewable energy adoption. Increasing the current CSP service temperature provides performance benefits when turning heat into electricity in the power block by employing more efficient thermodynamic cycles or mixed cycles using exhausted heat from the topping cycle as input for one or more bottom cycles. The particle route to high-temperature CSP is likely to develop into commercialization in the following years. So far, the advancement in solid particle technology has proved to be viable in specific components such as solar particle receivers and heat exchangers.

Nonetheless, further study on conveyance and particle thermal energy storage systems is needed. Materials issues must also be considered because various components of the power plant are subject to different interactions with the particle media. Issues such as erosion, wear, and improving heat transfer on the particle side are among the considerations for the future development of this technology. Finally, once the best candidate materials have been thoroughly investigated, material combinations or material augmentation might be explored to mitigate the potential disadvantages of the material choice.

REFERENCES

Aremco (2000) 'High temperature inorganic binders technical bulletin A11', *Aremco Products*, 3200(1760).
Aremco (2015) 'High temperature ceramic & graphite adhesives', *Technical Bulletin*, A2(845), pp. 1–6.
Aremco (2023) 'High temperature refractory coatings technical bulletin A5-S5 Aremco's refractory coatings offer the ultimate protection of high temperature components and structures used in the processing of metals, glasses and plastics.' *Aremco Products*. Available at: https://www.aremco.com/high-temp-refractory-coatings/ Accessed 10 Oct-2023.
Aremco Products Incorporated (2018) 'Technical bulletin A5-S2: High temperature high emissivity coatings', *Aremco Products*, (845), pp. 1–2. Available at: www.aremco.com/wp-content/uploads/2018/05/A05_S2_18_Emissivity.pdf.
Breitenstein, O., Warta, W. and Langenkamp, M. (2010) *Lock-in Thermography* (Springer Series in Advanced Microelectronics). Springer, Berlin, Heidelberg. Available at: https://doi.org/10.1007/978-3-642-02417-7.

Calderón, A. *et al.* (2019) 'Review of solid particle materials for heat transfer fluid and thermal energy storage in solar thermal power plants', *Energy Storage*, 1(4), pp. 1–20. Available at: https://doi.org/10.1002/est2.63.

Corona, J. *et al.* (2022) 'Characterizing particle-based thermal storage performance using optical methods for use in next generation concentrating solar power plants', in L. Jiang, R. Winston, and R.J. Koshel (eds) *Nonimaging Optics: Efficient Design for Illumination and Solar Concentration XVIII.* SPIE, p. 21. Available at: https://doi.org/10.1117/12.2632268.

Diago, M. *et al.* (2018) 'Characterization of desert sand to be used as a high-temperature thermal energy storage medium in particle solar receiver technology', *Applied Energy*, 216, pp. 402–413. Available at: https://doi.org/10.1016/J.APENERGY.2018.02.106.

Flamant, G. *et al.* (2014) 'A new heat transfer fluid for concentrating solar systems: Particle flow in tubes', *Energy Procedia*, 49, pp. 617–626. Available at: https://doi.org/10.1016/j.egypro.2014.03.067.

Ho, C.K. (2016) 'A review of high-temperature particle receivers for concentrating solar power', *Applied Thermal Engineering*, 109, pp. 958–969. Available at: https://doi.org/10.1016/j.applthermaleng.2016.04.103.

Ho, C.K. (2017) 'Advances in central receivers for concentrating solar applications', *Solar Energy*, 152, pp. 38–56. Available at: https://doi.org/10.1016/j.solener.2017.03.048.

Ho, C.K. *et al.* (2014) 'Characterization of pyromark 2500 paint for high-temperature solar receivers', *Journal of Solar Energy Engineering*, 136(1). Available at: https://doi.org/10.1115/1.4024031.

Ho, C.K. *et al.* (2017) 'Highlights of the high-temperature falling particle receiver project: 2012–2016', *AIP Conference Proceedings*, 1850, pp. 2012–2016. Available at: https://doi.org/10.1063/1.4984370.

Hruby, J.M. (1986) *Technical Feasibility Study of a Solid Particle Solar Central Receiver for High Temperature Applications.* Sandia National Labs., Livermore, CA.

Kondaiah, P. and Pitchumani, R. (2021) 'Fractal textured surfaces for high temperature corrosion mitigation in molten salts', *Solar Energy Materials and Solar Cells*, 230(March), p. 111281. Available at: https://doi.org/10.1016/j.solmat.2021.111281.

Kondaiah, P. and Pitchumani, R. (2022a) 'Fractal coatings of Ni and NiYSZ for high-temperature corrosion mitigation in solar salt', *Corrosion Science*, 201, p. 110283. Available at: https://doi.org/10.1016/j.corsci.2022.110283.

Kondaiah, P. and Pitchumani, R. (2022b) 'Novel textured surfaces for superior corrosion mitigation in molten carbonate salts for concentrating solar power', *Renewable and Sustainable Energy Reviews*, 170, p. 112961. Available at: https://doi.org/10.1016/j.rser.2022.112961.

Ma, Z. *et al.* (2015) 'Development of a concentrating solar power system using fluidized-bed technology for thermal energy conversion and solid particles for thermal energy storage', *Energy Procedia*, 69, pp. 1349–1359. Available at: https://doi.org/10.1016/j.egypro.2015.03.136.

Ma, Z., Davenport, P. and Zhang, R. (2020) 'Design analysis of a particle-based thermal energy storage system for concentrating solar power or grid energy storage', *Journal of Energy Storage*, 29(1), p. 101382. Available at: https://doi.org/10.1016/j.est.2020.101382.

Ma, Z., Glatzmaier, G. and Mehos, M. (2014) 'Fluidized bed technology for concentrating solar power with thermal energy storage', *Journal of Solar Energy Engineering*, 136(3). Available at: https://doi.org/10.1115/1.4027262.

Ma, Z., Zhang, R. and Sawaged, F. (2017) 'Design of particle-based thermal energy storage for a concentrating solar power system', in *ASME 2017 11th International Conference on Energy Sustainability.* American Society of Mechanical Engineers, pp. 1–8. Available at: https://doi.org/10.1115/ES2017-3099.

Nithyanandam, K. and Pitchumani, R. (2014) 'Cost and performance analysis of concentrating solar power systems with integrated latent thermal energy storage', *Energy*, 64, pp. 793–810. Available at: https://doi.org/10.1016/j.energy.2013.10.095.

NorPro, S.-G. (2000) *Denstone® 2000 Support Balls Supreme Reliability and Survival from Denstone®*. Saint-Gobain. Available at: file:///C:/Users/k1091/Downloads/sgnorpro-denstone-2000-support-media.pdf.

Reyes-Belmonte, M.A. *et al.* (2018) 'Particles-based thermal energy storage systems for concentrated solar power', *AIP Conference Proceedings*, 2033(November). Available at: https://doi.org/10.1063/1.5067215.

Rodríguez-Sánchez, M.R. *et al.* (2014) 'Thermal design guidelines of solar power towers', *Applied Thermal Engineering*, 63(1), pp. 428–438. Available at: https://doi.org/10.1016/j.applthermaleng.2013.11.014.

Rovense, F. *et al.* (2019) 'Flexible electricity dispatch for CSP plant using un-fired closed air Brayton cycle with particles based thermal energy storage system', *Energy*, 173, pp. 971–984. Available at: https://doi.org/10.1016/J.ENERGY.2019.02.135.

Siegel, N. *et al.* (2014) 'Physical properties of solid particle thermal energy storage media for concentrating solar power applications', *Energy Procedia*, 49, pp. 1015–1023. Available at: https://doi.org/10.1016/j.egypro.2014.03.109.

Sment, J.N. *et al.* (2020) 'Testing and simulations of spatial and temporal temperature variations in a particle-based thermal energy storage bin', in *ASME 2020 14th International Conference on Energy Sustainability*. American Society of Mechanical Engineers. Available at: https://doi.org/10.1115/ES2020-1660.

Sment, J.N. *et al.* (2022) 'Design considerations for commercial scale particle-based thermal energy storage systems', *AIP Conference Proceedings*, 2445(May). Available at: https://doi.org/10.1063/5.0086995.

Stekli, J., Irwin, L. and Pitchumani, R. (2013) 'Technical challenges and opportunities for concentrating solar power with thermal energy storage', *Journal of Thermal Science and Engineering Applications*, 5(2). Available at: https://doi.org/10.1115/1.4024143.

Strength, B. (2023) 'High bond strength, thermal shock and impact resistance 2000°F, metallic adhesives'. Available at: https://www.masterbond.com/properties/thermal-shock-resistant-adhesives.

Sulzgruber, V., Wünsch, D., Haider, M. *et al.* (2020) 'Numerical investigation on the flow behavior of a novel fluidization based particle thermal energy storage (FP-TES)', *Energy*, 200, p. 117528. Available at: https://doi.org/10.1016/J.ENERGY.2020.117528.

Sulzgruber, V., Wünsch, D., Walter, H. *et al.* (2020) 'FP-TES: Fluidization based particle thermal energy storage, part II: Experimental investigations', *Energies*, 13(17), p. 4302. Available at: https://doi.org/10.3390/en13174302.

Trevisan, S., Wang, W. and Laumert, B. (2022) 'A high-temperature thermal stability and optical property study of inorganic coatings on ceramic particles for potential thermal energy storage applications', *Solar Energy Materials and Solar Cells*, 239, p. 111679. Available at: https://doi.org/10.1016/j.solmat.2022.111679.

Turchi, C.S. *et al.* (2013) 'Thermodynamic study of advanced supercritical carbon dioxide power cycles for concentrating solar power systems', *Journal of Solar Energy Engineering*, 135(4). Available at: https://doi.org/10.1115/1.4024030.

Wu, W. *et al.* (2015) 'Prototype testing of a centrifugal particle receiver for high-temperature concentrating solar applications', *Journal of Solar Energy Engineering, Transactions of the ASME*, 137(4), pp. 1–7. Available at: https://doi.org/10.1115/1.4030657.

Wünsch, D. *et al.* (2020) 'FP-TES: A fluidisation-based particle thermal energy storage, part I: Numerical investigations and bulk heat conductivity', *Energies*, 13(17), p. 4298. Available at: https://doi.org/10.3390/en13174298.

14 Thermal Management of Batteries by Thermal Energy Storage Materials

Fathimathul Faseena A M and A. Sreekumar

14.1 INTRODUCTION

The expeditious rise in energy demand from the ever-growing overall inhabitants has encouraged the demand for extra significant and more efficient energy storage technologies. Currently, the entire world depends significantly on fossil fuels, which contribute nearly 80% of world energy needs. New batteries have emerged in recent years that provide potential technology for storing energy. Factors such as uneven temperature dispersion among battery cells, temperature rise or fall, and overheating can negatively affect battery safety, performance, and cycle lifetime.

The temperature fluctuations in batteries are normally unavoidable, as they release heat through chemical reactions during charging and discharging with the environmental conditions. The battery function is highly correlated to its operating temperature because it is a temperature-sensitive device. During the charging and discharging cycles, untimed heat dissipation of batteries will lead to overheating of the battery, adversely affecting its smooth functioning. As a result, the lack of a suitable thermal management system raises the battery temperature, having a negative impact on battery performance. To prevent temperature-related adverse effects, the development of an efficient and safe battery thermal management system (BTMS) is needed to minimize the temperature gradient and maintain a proper range. Furthermore, proper thermal management is vital for public adoption of new technologies, e.g., electric vehicles (EVs), hybrid electric vehicles (HEVs), and battery electric vehicles (BEVs). This chapter discusses various thermal management strategies and how a thermal energy storage system helps in addressing the problem of thermal management in batteries.

14.2 TYPES OF BATTERIES

An electrochemical device that can produce a specific amount of energy wherever it is required is termed a battery. In order to achieve the voltage and power requirements, a battery pack comprises numerous individual cells. The cell comprises four elements: two electrodes—cathode and anode, an electrolyte for ion transportation, and a separator to provide a barrier between the anode and the cathode during the exchange of ions (Goodenough & Park, 2013; Alaswad et al., 2016). Batteries are

DOI: 10.1201/9781003345558-17

TABLE 14.1

Some Features of Rechargeable Batteries

Battery Type	Specific Power (kW/kg)	Specific Energy (Wh/kg)	Self-Discharge Rate (loss per month, %)	Life Cycle (up to 80% capacity)	Charge Time (hour)
Lead-acid	150–250	25–40	5	200–700	8
Nickel-cadmium	200	45–80	20	500–2000	1
Nickel-metal hydride	200	60–120	30	500–1000	1
Lithium-ion	300	110–180	10	> 1000	2–3
Lithium-ion polymer	300	100–130	10	300–500	2–4

Source: Adapted from (Larminie & Lowry, n.d.).

commonly utilized in various appliances such as mobile phones, cameras, and vehicles (Jouhara et al., 2019).

Batteries are generally categorized as primary or secondary, based on whether they are rechargeable or not. The electron flow in the battery is caused by a reversible redox process, which takes place between various electrodes and electrolytes. Some of the specifications of rechargeable batteries that are commonly used for commercial applications are depicted in Table 14.1.

14.3 BATTERY THERMAL PERFORMANCES

14.3.1 Effect of Temperature on Batteries

Temperature is the most significant state variable as it directly impacts the redox reactions that take place in the course of discharging and charging in batteries. Temperature also affects the reaction rate, local resistance, and heat generation. Based on the material utilized for manufacturing the cells and type of battery, variable temperature ranges and hence a distinct thermal management system would have adopted. Even though, for most common batteries, the optimum temperature should be between 15–35°C (Khan et al., 2017).

Temperature of the cell is influenced by the internal structure and load profile of the battery, along with the ambient conditions, due to the endothermic and exothermic reactions that come along with the charging and discharging operation. Thus, temperature within the cell enormously varies with time, affecting nearly all of its characteristics transiently (Sato, 2000). Since the exponential Arrhenius equation (1) can be utilized to determine the electron transfer reaction rate, even minor temperature changes greatly impact the amount of the reaction.

$$k = k_0 exp\left(-\frac{E_A}{R \cdot T}\right) \quad (1)$$

Since the molar gas constant R (8.314 J mol^{-1} K^{-1}) and the reaction constant K_0 (unitless) are always constant, and the activation energy E_A (J mol^{-1}) has only a minimal temperature dependence, it is clear that the temperature T (K) is the most essential factor ("Encyclopedia of Sustainability Science and Technology," 2012).

14.3.2 THERMAL ISSUES IN BATTERIES

Heat production within the battery is primarily caused by chemical reactions and charge transmitted during the normal charging/discharging process. Furthermore, heat can also emerge from uncontrolled side reactions at increased temperature. The positive electrode produces the majority of heat in the battery, which is higher than that generated from the remaining parts (Selman et al., n.d.). Thus, the overall heat output consists of reversible and irreversible heat, as shown in Figure 14.1.

Generally, at cold conditions, battery performance is usually diminished. It can impact battery lifespan, charge acceptance, round-trip efficiency, power capacity, etc. (Burow et al., 2016). At extreme temperatures, battery performance is negatively impacted in a number of ways, including power/capacity loss, as well as it can induce autonomy loss at self-discharge. These effects can cause a significant loss in available energy (Wright et al., 2003). Increased temperature can also reduce the discharge rate and cycle life, as well as reduce overall power density and capacity. Whenever an internal cell breaks, the neighboring cells rapidly receive an influx of energy from the failing cell.

However, apart from far and near temperature performance, a uniform temperature diffusion is crucial in ensuring that batteries work efficiently. Temperature maldistribution could vary electrochemical performances and charging or discharging behaviors. Since the morphological features and thermophysical properties of materials within the cell varies, heat production in various elements and its transmission in overall directions differ widely. Consequently, a temperature gradient is developed

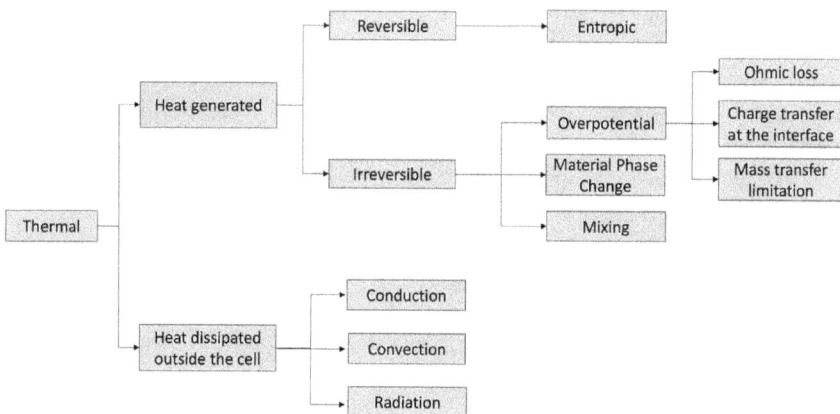

FIGURE 14.1 Battery thermal issues.

Source: Reproduced from (Abada et al., 2016).

278

within the cell (Wu et al., 2018). Compared to other regions of the cell, the temperature is higher adjacent to the electrode. Similar unequal distribution of temperature will cause uneven reaction rates of the electrode, which would lower the cell efficiency and lifetime.

Exothermic reactions within the cells can emit excess heat that can be dispersed to the surroundings at a specific starting temperature threshold, resulting in thermal runaway, an irreversible self-accelerating heating. An improper charging or discharging and short circuit are more often responsible for thermal runaway (Lamb et al., 2015). When the temperature rises, the rate of reaction gradually gets increased. Hence, a tremendous temperature rise would possibly lead to an explosion (Spotnitz et al., 2007).

Thermal runaway is initiated and accelerated by the following processes (Q. Wang et al., 2012):

- Interior shorts induced by separator melting
- Electrolyte oxidation by the cathode
- Electrolyte reduction by the anode
- Electrolyte thermal degradation
- Thermal decomposition of the anode or cathode

14.4 BATTERY THERMAL MANAGEMENT SYSTEM (BTMS)

Thermal management entails monitoring and controlling battery temperature to ensure that the battery is not damaged by low or high temperatures. Unlike state of charge (SOC) and state of health (SOH), which have been estimated using multiple parameters, temperature estimation is solely based on individual cell temperature measurements.

A battery should perform in its optimal temperature range for maximum efficiency and longevity. For example, the operating temperature specified by the manufacturer for a Li-ion battery ranges from −20°C to 60°C, with 30°C being its optimum temperature (Malik et al., 2016). A highly effective thermal controlling system is vital to retain the temperature under the operating range and beneath the threshold of the cell in order to control, monitor, and manage the battery performance and to improve the lifespan. A BTMS is a functional mechanism that employs a monitoring technique for maintaining the battery under the optimum working temperature (Malik et al., 2016).

BTMS should have some properties to optimize the battery performance (Siddique et al., 2018):

- Low cost
- Rigidity
- Compact and lightweight
- Easy to operate
- Easily packaged
- Low maintenance
- Safe and reliable
- No harmful gas emissions

14.5 THERMAL MANAGEMENT TECHNIQUES FOR BATTERY COOLING

The two forms of heat management for batteries are heating and cooling (An et al., 2017). Thermal management of a battery suffering from low-temperature effects is relatively simple. This is easily accomplished by utilizing the thermal energy produced in the interior of the battery using selective heaters or by constantly charging and discharging the battery, which generates internal heat continuously (Ghadbeigi et al., 2018). However, the techniques used for cooling the batteries are more complex than those for heating.

The primary benefits of cooling techniques include:

• Overheating protection
• Heat dissipation
• Heat distribution

Some technologies can be used to avoid excessive temperature rise in batteries and to implement heat management of the same. To mitigate the generated heat, two strategies can be employed (R. Zhao et al., 2015):

• Interior modifications in the chemical composition and components of the battery to minimize heat production, such as modification of electrode thickness.
• External heat management technique to improve heat elimination from the battery includes coolant circuits, air circuits, heat pipes, or employing PCMs.

Currently, the exterior thermal management system is extensively used. The key concerns with these techniques are their complicated installation, relatively high expense, unreliability, and thorough power consumption (Ianniciello et al., 2018). During the design process of BTMS, the system reliability, capability, energy consumption, and weight have to be considered (J. Chen et al., 2019). Rapid development is expected to continue for BTMS, with estimates indicating up to 39% growth between 2020 and 2024 (Murali et al., 2021). The production and analysis of EVs, which require this technology to ensure reliable operation, are two main areas where it is in high demand.

Based on whether a power drive is present in the heat transfer medium, cooling techniques are categorized into two types:

• Active cooling: The cooling method that consumes energy
• Passive cooling: The cooling method that doesn't use energy

The application determines the selection between active or passive cooling. Figure 14.2 shows the classification of cooling methods. The benefits and drawbacks of active and passive BTMS are illustrated in Table 14.2.

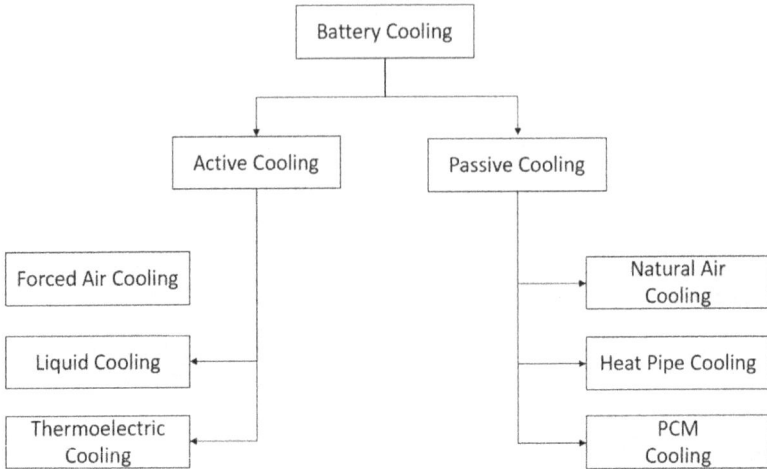

FIGURE 14.2   Flowchart showing different techniques for battery cooling.

TABLE 14.2

Advantages and Disadvantages of Active and Passive Cooling
(Al-Zareer et al., 2018; Shen et al., 2021; Siddique et al., 2018)

BTMS	Types	Advantages	Disadvantages
Active system	Forced air cooling	Low maintenance Low initial cost Commonly used	Low specific heat Low efficiency Low heat transfer coefficient Insufficient for extreme conditions
	Liquid cooling	High cooling capacity High efficiency Commonly commercialized More uniform temperature distribution Higher specific heat capacity	Complex structure Expensive Risk of leakage Maintenance difficulty High operative cost Short lifespan Insufficient for extreme conditions
	Thermoelectric cooling	Noise-free Low maintenance cost Low cost Reliable and long lifespan	Consuming power Low cooling efficiency Low coefficient of performance
Passive system	Natural convection	Lightweight Simple configuration Low operative cost Easy to maintain	Low specific heat Low efficiency High temperature gradient Unsuitable for fast charging Low heat transfer coefficient

TABLE 14.2 (*Continued*)
Advantages and Disadvantages of Active and Passive Cooling
(Al-Zareer et al., 2018; Shen et al., 2021; Siddique et al., 2018)

BTMS	Types	Advantages	Disadvantages
	Heat pipe cooling	High efficiency	Complex structure
		Flexible shape design	Expensive
		High thermal conductivity	High operative cost
			Electricity consumption
			Limited thermal contact area
	PCM cooling	More uniform temperature distribution	Low thermal conductivity
			Still under laboratory study
		Fast temperature response	Volume expansion
		Compact system design	Risk of leakage
		Low maintenance cost	PCM restoration
		Higher efficiency	supercooling
		Able to function in extreme conditions	

14.5.1 Classification of BTMS

14.5.1.1 Air Cooling

Air cooling is the process whereby air is propagated through the inner channels of a battery pack. Heat transmission occurs on the battery surface, thus the heat goes out with the air. Here, the air inlet velocities determine the heat transfer rate (Behi et al., 2020).

It is categorized into two types: natural air convection (passive cooling) and forced air convection (active cooling). In natural convection, airflow occurs without any power consumption. However, for forced air convection, a dedicated energy-consuming system forces the airflow into the battery system.

Forced convection is not needed when the C-rate is less than 0.5C because natural convection is sufficient to maintain a safe temperature. However, as the C-rate increases, the current quickly rises, which causes a proportional increase in the development of heat within the cell. The cooling effect of any system is monitored by various elements such as cell spacing, air flow direction, inlet air temperature, fan location, design, etc. (K. Chen et al., 2017; Yu et al., 2019).

Air-based systems stabilize the battery cells using conditioned ambient air as a medium. The heat removal rate in an active cooling system is more significant than that in a passive cooling system. All air-based systems benefit significantly from the shortage of an insulator between the battery and the air, leading to configurations that are comparatively simple and extremely stable.

Air cooling is widely used in vehicle applications, computer cooling, or residential applications, etc. (Fathabadi, 2014). Higher velocity air cooling reduces the time needed for the cell to acquire a steady state while also minimizing overheating and thermal runaway effects.

14.5.1.2 Liquid Cooling

Liquid cooling can significantly reduce the battery temperature owing to the high heat transfer coefficient of commonly used heat transfer fluids (HTF). Hence, it has superior efficiency over air cooling. The liquid typically has a thinner liquid boundary layer and a much greater heat transfer coefficient than air, resulting in a higher effective thermal conductivity than air. Depending on whether or not the liquid makes direct contact with the battery, it can be distinguished into two types:

1. Direct-contact cooling: HTF has direct contact with the surface of battery.
2. Indirect-contact cooling: Use of a jacket, cooling plate, or distinct piping, beside each battery cell.

Insulating mineral oil with a high viscosity is served as the heat transfer medium in direct-contact cooling. To prevent electrical shorts, HTF for cell cooling would be dielectric, like mineral oils, silicon-based oils, or deionized water.

The indirect system is relatively easier to develop and maintain. Compared to the dielectric fluids used in direct cooling, the cooling media in indirect cooling has low viscosity, resulting in a significantly higher flow rate. Indirect cooling is further divided into three types:

- Jacket cooling
- Tube cooling
- Cold-plate cooling

Due to the risk of electrical short, water is not directly used for cooling. Also, water has the potential to freeze in cold-temperature regions. Hence, a blend of water and ethylene glycol is commonly used as HTF in indirect-contact cooling. Furthermore, liquid metal can be used as a battery coolant as it has a low T_{max} and a good temperature uniformity compared to water (Jiaqiang et al., 2018; Yang et al., 2016).

The channel counts, width, height, and coolant flowrate have a strong effect on the cooling capacity of this method. The flowrate and channel numbers are the primary elements and have a similar effect, while channel width and height are secondary. This technique is more efficacious than other cooling strategies since it has a compact structure and massive specific heat capacity.

14.5.1.3 Heat-Pipe Cooling

A heat pipe can operate on its own without being pumped by an external power source. It transmits a large amount of heat energy at high speeds over long distances using phase change heat transfer despite minimal temperature variations. Simply, it is a heat transfer device that contains a little quantity of fluid that vaporizes and produces heat. The three main parts of a heat pipe include: an evaporator portion, an adiabatic portion, and a condenser portion.

Heating the evaporator portion leads to the vaporization of the working fluid, where it transmits the heat from the battery to the condenser portion. Then, the vapor fluid discharges heat through the condenser into the surroundings and condenses into a liquid form (Ling et al., 2018; Ye et al., 2015). The evaporator part is connected to a heating element that requires cooling. The fluid in the heat pipe vaporizes by absorbing heat from the heating element, and then further it moves to the condenser part. Due to the external heat exchange, fluid gets condensed, which becomes liquid and goes back to the vaporizing part. Depending on the surging force, there are various methods to return the fluid to the evaporating section from the condenser section.

14.5.1.4 Thermoelectric Cooling (TEC)

A TEC can transform electrical energy into thermal energy to heat and cool a wide range of objects, utilizing the Peltier effect (D. Zhao & Tan, 2014). A TEC is mainly composed of alternating n- and p-type semiconductors sandwiched between two thin ceramic wafers. Simply, a portion of TEC is maintained in connection with the battery, whereas the remaining portion is attached to the cooling channel. The heat produced is transmitted to different portions of the TEC, if a direct current flows through the system, and thereby it cools the battery. The primary benefit of thermoelectric cooling is that it is able to heat the battery by simply inverting the flow of current.

14.5.1.5 Phase Change Material (PCM) Cooling

The ability of PCM to operate both as a passive and semi-passive system makes it an enticing alternative to traditional cooling systems. Researchers have become increasingly interested in PCM-based BTMS in recent years due to its better temperature-regulating ability, low energy requirement, and simple structure.

A PCM is capable of storing thermal energy in both latent and sensible forms and then releasing it through the inverse action. As the temperature elevates, PCM absorbs and stores energy until it arrives at its melting point, and then the temperature falls under the melting point, thereby the PCM reverts back into the previous phase and releases the stored energy. Thermal management using PCM as an energy storage medium is discussed in detail in the following sections.

14.6 APPLICATION OF LATENT HEAT STORAGE MATERIALS IN BTMS

14.6.1 Thermal Energy Storage (TES)

Various TES technologies are introduced to improve thermal energy utilization efficiency, allowing excess thermal energy to be stored and used later at larger scales. TES applications employ a wide range of materials. The TES systems are mainly classified as three categories according to whether the material is cold or hot storage (Figure 14.3). Three basic approaches to store thermal energy can be distinguished. The following sections discuss the fundamental processes and their suitability for battery application.

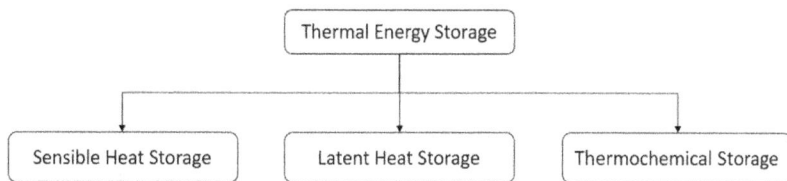

FIGURE 14.3 Classification of TES.

14.6.2 Classification and Properties

14.6.2.1 Sensible Heat Storage (SHS)

Storing sensible heat inside a material increases the energy and causes a temperature gradient. An SHS system uses the temperature difference and thermal capacity of the material, during the charge-discharge process.

The quantity of heat stored (Q) in a given condition of materials at a certain temperature change (T_f – final temperature, T_i – initial temperature) can be calculated from the amount of storage material (m) and temperature-dependent specific heat capacity (C_p) using Equations 2 and 3 (Raam Dheep & Sreekumar, 2014):

$$Q = \int_{T_i}^{T_f} mC_p dT \tag{2}$$

$$= mC_p \left(T_f - T_i \right) \tag{3}$$

The given relation can be applicable for temperature levels in which the entire material is in a homogeneous and defined gaseous, liquid, or solid phase. In addition, specific offsets should be maintained towards the phase transition temperature regions. When approaching phase change temperatures, additional latent heat correlations must be considered. The thermal inertia caused by sensible heat storage can help mitigate internal or external thermal peaks, depending on the battery mass (Furbo, 2015).

14.6.2.2 Thermochemical Storage (TCS)

Thermochemical systems use thermochemical materials that absorb and release heat via splitting and remaking molecular bonds through a totally reversible endothermic and exothermic process. Briefly, to store and retrieve heat, reversible reactions and sorption methods should be used. As shown in Equation 4, the thermochemical substance AB, which is utilized to store thermochemical energy, dissociates into A and B via an endothermic reaction, whereas the reverse product is formed through an exothermic reaction:

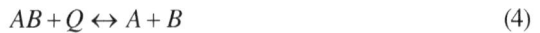

$$AB + Q \leftrightarrow A + B \tag{4}$$

[Q – amount of heat required to dissociate A and B]

An endothermic reaction occurs during charging, causing the substance to absorb heat and split into reactants. The resulting reaction products should be kept off-site. To regain the thermal energy stored, reactants are mixed at suitable conditions; an exothermic reaction occurs, and hence energy is released (Sarbu & Sebarchievici, 2018). The amount of heat stored (Q) is estimated by the given Equation 5:

$$Q = a_r m \Delta h_r \tag{5}$$

[a_r – fraction melted, m – amount of storage material (kg), Δh_r – endothermic heat reaction] (Sharma et al., 2009). TCS possess more energy density than any other

TES, while the reaction efficacy reduces gradually. Meanwhile, the complexity of handling, particularly separating/combining the liquid components, prevents it from being used in almost all battery applications.

14.6.2.3 Latent Heat Storage (LHS)

In the LHS system, thermal energy is stored in the latent heat of TES materials during a continuous process like phase transformation. Latent heat is the heat that a material stores when it goes through a phase transition. Due to its isothermal storage mechanism, the PCM-based LHS system provides an efficient method for storing thermal energy.

The amount of energy stored in a LHS substrate can be calculated as follows in Equations 6 and 7:

$$Q = \int_{T_i}^{T_m} mC_p dT + ma_m \Delta h_m + \int_{T_m}^{T_f} mC_p dT \tag{6}$$

$$Q = m \left[C_{sp} \left(T_m - T_i \right) + a_m \Delta h_m + C_{lp} \left(T_f - T_m \right) \right] \tag{7}$$

[T_i – initial temperature (°C), T_m – melting temperature (°C), m – mass of heat storage medium (kg), C_p – specific heat (kJ/kg K), a_m – melted fraction, h_m – heat of fusion per unit mass (kJ/kg), T_f – final temperature (°C), C_{sp} – average specific heat between T_i and T_m (kJ/kg K), C_{lp} – average specific heat between T_m and T_f (kJ/kg K)]

Phase changes comprise processes that result in a change in the aggregation state: crystallization and melting, evaporation and condensation, as well as processes that occur in the liquid or solid phase. Depending on the properties within the chosen material, it can store a substantial quantity of heat or cold. The phase change occurs from solid to liquid with only a slight volume change. All of this causes the stored material to melt and solidify at a fixed temperature, which is termed as the phase change temperature. When the melting process is complete, additional heat transfer causes sensible heat to be stored again. At the melting point, the temperature remains nearly constant until the phase transition of the entire mass has completed. The primary benefit of using LHS over SHS is its ability to store heat at nearly constant temperatures. At first, this material functions as SHS material, where the temperature increases proportionally with the energy of the system; but even if, there upon a change in phase, heat is absorbed or released under nearly isothermal conditions.

Hence, a group of materials known as PCMs has emerged as the best suitable for storing latent heat for various applications. Their characteristics, as well as the benefits of LHS, made them intriguing options for battery applications.

14.6.3 PHASE CHANGE MATERIALS

14.6.3.1 Generalities About PCM

PCM is extensively utilized in thermal management applications due to its unique properties (Farid et al., 2004). According to the phase changing abilities by absorption and desorption of heat, LHS materials are categorized as solid-solid, solid-liquid, or liquid-gas.

Latent heat thermal energy storage (LHTES) using PCMs is an effective method for achieving passive thermal management and ensuring the $T_{battery}$ remains within the specified limit. When the solid PCM reaches its melting point, it melts and stores energy. Similarly, if the liquid material cools below its melting point, it solidifies and releases energy.

14.6.3.2 Classification of Phase Change Material

There are numerous classification standards for PCMs. Figure 14.4 depicts an explicit classification. On the basis of phase transformation, PCMs are divided into liquid-gas, solid-liquid, and solid-solid. Among them, solid-liquid PCMs have the characteristics of minor change in volume during phase change and large latent heat capacity. As solid-liquid PCMs can be generalized and classified as organic PCMs, inorganic PCMs, and eutectic PCMs. The phase transition temperature and specific latent heat storage capacity are the key characteristics of PCMs for technical applications.

The desired operating temperature range is predefined in most technical applications, particularly battery thermal management. Thereupon, the melting point should be the first preference for choosing a PCM, which must be within the application's limit. A wide temperature range can be covered due to the variety of PCM. The range of storage density for the material is as broad as the range of melting temperatures. Generally, the material would have large specific latent energy density to be selected for battery applications. Moreover, there are extensive features unique to these types of materials that affect their suitability for use.

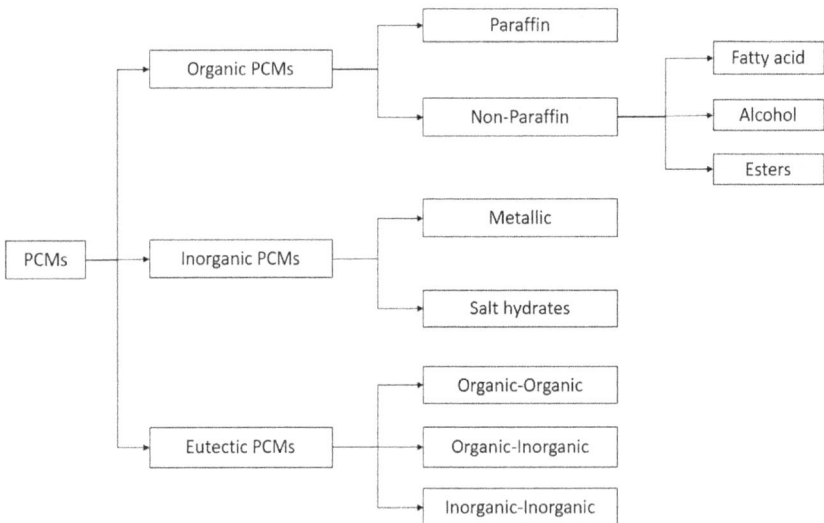

FIGURE 14.4   Classification of PCMs.

14.6.3.2.1 Organic PCMs

An organic compound is distinguished with carbon atoms in its composition. Organic materials have the potential to undergo concurrent melting with no phase separation about a multitude of cycles while degrading the latent heat of fusion and are neither affected by supercooling. These substances are accessible with a variety of melting points.

Organic PCMs have been the primary raw material source for PCMs due to their stable phase change temperature, less corrosive nature, non-toxicity, superior chemical endurance, lack of supercooling (self-nucleation), and subsequent degradation of latent heat of fusion. Apart from the numerous advantages listed, the major disadvantages are low thermal conductivity, a supercooling effect, and PCM spilling in chambers. Organic PCMs can also be subdivided into paraffins and non-paraffins.

14.6.3.2.1.1 Paraffins

Paraffins are straight-chain n-alkanes having the general formula C_nH_{2n+2}. The blending of (CH_3)-chain dissipates a substantial fraction of latent heat. Paraffin chains can be even (n-paraffin) or odd (iso-paraffin) depending on the length of carbon chain (Abhat, 1983).

The melting point (T_m) range of paraffins are between 5.5–76°C, where the T_m and heat of fusion are linearly proportional to the carbon chain length. The increased melting temperature is due to the increased attraction between n-alkane chains (Sarier & Onder, 2012). Because kinds of paraffin are synthesized by the oil refineries, their cost varies with the crude oil prices. So, due to cost constraints, technical-grade paraffin is the only type that can be employed as PCMs in LHS systems. Meanwhile, paraffins are non-corrosive, reliable, non-toxic, and easy to recycle, making them convenient for a variety of LHS applications (Lane, 1983).

They are stable and chemically inert below 500°C and exhibit minimal volume change when melted. Hence, a system using paraffin typically has a longer freeze-melt cycle for these properties. In addition to a number of advantageous properties such as congruent melting and good nucleating properties, they have some flaws such as:

1. Moderately flammable
2. Low thermal conductivity
3. Noncompatible for plastic chambers

All of these drawbacks are being limited to some extent by altering the storage system and the wax.

14.6.3.2.1.2 Non-Paraffins

Non-paraffin PCMs are the largest category available among PCMs with diverse characteristics, unlike paraffins, that have extremely comparable features. It is the most extensive group of phase change storage candidates. These organic materials have the following characteristics:

1. Unstable at high temperatures
2. Variable toxic effects

3. Low flashpoints
4. High latent heat of fusion
5. Inflammable
6. Low thermal conductivity

This category includes fatty acids, esters, alcohols, and glycols.

14.6.3.2.1.3 Fatty Acids

Fatty acids are primarily obtained from vegetables and animals, enabling a non-polluting supply. These are classified as follows: stearic, caprylic, lauric, capric, myristic, and palmitic (Sarier & Onder, 2012). When compared to paraffins, these substances have exceptional phase transition features and also freeze without supercooling. Their primary disadvantage is the price, which is approximately three times more expensive than that of technical-grade paraffin. Fatty acids possess excellent characteristics such as good thermal and chemical stabilities, non-toxicity, biodegradability, and a range of melting points that is suitable for a wide range LHS applications. It can withstand about thousands of thermal cycles with no deterioration in the thermal properties (Feldman et al., 1991).

14.6.3.2.1.4 Esters

Esters are formed by replacing one hydroxyl group with one alkyl group in an acid. Fatty acid esters exhibit a small solid-liquid transition range. Therefore, they can generate eutectics with little subcooling (Nkwetta & Haghighat, 2014). Such materials are widely available in huge amounts for commercial use.

14.6.3.2.1.5 Sugar Alcohol

They are commonly called polyalcohols with a moderate melting range (90–200°C). Prior research has shown that alcohols such as mannitol, erythritol, and xylitol have significantly better latent heat potential than remaining compounds in the group. Among sugar alcohols, erythritol has proven to be an excellent thermal energy storage material. PCMs at low temperatures, sugar alcohol displayed a significantly high rate of supercooling, which should impair thermal energy storage efficiency. During melting, these materials experience a 10–15% increase in volume (Kaizawa et al., 2008).

14.6.3.2.1.6 Polyethylene Glycols

Polyethylene glycols (PEGs), consists of dimethyl ether chains. Since they have a terminal hydroxyl group, they are soluble in both organic substances and water. They are available in various grades, and they have some advantages such as being thermally and chemically stable, less corrosive, non-flammable, safe, and affordable (Meng & Hu, 2008). With the increasing molecular weight of the PEGs, the T_m and latent heat of fusion are also enhanced. PEGs also possess low heat conduction, as like other PCMs. There have been extensive numerical and experimental researches performed to improve this property.

14.6.3.2.2 Inorganic PCMs

Metals, salts, and salt hydrates are the common inorganic substrates used for latent heat storage applications, as they can store a significant amount of energy due to their higher phase change enthalpy. However, at elevated temperatures a phase change takes place, particularly for salts and metals, preventing a wide range of technical applications. Salts are commonly used for commercial uses, but there are not known battery applications (Mathur et al., 2014). Salt hydrates and metallics are the main two categories.

14.6.3.2.2.1 Salt Hydrate

They are blends of water and inorganic salts that produce crystalline solids having the formula A.nH$_2$O (A – salt, n – number of molecules, H$_2$O – water molecule) (Su et al., 2015). Salt hydrates, which are formed by anhydrous salt at room temperature, possess a phase change enthalpy that varies with bond strength between the salt and the water molecules. The major limiting factor restricting their application in LHTES systems are supercooling during crystallization, corrosion, and phase segregation.

The transition of salt hydrates from solid to liquid is a deterioration of the salt. The hydrate crystals disintegrate at its melting temperature into anhydrous salt and water or lower hydrate and water. There are distinct behaviors shown by molten salts such as congruent, semi-congruent, and incongruent melting. At the phase transition temperature:

- The salt is fully soluble in its water of hydration, referred to as *congruent melting*.
- When the salt is fully insoluble in its water of hydration, *incongruent melting* occurs.
- *Semi-congruent melting* occurs when the solid and liquid phases are in equilibrium.

Some of the intriguing characteristics of salt hydrates are as follows:

1. Minimal volume variations during melting
2. Relatively high thermal conductivity
3. High latent heat of fusion per unit volume
4. Low corrosiveness
5. Compatible with plastic and less toxic

14.6.3.2.2.2 Metallic

This category includes metals and their alloys, having low melting temperature, high heat of fusion per unit volume, superior thermal conductivity over other PCMs, and minimal volume change during phase transition. However, these are rarely considered for latent heat storage applications because of its relatively low heat of fusion per unit weight (Ge et al., 2013). Some of the characteristics of these materials are as follows (Su et al., 2015):

1. Low specific heat
2. High heat of fusion per unit volume
3. Relatively low vapor pressure
4. Low specific heat of fusion

5. High thermal conductivity
6. Small volume change during phase transition

Hence, when a specific metal is employed as a PCM, the heat transfer ability is considerably enhanced when compared to traditional PCMs.

14.6.3.2.3 Eutectic PCMs

Eutectic mixtures are made by two or more low melting materials, such as organic, inorganic, or both inorganic and organic compounds, with almost similar freezing and melting points. They are crystal mixtures of multiple soluble components that exhibit simultaneous melting and solidification, and their freezing and melting points are usually lower than those of individual components.

In contrast to other PCMs, eutectics usually melt and freeze without segmentation. When compared to other PCMs, eutectic PCMs possess high thermal conductivity and density (Sharma et al., 2009). Variations in the T_m of the resulting eutectic mixture could be achieved by changing the mass of the individual substrate. Due to the abundance of prevailing organic compounds and inorganic salts, the eutectic types are expanding with different combinations.

14.6.3.3 Selection Criteria of PCM for Thermal Management

The efficiency of thermal management is significantly dependent on proper PCM. The following are the primary criteria for selecting a PCM for BTMS:

- Melting point in the desired operating temperature
- High specific heat
- Minor volume change in phase transition
- No supercooling
- Long-term chemical endurance
- Non-poisonous
- High latent heat
- Non-corrosive
- Non-flammable and non-explosive
- Easily available
- High thermal conductivity
- Compatibility with materials of construction

14.7 PCM-BASED THERMAL MANAGEMENT

Battery thermal management using PCM, first demonstrated and patented by Al Hallaj and Selman, outperformed the conventional systems (al Hallaj & Selman, 2000). This study analyzed a thermal management system for EV batteries. In order to evaluate the temperature profile for a 100 Ah battery cell, they performed a numerical simulation. They experimentally investigated a battery pack with paraffin wax (T_m –56°C) between eight cylindrical cells. Hence, they tested thermal performance at a 1C discharge rate from the experimental prototype developed, observing a decrease in T_{max}.

The required mass of PCM for the battery thermal management is calculated as follows in Equation 8 (Shen et al., 2021):

$$M_{PCM} = \frac{Q_{dis}}{C_p\left(T_m - T_i\right) + H} \qquad (8)$$

[Q_{dis} (J) – heat released by the battery, M_{PCM} (kg) – quantity of PCM, C_p (J kg^{-1} K^{-1}) – specific heat of PCMs, T_m ($^{\circ}C$) – melting point, T_i ($^{\circ}C$) – initial temperature of PCMs, H (J kg^{-1}) – latent heat of PCM]

PCM absorbs heat from the battery through the phase change by latent heat of the material, and inhibits the $T_{battery}$ from rising too quickly. It can not only minimize the temperature gradient between moieties but also store energy in cold environments and transfer energy into batteries to attain thermal insulation and increase battery performance in cold environments.

Conversely, some researches are concentrated on the thermal or mechanical properties of PCM. Rao et al. (2011) studied the impact of phase transition temperature range and discovered that a lower T_m of PCM might be advantageous for cooling performance.

Duan and Naterer (2010) performed an investigation experimentally on the usage of PCM for thermal control of an Li-ion battery in 2010. Two different PCM designs are being investigated: one consists of a heater surrounded by PCM cylinders, while the other consists of a heater wrapped by PCM jackets. The experimental results show that two models are capable of keeping the $T_{battery}$ within the defined limits.

Ramandi et al. (2011) examined the exergy analysis using four various PCMS having a melting point ranging from 30–40°C: capric acid, $Na_2(SO_4)10H_2O$, eicosane, and $Zn(NO_3)_26H_2O$. BTMS using a single layer of capric acid had a maximum exergy efficiency of 46%, while in double-layer PCM, a blend of capric acid and $Zn(NO_3)_26H_2O$ has been observed as more effective. According to the results, the double-layer configuration has higher efficiency than the single-layer configuration. PCM-based BTMS is significantly influenced by the geometric characteristics.

The effect of PCM plate thickness on $LiFePO_4$ battery performance is investigated by Malik et al. (2017). They observed that $T_{battery}$ is lowered by increasing PCM thickness at low discharge rates, while no substantial progress is noticed by increasing PCM plate thickness at high discharge rates. Javani et al. (2014) also analyzed the influence of PCM layer thickness on the heat transfer performance of the battery. With increased thickness, temperature uniformity was attained; however, there was no noticeable drop in T_{max}.

The uncertain phase transition thermal energy transfer of an Li-ion battery employed in solar automobiles with volumetric heat production is numerically studied by Moraga et al. (2016). They examined a thermal control system incorporating several PCMs with varying thickness in one or three layers wrapping the battery. Three successive layers of PCM reduced the temperature more effectively than a single-layer PCM. They also found that for better performance, the PCM having low thermal conductivity must be inserted into the exterior wall, and the PCM possessing high thermal conductivity should be organized close to the battery.

When compared to conventional systems, PCM can attain lower T_{max} and good temperature uniformity (Wu, Wu, et al., 2017).

14.8 PERFORMANCE ENHANCEMENT OF PCM-BASED BTMS

PCMs, which integrate into the battery, absorb heat and control battery temperature during charge and discharge cycles. Reduced heat transfer rate on account of low thermal conductivity and limited thermal storage capacity of PCMs are the most significant drawbacks of the PCM-based BTMS (Al-Zareer et al., 2019; Li et al., 2021). PCMs with a lower Q would not absorb the battery heat rapidly, since a significant fraction of PCM stayed as unmelted, resulting in the elevated rate of $T_{battery}$ (Barnes & Li, 2020).

Several techniques were proposed to elevate the performance of PCM-based BTMS. These are categorized as follows:

- Increasing heat transfer rate
  For enhancing Q, two commonly adopted strategies are extending the heat transfer area A_h and increasing thermal conductivity λ_{PCM}, which allows battery heat to be dispersed gradually, hence preventing the accumulation of heat in batteries.
- Retrieving thermal storage capacity of PCM
  For recovering H_{PCM}, coupling with other cooling methods removes the heat contained in PCMs.
- Reduction of PCM cooling load
  Hybridization of cooling techniques with PCM is developed to reduce the PCM cooling load. Here, the PCM absorbs only a fraction of heat from the battery, and the rest is dispersed by other cooling techniques.

14.8.1 IMPROVING HEAT TRANSFER RATE

A generic equation for estimating the heat transfer rate, Q, between PCMs and batteries should be stated as follows in Equation 9:

$$Q = kA_h \left(T_{battery} - T_{pcm} \right) \tag{9}$$

[k – heat transfer coefficient, A_h – heat transfer area, and T_{PCM} – melting point of PCM]

Hence, $T_{battery}$ is affected by three key parameters: k, A_h, and T_{PCM}, in accordance with Equation 9. Since the intended working temperature of batteries is 15–35°C and other thermophysical characteristics of PCM need to be taken into account concurrently, choosing a PCM with low T_m to increase Q is practically impossible.

Therefore, two ways to improve the heat transfer rate are increasing thermal conductivity and broadening the heat transfer area of PCM.

14.8.1.1 Enhancing PCM Thermal Conductivity

As PCMs have low thermal conductivity, the rate at which heat may be transferred to the materials is also limited. Fillers having high heat conduction ability, such as porous materials and nanomaterials, are added to increase the thermal conductivity. After comparing various methods and numerous experiments to increase this property, carbon-based additive particles outperform metal-based additives (Singh et al., 2019).

The PCM might fully melt in summer weather or by multiple charge-discharge cycles, while the poor heat conduction will become a barrier for transmission of thermal energy. According to the studies on PCM cooling BTMS, the primary concern is its poor thermal conductivity (Farid et al., 2004). Herein are presented some works based on strategies for enhancing PCM thermal conductivity.

14.8.1.1.1 Adding Porous Materials

The composite PCMs are prepared to enhance the ability of pure PCM to conduct heat, as well as a metal foam combination has gained the most attention. The porous material must have high porosity, low heat transfer resistance, and high thermal conductivity to be used as fillers. Aluminium foam, nickel foam, and copper foam are usually mixed with the PCM.

Hussain et al. (2016) performed a study by designing an effective thermal control system for a lithium-ion battery with a nickel foam-paraffin wax composite PCM. Compared to the cooling by pure PCM and natural air, a drop in the battery surface temperature ($T_{surface}$) of 31% and 24% is obtained with the composite PCM under a 2C discharge rate. With nickel foam, the heat transfer rate of PCM was raised by a factor of 4.8.

This study reveals the impacts of the geometrical properties of the foam on the $T_{surface}$ of the battery. The battery $T_{surface}$ diminishes with the porosity of metal foam and pore density. The discharging efficiency, in contrast, expands with porosity but lowers with pore density. Moreover, nickel foam provides an economic benefit since it costs less than most other metal foams. The schematics pack surrounded by CPCM is shown in Figure 14.5.

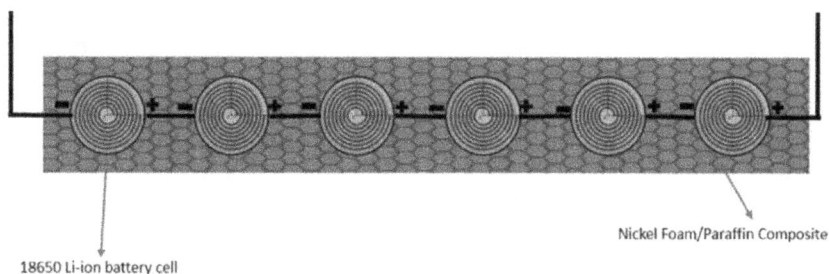

Nickel Foam/Paraffin Composite

18650 Li-ion battery cell

FIGURE 14.5   Battery pack wrapped by nickel foam-paraffin composite.

Source: Reproduced from (*Hussain et al., 2016*).

X. Wang et al. (2018) developed a passive technique by utilizing copper foam-paraffin composite PCM, where the PCM has the indirect interaction with the cell. In this work, they numerically and experimentally investigated the temperature gradient with composite PCM and air cooling techniques. Here, the $T_{battery}$ is managed within the intended range even at a high discharge rate.

14.8.1.1.2 Nanomaterial Additives

Nano-enhanced PCMs are produced via uniform diffusion of nanomaterial fillers in the PCM. To ensure the stability of composite PCM, nanomaterial additives must possess high chemical endurance and high thermal conductivity. Metal nanoparticles and carbon nanomaterials are commonly used as fillers.

The heat transfer characteristics of a battery-like prototype are investigated by Karimi et al. (2016) using three different metal nanoparticles: Cu, Ag, and Fe_2O_3. The findings demonstrated that the improvement in heat transfer potential of the system differed with nanoparticle, in which the composites containing Ag nanoparticles have the highest potential.

Lin et al. (2015) studied the BTMS with an EG matrix/PCM cooling system experimentally and numerically. An expanded graphite matrix mended with PCM has been utilized to maximize the heat conduction and temperature uniformity. The T_{max} is lowered by 32% and 37% at 1C and 2C discharge rates while maintaining the ΔT_{max} under 5°C. Hence, the studied composite PCM is suitable for EV/BEV vehicles with high-capacity rectangular batteries.

PCM blended with carbon-based materials have shown superior physical, chemical, thermal, and electrical properties. Samimi et al. (2016) proposed carbon fiber PCM composites for an Li-ion battery cell. Carbon fiber-enhanced PCM boosts the heat transfer ability of the paraffin mixture significantly. This study shows that CPCM having a higher percentage of carbon fibers exhibit excellent temperature uniformity. Therefore, this composite PCM is viable for using in BTMS. Several other numerical and experimental studies have been conducted in nanomaterial addition (Ghadbeigi et al., 2018; X. Zhang et al., 2018).

14.8.1.2 Expanding Heat Transfer Area (A_h)

14.8.1.2.1 Adding Fins
The incorporation of fins to the PCM cooling system raises the transmission of thermal energy. Copper and aluminium are widely used as fin materials, because of their higher thermal conductivity and specific heat capacity relative to other materials.

A novel PCM-fin structure BTMS proposed by Ping et al. (2018) determines the heat dissipation effect and the thermal response of a prismatic $LiFePO_4$ battery module during the discharging process. Three different paraffin waxes having close physical properties were used as PCMs: n-pentacosane, n-tetracosane, and n-docosane. Longitudinal aluminum fins are mounted between batteries, as demonstrated in Figure 14.6. The study found that BTMSs with fin structure could keep the T_{max} of the battery surface below 51°C, even at a 3C discharge rate. Moreover, apart from heat conduction, it also improves natural convection. In this work, the effects of PCM thickness, fin thickness, and fin spacing are numerically analyzed. Reducing fin thickness could decrease the T_{max} and ΔT_{max} of the battery module by raising

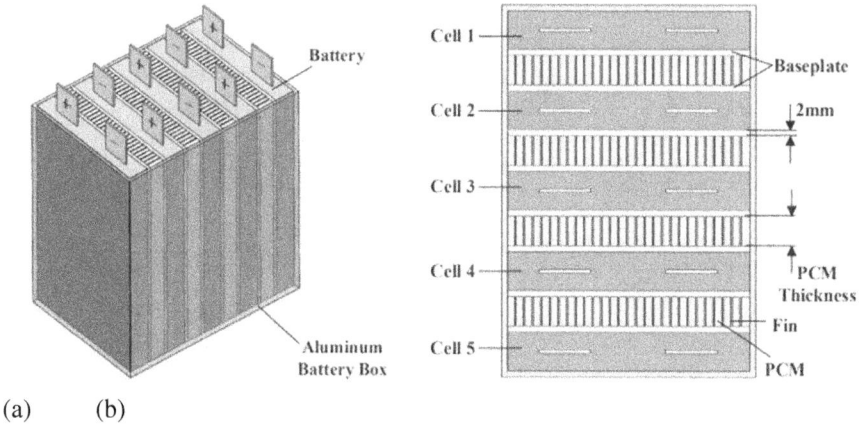

(a)   (b)

FIGURE 14.6 Schematic diagram of closely placed PCM fin structure and prismatic cells: (a) side view, (b) top view.

Source: Reproduced from (*Ping et al., 2018*).

the heat exchange area with the PCM. Also, close arrangement of the fins would contribute to a negative impact on lowering PCM volume. Increasing PCM thickness can improve thermal performance without exceeding the critical value.

Z. Wang et al. (2017) investigated BTMSs with PCM-fin structures using bench testing and simulation. Both the finned and pure PCM casings are subjected to a heat transfer analysis. The results show that when 8-fin was incorporated, the heat transfer area increased by 170%. Also, $T_{battery}$ can be maintained under 50°C at a 5C discharge rate when contrasted to pure PCM-based BTMS.

Although the potency and the dominance of thermal management with PCM-fin structures have been demonstrated, researches on extending the heat transfer area are far limited in comparison to maximizing the thermal conductivity of PCM.

14.8.2 Retrieving Thermal Storage Capacity (H_{PCM})

During the charging and discharging process, PCM absorbs battery heat. Thereupon, liquid fraction increases by absorbing heat. As the phase change occurs in an isothermal condition, $T_{battery}$ is precisely controlled prior to the PCM melting entirely. Owing to the restricted H_{PCM}, PCMs, however, will undoubtedly undergo a complete meltdown with continuous discharge. $T_{battery}$ will drastically increase once fully melted, exceeding the maximum value and triggering thermal runaway. Coupling with other cooling methods such as forced liquid cooling, air cooling, thermoelectric cooling, and HP cooling that remove heat from the battery is one of the popular methods for recovering H_{PCM}. Here, the cooling methods are used to remove the heat occupied in the PCM.

As demonstrated in Figure 14.7a, the air is forcefully passed through the PCM casing, thereby dissipating PCM-absorbed heat. Since air convection has a low heat

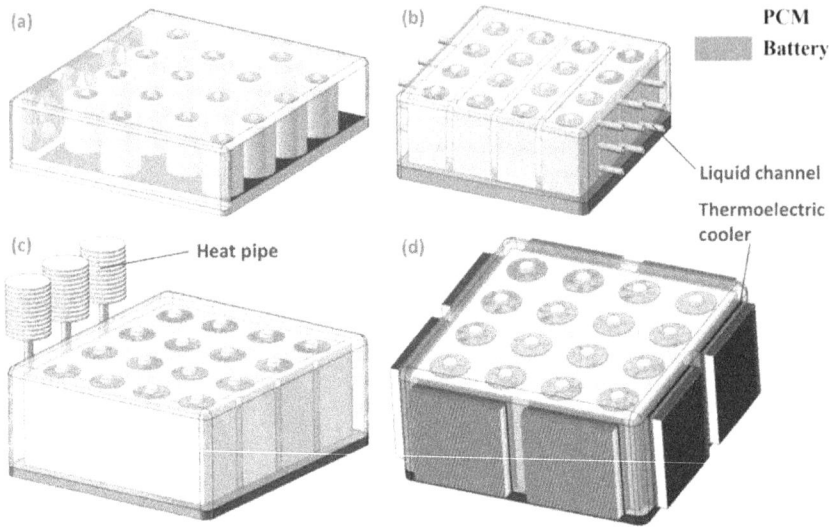

FIGURE 14.7 PCM coupled with (a) forced air, (b) forced liquid, (c) HP, and (d) TEC cooling.

Source: Reproduced from (Shen et al., 2021).

transfer coefficient, PCM coupled with liquid cooling has been proposed, as depicted in Figure 14.7b. As illustrated in Figure 14.7c, heat pipes, having a higher heat transfer coefficient and a low temperature gradient, have been chosen to expedite heat absorption from the phase change material. The evaporator part is encased within PCMs, while the condenser part is opened to the surrounding. Furthermore, to dissipate heat in the PCMS, Peltier effect-based thermoelectric coolers (TECs) have been utilized, as shown in Figure 14.7d. PCMs came in contact with the colder side of TECs. The current flowing through TECs primarily controls the cold side temperature.

14.8.2.1 Coupled BTMS with PCM and Forced Air Cooling

R. D. Jilte et al. (2019) recreated a battery system to investigate the temperature controlling potency of the coupled system with active air cooling and PCM. Here, the conventional battery system, where the basic units upon the whole battery are placed in one PCM, is altered to stimulate passive and active cooling for every battery cell. Each module contains six cells packed in a 1S6P pattern, where the cells are encased in a 4 mm cylindrical space loaded with lauric acid (melting point –43.5 °C to –48.2°C) PCM, as illustrated in Figure 14.8. It demonstrates that PCM usage in battery aids in maintaining the uniform temperature under 0.05°C for the entire battery working at a 2C discharge rate. The maximum increase in cell temperature with the suggested cooling design for individual cells are below 5°C, even at high ambient temperatures.

The study performed by Ling et al. (2015) deduced that, in a PCM-based conventional system, after two cycles of the 5S4P pack with 1C charge rate and 1.5C/2C

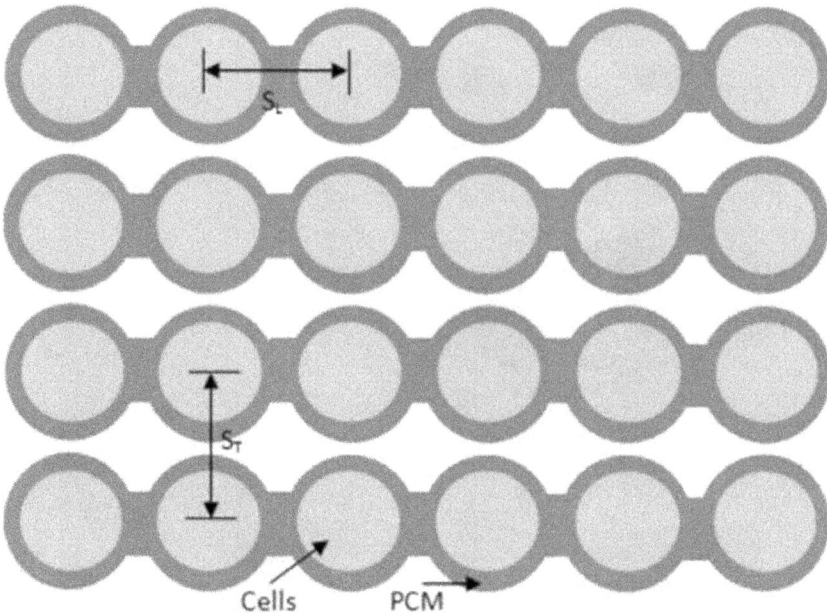

FIGURE 14.8 BTMS with enhanced PCM cooling system and natural ventilation.

Source: Reproduced from (R. D. *Jilte et al., 2019*).

discharge rate at a room temperature of 25°C, the T_{max} surpassed the safety limit of 60°C. Forced air convection is used to evacuate the occupied heat from the PCM, as illustrated in Figure 14.9. This hybrid system succeeded in keeping the T_{max} under 50°C in cycles at less than 2C, with a 7°C rise in $T_{ambient}$. The investigation on air-speed effect demonstrates that the T_{max} and ΔT_{max} in the battery are independent of airspeed, whether the PCMs possess sufficient thermal energy storage potential. Therefore, the roles handled by PCMs and forced air convection seem to be distinct. PCMs regulate the T_{max} and ΔT_{max}, whereas forced air convection cools the PCMs between two discharges, assisting in the recovery of PCM latent heat.

The experiment agrees reasonably well with the numerical model, having a mean temperature gradient below 2.3°C. The numerical results clearly show that the fraction of PCMs that have melted at the completion of each cycle rises in the purely passive BTMS, whereas in the hybrid system, it remains zero, through which heat deposition in PCMs can be explained.

A novel hybrid structure proposed by Bamdezh and Molaeimanesh (2020) comprising an air cooling system and PCM wrapped around 18650 lithium cells studied the geometry of coupled PCM and forced air cooling. These authors investigated the impacts of major structural factors on performance with the simulation of two arrangements of cells, three different distances between adjacent cells, and three different PCM thicknesses. The results showed that the geometrical factors significantly

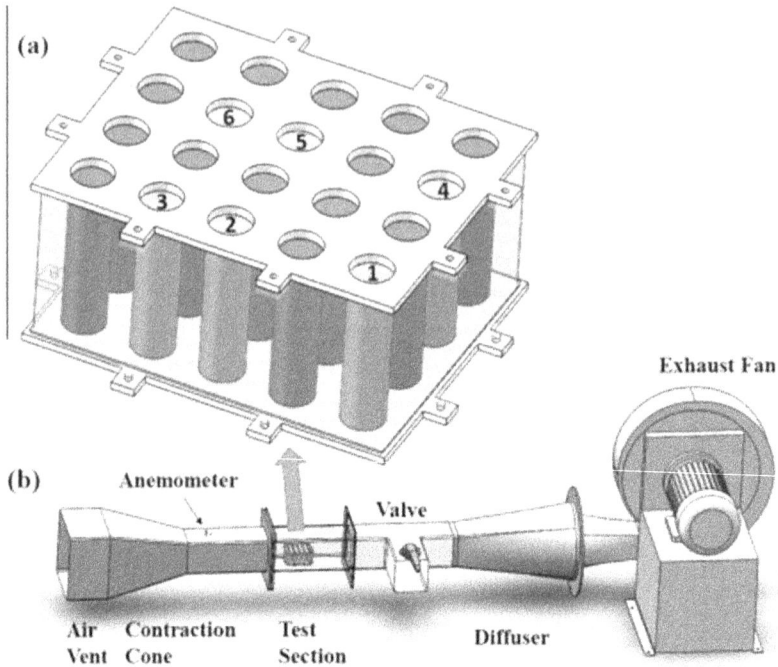

FIGURE 14.9   (a) 5S4P battery pack diagram; (b) air channel structure.

Source: Reproduced from (*Ling et al., 2015*).

impacted the system efficiency. With increasing PCM thickness, $T_{battery}$ decreased while the maximum temperature difference increased.

14.8.2.2 Coupled BTMS with PCM and Liquid Cooling

R. Jilte et al. (2021) proposed and analyzed a novel design for coupling liquid passages to a PCM chamber in a battery system with liquid passages enveloped on the exterior of the PCM casing. Heat can be dissipated from the PCM chamber to both a convecting air and circulation medium using the channelling method. Moreover, the design enables the cooling system to operate without the circulation of liquid media when the $T_{ambient}$ is low. Even at an incredibly high atmospheric temperature of 40°C, the coupling of liquid passages to phase change chambers did result in a $T_{battery}$ under 41.2°C.

14.8.2.3 Coupled BTMS with PCM and Heat Pipe Cooling

An HP-assisted PCM-based BTMS for EVs and HEVs was designed by Wu, Yang, et al. (2017). Experiments show that heat pipes can influence the temperature distribution of a battery module when activated at high discharge rates of the batteries. The schematic diagram is shown in Figure 14.10. With active convection of air elevating the heat dispersion of the condenser portion, the temperature rise has been

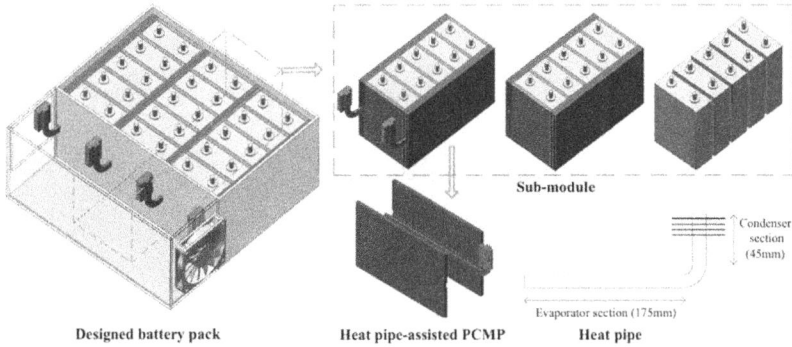

Designed battery pack   **Heat pipe-assisted PCMP**   **Heat pipe**

FIGURE 14.10 Schematic diagram of battery pack and sub-modules.

Source: Reproduced from (Wu, Yang, et al., 2017).

maintained under 50°C even at a high discharge rate of 5C. Hence it results in an extra stable and minimized temperature gradient during the cycling process. However, elevating the air velocity has minimal effectiveness when it reaches a crucial point during the PCM phase change process.

14.8.2.4 Coupled BTMS with PCM and Thermoelectric Cooling

Song et al. (2018) analyzed the functioning of PCM and semiconductor thermoelectric cooling (TEC) device-coupled BTM systems. The effect of various factors, particularly cooling and heating power, thermal management temperature range, and TEC device arrangement, was numerically investigated. The exterior portion of the aluminium casing that enclosed the PCMs was equipped with TECs. As illustrated in Figure 14.11a, four TEC configurations were analyzed.

Semiconductor thermoelectric devices are distributed uniformly beneath the battery module in case 1. It is mounted on the two opposing faces of the battery module in cases 2 and 3. Likewise, it is arranged along each of the four flanks of the pack in case 4. In the context of ΔT_{max}, case 2 was the highest and case 3 was the lowest.

14.8.3 REDUCTION OF PCM COOLING LOAD

In this method, PCMs absorb only a fraction of the heat generated, with the remainder being removed by other cooling strategies including forced air cooling, liquid cooling, and HP cooling.

Here, cooling strategies are employed in order to eliminate the heat directly from the batteries or concurrently from the batteries and the PCMs (Molaeimanesh et al., 2020). Because of the decreasing thermal resistance among the batteries and the ambient, the heat emission rate for this method is greater than that of retrieving H_{PCM}.

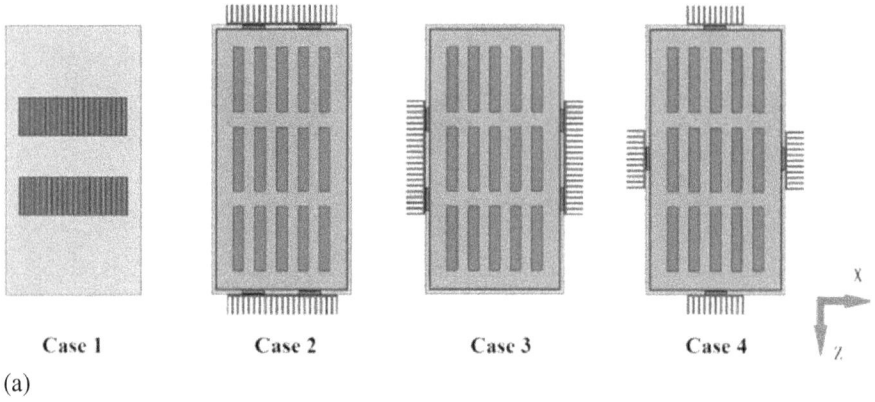

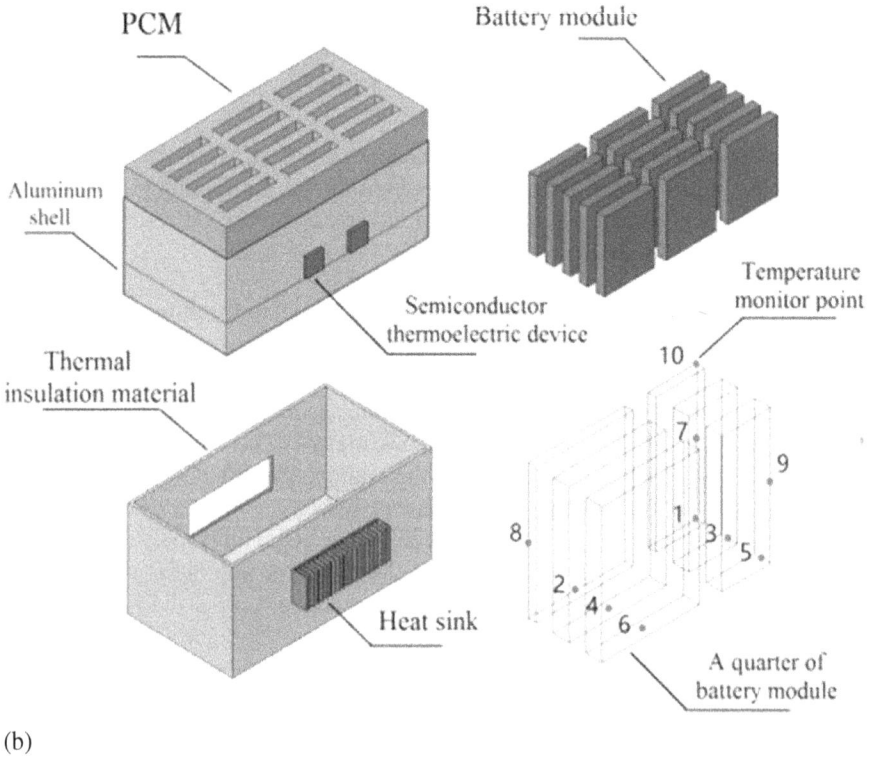

FIGURE 14.11 (a) Layout of semiconductor thermoelectric devices; (b) battery pack with PCMs and TEC.

Source: Reproduced from (Song et al., 2018).

14.8.3.1 Hybrid BTMS with PCM and Forced Air Cooling

Lv et al. (2020) proposed a hybridized configuration in which the serpentine CPCM (S-CPCM) plates replaced the common block-shaped CPCM (B-CPCM) module. This novel structure effectively minimizes the mass of the CPCM module by

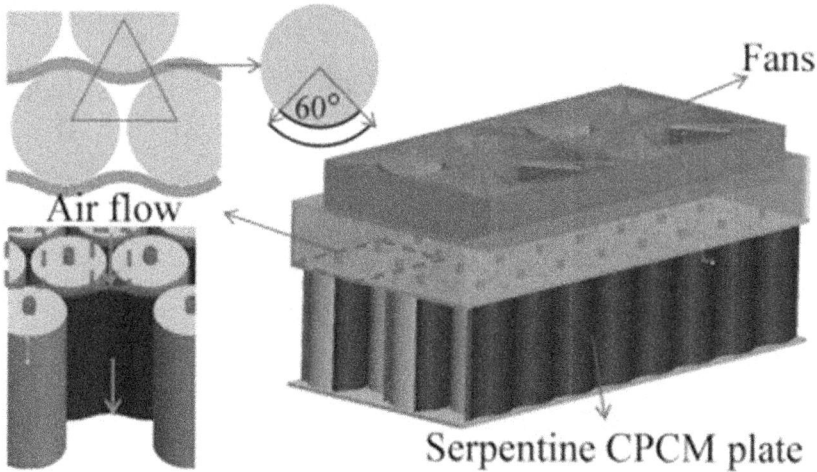

FIGURE 14.12   Serpentine composite PCM modules.

Source: Reproduced from (Lv et al., 2020).

approximately 70%, enhancing the energy density of the battery pack from 107.7 to 121.6 Wh kg^{-1}. Under forced air convection, the S-CPCM module outperforms the traditional bulky CPCM module in terms of heat dissipation, as the S-CPCM plate with good stability provides multiple air convection channels and a larger surface area for heat transfer. A schematic diagram is shown in Figure 14.12.

14.8.3.2 Hybrid BTMS with PCM and Liquid Cooling

Cao et al. (2020) developed a hybridized thermal control system having a combination of liquid cooling and PCMs for lithium-ion cylindrical cells. A cooling plate was kept at the middle portion of batteries with three liquid channels through by water that goes into a PCM matrix. A combination of expanded graphite and RT44HC composites is used here. A schematic representation is illustrated in Figure 14.13. Due to the contact with PCMs, H_{PCM} was recovered simultaneously by liquid via absorbing the stored heat in the PCMs. The reduction of maximum temperature and planar $\Delta T_{battery}$ was restricted, whereas the axial $\Delta T_{battery}$ increased substantially. Hence, a low flowrate was favorable. Therefore, a managing approach was proposed that would limit the liquid temperature under 40°C and relatively closer to room temperature as possible.

14.8.3.3 Hybrid BTMS with PCM and HP Cooling

Putra et al. (2020) proposed and tested the performance of hybridized HP and PCM cooling with battery simulators. The integrated PCM and finned L-shaped HP test cell consists of a PCM container, two finned L-shaped HPs, two heater-equipped battery simulators, and a stainless-steel conductor. Each battery simulator was equipped with a heater to resemble the heat produced by an actual Li-ion battery. Consequently, a voltage regulator could be used to manage the heat load.

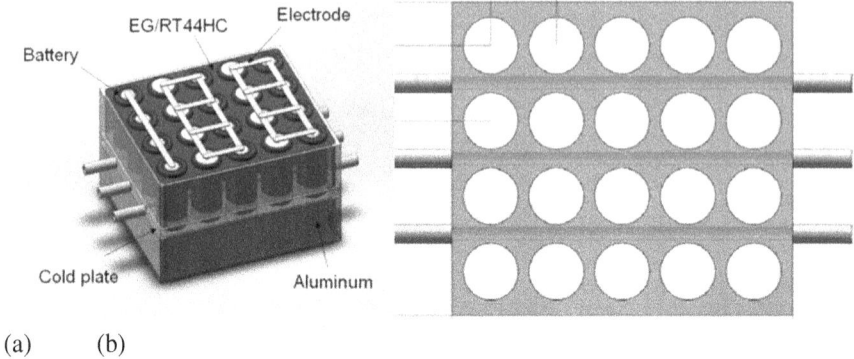

(a)   (b)

FIGURE 14.13   (a) battery pack; (b) model diagram of the cold plate.

Source: Adapted from (Cao et al., 2020).

For this study, two battery simulators and cables with heating elements are integrated into an air duct. The heat gets transferred to the battery simulator from the heater, raising its temperature. Between the battery simulators, the HPs are placed. The PCM and HP were used here to control the battery simulator temperature. The experimental outcomes demonstrate that the HP can lower $T_{battery}$ for various heat energies, relative to a battery with no passive cooling mechanism.

In this study, two PCMs were considered: RT44HC, where the T_m is around the defined temperature of the battery, and beeswax, where the T_m surpassed the specified temperature of the battery. Two parts of a heat pipe are an evaporator and a condenser part. In the evaporator, the fluid is in the liquid form, while it is in the vapor phase in the condenser portion, as shown in Figure 14.14. In comparison to the HP system, the PCM-HP system can minimize the $T_{surface}$ by 31.9°C or 33.2°C.

14.9 CONCLUSION

BTMS has a critical role in removing thermal impacts from batteries, which improves the temperature uniformity, reduces the T_{max}, and extends the lifespan, safety, and performance of the battery pack. Effective BTMS is crucial to improve the safety and efficiency of batteries through cooling and heating. A battery with low-temperature effects can be thermally managed quite easily. However, strategies used for battery cooling can be more complicated than heating methods. Battery cooling systems were designed to actively or passively remove heat generated by batteries using various heat transfer principles. The main benefits of cooling techniques were demonstrated to be heat dissipation, uniform heat distribution through the battery pack, and protection from overheating. Thermal management methods can be active or passive, as well as categorized according to the medium: air, liquid, heat pipe, PCM, and thermoelectric cooler.

PCM-based BTMS is considered as the latest novel cooling technology. In a phase change process, the PCM cooling systems could absorb a substantial fraction

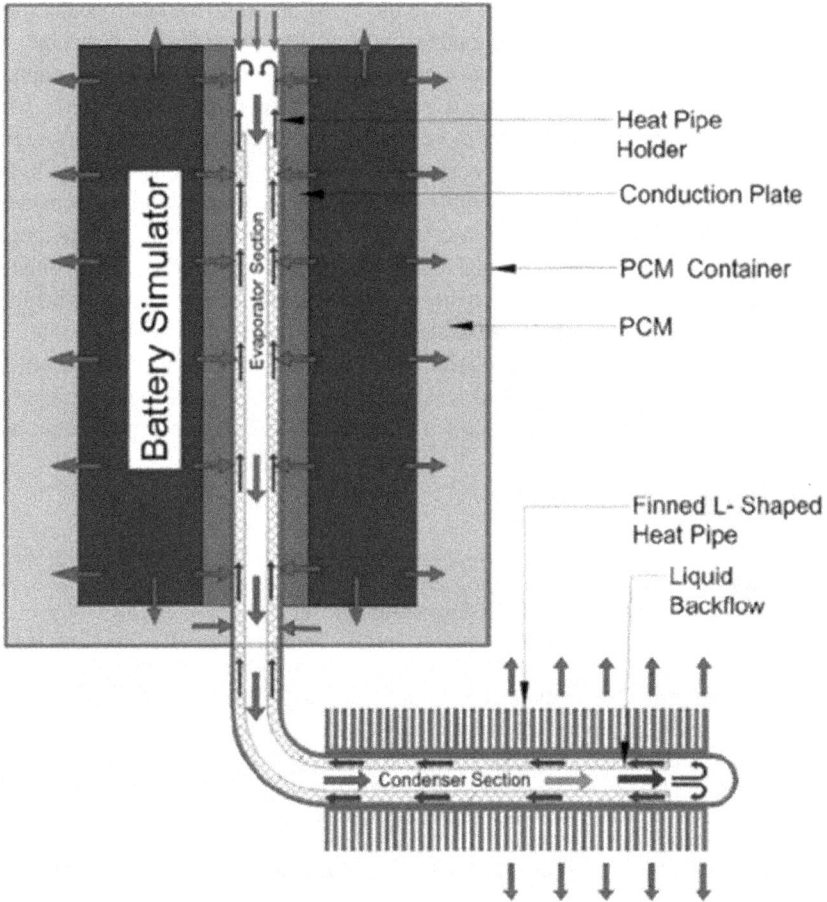

FIGURE 14.14 A diagrammatic representation showing the combined PCM and heat pipe system (Putra et al., 2020).

of heat at the constant temperature while consuming little or no energy. With the growing trend of rapid charging and discharging and large capacity, the performance improvement of PCM cooling technology is crucial. The efficiency of PCM cooling is significantly affected by the phase transition point, mass, thermal conductivity, and thickness of PCM. In addition, temperature effects, type of batteries, heat generation sources, and thermal parameters should be considered and optimized to achieve better cooling performance.

Currently, BTMS research is focusing on improving cooling efficiency by extending heat transfer area, enhancing thermal conductivity, and recovering the heat storage capacity of PCMs. Several researches have been conducted to increase the thermal conductivity of PCMs by adding metal meshes, porous media, and nanomaterials to the pure PCM. As a common strategy for extending the heat transfer area, inserting fins can promote heat transmission inside the PCM and reduces heat occupation at the bottom of the battery. Cooling methods such as forced air, liquid, heat pipe, and thermoelectric coupled with PCM are commonly used to recover the storage capacity of PCM. Hybrid cooling methods are used to reduce the cooling load of PCMs, while researches on hybrid systems are far more restricted as compared to coupling systems. PCM with air, liquid, and HP are commonly studied in the recent work on hybrid PCM-based BTMS. Hybridizing with other cooling techniques can be considered in the future. Furthermore, paraffin wax is the extensively used PCM, due to its ease of availability and low cost. Although composite PCMs are widely used to enhance thermal management performance, many problems, such as design of hybridized systems, additive filling strategy, large-scale studies, and more still need to be fulfilled before commercialization. Even so, significant hurdles persist with this technology, such as low structural strength and leakage of melted PCM, minimal surface heat transfer coefficient, low thermal conductivity, and exhaustion of potential latent heat.

REFERENCES

Abada, S., Marlair, G., Lecocq, A., Petit, M., Sauvant-Moynot, V., & Huet, F. (2016). Safety focused modeling of lithium-ion batteries: A review. *Journal of Power Sources*, *306*, 178–192. https://doi.org/10.1016/j.jpowsour.2015.11.100

Abhat, A. (1983). Low temperature latent heat thermal energy storage: Heat storage materials. *Solar Energy*, *30*(4), 313–332. https://doi.org/10.1016/0038-092X(83)90186-X

al Hallaj, S., & Selman, J. R. (2000). A novel thermal management system for electric vehicle batteries using phase-change material. *Journal of the Electrochemical Society*, *147*(9).

Alaswad, A., Baroutaji, A., Achour, H., Carton, J., al Makky, A., & Olabi, A. G. (2016). Developments in fuel cell technologies in the transport sector. *International Journal of Hydrogen Energy*, *41*(37), 16499–16508. https://doi.org/10.1016/j.ijhydene.2016.03.164

Al-Zareer, M., Dincer, I., & Rosen, M. A. (2018). A review of novel thermal management systems for batteries. *International Journal of Energy Research*, *42*(10), 3182–3205. https://doi.org/10.1002/er.4095

Al-Zareer, M., Dincer, I., & Rosen, M. A. (2019). A novel approach for performance improvement of liquid to vapor based battery cooling systems. *Energy Conversion and Management*, *187*, 191–204. https://doi.org/10.1016/J.ENCONMAN.2019.02.063

An, Z., Jia, L., Ding, Y., Dang, C., & Li, X. (2017). A review on lithium-ion power battery thermal management technologies and thermal safety. *Journal of Thermal Science*, *26*(5), 391–412. https://doi.org/10.1007/s11630-017-0955-2

Bamdezh, M. A., & Molaeimanesh, G. R. (2020). Impact of system structure on the performance of a hybrid thermal management system for a Li-ion battery module. *Journal of Power Sources*, *457*. https://doi.org/10.1016/j.jpowsour.2020.227993

Barnes, D., & Li, X. (2020). Battery thermal management using phase change material-metal foam composite materials at various environmental temperatures. *Nutrition Today*, *17*(2). https://doi.org/10.1115/1.4045326

Behi, H., Karimi, D., Behi, M., Ghanbarpour, M., Jaguemont, J., Sokkeh, M. A., Gandoman, F. H., Berecibar, M., & van Mierlo, J. (2020). A new concept of thermal management system in Li-ion battery using air cooling and heat pipe for electric vehicles. *Applied Thermal Engineering*, *174*, 115280. https://doi.org/10.1016/J.APPLTHERMALENG.2020.115280

Burow, D., Sergeeva, K., Calles, S., Schorb, K., Börger, A., Roth, C., & Heitjans, P. (2016). Inhomogeneous degradation of graphite anodes in automotive lithium ion batteries under low-temperature pulse cycling conditions. *Journal of Power Sources, 307*, 806–814. https://doi.org/10.1016/j.jpowsour.2016.01.033

Cao, J., Luo, M., Fang, X., Ling, Z., & Zhang, Z. (2020). Liquid cooling with phase change materials for cylindrical Li-ion batteries: An experimental and numerical study. *Energy, 191*. https://doi.org/10.1016/j.energy.2019.116565

Chen, J., Kang, S., Jiaqiang, E., Huang, Z., Wei, K., Zhang, B., Zhu, H., Deng, Y., Zhang, F., & Liao, G. (2019). Effects of different phase change material thermal management strategies on the cooling performance of the power lithium ion batteries: A review. *Journal of Power Sources, 442*. https://doi.org/10.1016/j.jpowsour.2019.227228

Chen, K., Li, Z., Chen, Y., Long, S., Hou, J., Song, M., & Wang, S. (2017). Design of parallel air-cooled battery thermal management system through numerical study. *Energies, 10*(10). https://doi.org/10.3390/en10101677

Duan, X., & Naterer, G. F. (2010). Heat transfer in phase change materials for thermal management of electric vehicle battery modules. *International Journal of Heat and Mass Transfer, 53*(23–24), 5176–5182. https://doi.org/10.1016/j.ijheatmasstransfer.2010.07.044

Encyclopedia of Sustainability Science and Technology. (2012). *Encyclopedia of sustainability science and technology*. Springer. https://doi.org/10.1007/978-1-4419-0851-3

Farid, M. M., Khudhair, A. M., Razack, S. A. K., & Al-Hallaj, S. (2004). A review on phase change energy storage: Materials and applications. *Energy Conversion and Management, 45*(9–10), 1597–1615. https://doi.org/10.1016/j.enconman.2003.09.015

Fathabadi, H. (2014). A novel design including cooling media for Lithium-ion batteries pack used in hybrid and electric vehicles. *Journal of Power Sources, 245*, 495–500. https://doi.org/10.1016/J.JPOWSOUR.2013.06.160

Feldman, D., Banu, D., Hawes, D., & Ghanbari, E. (1991). Obtaining an energy storing building material by direct incorporation of an organic phase change material in gypsum wallboard. *Solar Energy Materials, 22*(2–3), 231–242. https://doi.org/10.1016/0165-1633(91)90021-C

Furbo, S. (2015). Using water for heat storage in thermal energy storage (TES) systems. In *Advances in thermal energy storage systems: Methods and applications* (pp. 31–47). https://doi.org/10.1533/9781782420965.1.31

Ge, H., Li, H., Mei, S., & Liu, J. (2013). Low melting point liquid metal as a new class of phase change material: An emerging frontier in energy area. *Renewable and Sustainable Energy Reviews, 21*, 331–346. https://doi.org/10.1016/J.RSER.2013.01.008

Ghadbeigi, L., Day, B., Lundgren, K., & Sparks, T. D. (2018). Cold temperature performance of phase change material based battery thermal management systems. *Energy Reports, 4*, 303–307. https://doi.org/10.1016/j.egyr.2018.04.001

Goodenough, J. B., & Park, K. S. (2013). The Li-ion rechargeable battery: A perspective. *Journal of the American Chemical Society, 135*(4), 1167–1176. https://doi.org/10.1021/ja3091438

Hussain, A., Tso, C. Y., & Chao, C. Y. H. (2016). Experimental investigation of a passive thermal management system for high-powered lithium ion batteries using nickel foam-paraffin composite. *Energy, 115*, 209–218. https://doi.org/10.1016/j.energy.2016.09.008

Ianniciello, L., Biwolé, P. H., & Achard, P. (2018). Electric vehicles batteries thermal management systems employing phase change materials. *Journal of Power Sources, 378*, 383–403. https://doi.org/10.1016/j.jpowsour.2017.12.071

Javani, N., Dincer, I., Naterer, G. F., & Yilbas, B. S. (2014). Heat transfer and thermal management with PCMs in a Li-ion battery cell for electric vehicles. *International Journal of Heat and Mass Transfer, 72*, 690–703. https://doi.org/10.1016/j.ijheatmasstransfer.2013.12.076

Jiaqiang, E., Han, D., Qiu, A., Zhu, H., Deng, Y., Chen, J., Zhao, X., Zuo, W., Wang, H., Chen, J., & Peng, Q. (2018). Orthogonal experimental design of liquid-cooling structure on the cooling effect of a liquid-cooled battery thermal management system. *Applied Thermal Engineering, 132*, 508–520. https://doi.org/10.1016/j.applthermaleng.2017.12.115

Jilte, R., Afzal, A., Islam, M. T., & Manokar, A. M. (2021). Hybrid cooling of cylindrical battery with liquid channels in phase change material. *International Journal of Energy Research, 45*(7), 11065–11083. https://doi.org/10.1002/er.6590

Jilte, R. D., Kumar, R., Ahmadi, M. H., & Chen, L. (2019). Battery thermal management system employing phase change material with cell-to-cell air cooling. *Applied Thermal Engineering, 161*, 114199. https://doi.org/10.1016/J.APPLTHERMALENG.2019.114199

Jouhara, H., Khordehgah, N., Serey, N., Almahmoud, S., Lester, S. P., Machen, D., & Wrobel, L. (2019). Applications and thermal management of rechargeable batteries for industrial applications. *Energy, 170*, 849–861. https://doi.org/10.1016/j.energy.2018.12.218

Kaizawa, A., Maruoka, N., Kawai, A., Kamano, H., Jozuka, T., Senda, T., & Akiyama, T. (2008). Thermophysical and heat transfer properties of phase change material candidate for waste heat transportation system. *Heat and Mass Transfer/Waerme-Und Stoffueber-tragung, 44*(7), 763–769. https://doi.org/10.1007/s00231-007-0311-2

Karimi, G., Azizi, M., & Babapoor, A. (2016). Experimental study of a cylindrical lithium ion battery thermal management using phase change material composites. *Journal of Energy Storage, 8*, 168–174. https://doi.org/10.1016/J.EST.2016.08.005

Khan, M. R., Swierczynski, M. J., & Kær, S. K. (2017). Towards an ultimate battery thermal management system: A review. *Batteries, 3*(1). https://doi.org/10.3390/batteries3010009

Lamb, J., Orendorff, C. J., Steele, L. A. M., & Spangler, S. W. (2015). Failure propagation in multi-cell lithium ion batteries. *Journal of Power Sources, 283*, 517–523. https://doi.org/10.1016/j.jpowsour.2014.10.081

Lane, G. A. (1983). *Solar heat storage: Latent heat materials*. CRC Press.

Larminie, James., & Lowry, J. (2012). *Electric vehicle technology explained*. John Wiley & Sons.

Li, W. Q., Guo, S. J., Tan, L., Liu, L. L., & Ao, W. (2021). Heat transfer enhancement of nano-encapsulated phase change material (NEPCM) using metal foam for thermal energy storage. *International Journal of Heat and Mass Transfer, 166*. https://doi.org/10.1016/j.ijheatmasstransfer.2020.120737

Lin, C., Xu, S., Chang, G., & Liu, J. (2015). Experiment and simulation of a LiFePO4 battery pack with a passive thermal management system using composite phase change material and graphite sheets. *Journal of Power Sources, 275*, 742–749. https://doi.org/10.1016/j.jpowsour.2014.11.068

Ling, Z., Cao, J., Zhang, W., Zhang, Z., Fang, X., & Gao, X. (2018). Compact liquid cooling strategy with phase change materials for Li-ion batteries optimized using response surface methodology. *Applied Energy, 228*, 777–788. https://doi.org/10.1016/J.APENERGY.2018.06.143

Ling, Z., Wang, F., Fang, X., Gao, X., & Zhang, Z. (2015). A hybrid thermal management system for lithium ion batteries combining phase change materials with forced-air cooling. *Applied Energy, 148*, 403–409. https://doi.org/10.1016/j.apenergy.2015.03.080

Lv, Y., Liu, G., Zhang, G., & Yang, X. (2020). A novel thermal management structure using serpentine phase change material coupled with forced air convection for cylindrical battery modules. *Journal of Power Sources, 468*. https://doi.org/10.1016/j.jpowsour.2020.228398

Malik, M., Dincer, I., & Rosen, M. A. (2016). Review on use of phase change materials in battery thermal management for electric and hybrid electric vehicles. *International Journal of Energy Research, 40*(8), 1011–1031. https://doi.org/10.1002/er.3496

Malik, M., Dincer, I., Rosen, M., & Fowler, M. (2017). Experimental investigation of a new passive thermal management system for a Li-Ion battery pack using phase change composite material. *Electrochimica Acta, 257,* 345–355. https://doi.org/10.1016/j.elect-acta.2017.10.051

Mathur, A., Kasetty, R., Oxley, J., Mendez, J., & Nithyanandam, K. (2014). Using encapsulated phase change salts for concentrated solar power plant. *Energy Procedia, 49,* 908–915. https://doi.org/10.1016/J.EGYPRO.2014.03.098

Meng, Q., & Hu, J. (2008). A poly(ethylene glycol)-based smart phase change material. *Solar Energy Materials and Solar Cells, 92*(10), 1260–1268. https://doi.org/10.1016/J.SOL-MAT.2008.04.026

Molaeimanesh, G. R., Mirfallah Nasiry, S. M., & Dahmardeh, M. (2020). Impact of configuration on the performance of a hybrid thermal management system including phase change material and water-cooling channels for Li-ion batteries. *Applied Thermal Engineering, 181,* 116028. https://doi.org/10.1016/J.APPLTHERMALENG.2020.116028

Moraga, N. O., Xamán, J. P., & Araya, R. H. (2016). Cooling Li-ion batteries of racing solar car by using multiple phase change materials. *Applied Thermal Engineering, 108,* 1041–1054. https://doi.org/10.1016/j.applthermaleng.2016.07.183

Murali, G., Sravya, G. S. N., Jaya, J., & Naga Vamsi, V. (2021). A review on hybrid thermal management of battery packs and it's cooling performance by enhanced PCM. *Renewable and Sustainable Energy Reviews, 150.* https://doi.org/10.1016/j.rser.2021.111513

Nkwetta, D. N., & Haghighat, F. (2014). Thermal energy storage with phase change material—A state-of-the art review. *Sustainable Cities and Society, 10,* 87–100. https://doi.org/10.1016/J.SCS.2013.05.007

Ping, P., Peng, R., Kong, D., Chen, G., & Wen, J. (2018). Investigation on thermal management performance of PCM-fin structure for Li-ion battery module in high-temperature environment. *Energy Conversion and Management, 176,* 131–146. https://doi.org/10.1016/j.enconman.2018.09.025

Putra, N., Sandi, A. F., Ariantara, B., Abdullah, N., & Indra Mahlia, T. M. (2020). Performance of beeswax phase change material (PCM) and heat pipe as passive battery cooling system for electric vehicles. *Case Studies in Thermal Engineering, 21.* https://doi.org/10.1016/j.csite.2020.100655

Raam Dheep, G., & Sreekumar, A. (2014). Influence of nanomaterials on properties of latent heat solar thermal energy storage materials – A review. *Energy Conversion and Management, 83,* 133–148. https://doi.org/10.1016/J.ENCONMAN.2014.03.058

Ramandi, M. Y., Dincer, I., & Naterer, G. F. (2011). Heat transfer and thermal management of electric vehicle batteries with phase change materials. *Heat and Mass Transfer/Waerme-Und Stoffuebertragung, 47*(7), 777–788. https://doi.org/10.1007/s00231-011-0766-z

Rao, Z., Wang, S., & Zhang, G. (2011). Simulation and experiment of thermal energy management with phase change material for ageing LiFePO4 power battery. *Energy Conversion and Management, 52*(12), 3408–3414. https://doi.org/10.1016/J.ENCON-MAN.2011.07.009

Samimi, F., Babapoor, A., Azizi, M., & Karimi, G. (2016). Thermal management analysis of a Li-ion battery cell using phase change material loaded with carbon fibers. *Energy, 96,* 355–371. https://doi.org/10.1016/j.energy.2015.12.064

Sarbu, I., & Sebarchievici, C. (2018). A comprehensive review of thermal energy storage. *Sustainability (Switzerland), 10*(1). https://doi.org/10.3390/su10010191

Sarier, N., & Onder, E. (2012). Organic phase change materials and their textile applications: An overview. *Thermochimica Acta, 540,* 7–60. https://doi.org/10.1016/J.TCA.2012.04.013

Sato, N. (2000). Thermal behavior analysis of lithium-ion batteries for electric and hybrid vehicles. *Journal of Power Sources, 99*(1–2), 70–77.

Selman, J. R., al Hallaj, S., Uchida, I., & Hirano, Y. (2001). Cooperative research on safety fundamentals of lithium batteries. *Journal of Power Sources, 97,* 726–732.

Sharma, A., Tyagi, V. V., Chen, C. R., & Buddhi, D. (2009). Review on thermal energy storage with phase change materials and applications. *Renewable and Sustainable Energy Reviews*, *13*(2), 318–345. https://doi.org/10.1016/J.RSER.2007.10.005

Shen, Z. G., Chen, S., Liu, X., & Chen, B. (2021). A review on thermal management performance enhancement of phase change materials for vehicle lithium-ion batteries. *Renewable and Sustainable Energy Reviews*, *148*, 111301. https://doi.org/10.1016/J.RSER.2021.111301

Siddique, A. R. M., Mahmud, S., & Van Heyst, B. (2018). A comprehensive review on a passive (phase change materials) and an active (thermoelectric cooler) battery thermal management system and their limitations. *Journal of Power Sources*, *401*, 224–237. https://doi.org/10.1016/j.jpowsour.2018.08.094

Singh, R., Sadeghi, S., & Shabani, B. (2019). Thermal conductivity enhancement of phase change materials for low-temperature thermal energy storage applications. *Energies*, *12*(1). https://doi.org/10.3390/en12010075

Song, W., Bai, F., Chen, M., Lin, S., Feng, Z., & Li, Y. (2018). Thermal management of standby battery for outdoor base station based on the semiconductor thermoelectric device and phase change materials. *Applied Thermal Engineering*, *137*, 203–217. https://doi.org/10.1016/j.applthermaleng.2018.03.072

Spotnitz, R. M., Weaver, J., Yeduvaka, G., Doughty, D. H., & Roth, E. P. (2007). Simulation of abuse tolerance of lithium-ion battery packs. *Journal of Power Sources*, *163*(2), 1080–1086. https://doi.org/10.1016/j.jpowsour.2006.10.013

Su, W., Darkwa, J., & Kokogiannakis, G. (2015). Review of solid–liquid phase change materials and their encapsulation technologies. *Renewable and Sustainable Energy Reviews*, *48*, 373–391. https://doi.org/10.1016/J.RSER.2015.04.044

Wang, Q., Ping, P., Zhao, X., Chu, G., Sun, J., & Chen, C. (2012). Thermal runaway caused fire and explosion of lithium ion battery. *Journal of Power Sources*, *208*, 210–224. https://doi.org/10.1016/j.jpowsour.2012.02.038

Wang, X., Xie, Y., Day, R., Wu, H., Hu, Z., Zhu, J., & Wen, D. (2018). Performance analysis of a novel thermal management system with composite phase change material for a lithium-ion battery pack. *Energy*, *156*, 154–168. https://doi.org/10.1016/j.energy.2018.05.104

Wang, Z., Zhang, H., & Xia, X. (2017). Experimental investigation on the thermal behavior of cylindrical battery with composite paraffin and fin structure. *International Journal of Heat and Mass Transfer*, *109*, 958–970. https://doi.org/10.1016/j.ijheatmasstransfer.2017.02.057

Wright, R. B., Christophersen, J. P., Motloch, C. G., Belt, J. R., Ho, C. D., Battaglia, V. S., Barnes, J. A., Duong, T. Q., & Sutula, R. A. (2003). Power fade and capacity fade resulting from cycle-life testing of advanced technology development program lithium-ion batteries. *Journal of Power Sources*, *119–121*, 865–869. https://doi.org/10.1016/S0378-7753(03)00190-3

Wu, W., Wu, W., & Wang, S. (2017). Thermal optimization of composite PCM based large-format lithium-ion battery modules under extreme operating conditions. *Energy Conversion and Management*, *153*, 22–33. https://doi.org/10.1016/J.ENCONMAN.2017.09.068

Wu, W., Wu, W., & Wang, S. (2018). Thermal management optimization of a prismatic battery with shape-stabilized phase change material. *International Journal of Heat and Mass Transfer*, *121*, 967–977. https://doi.org/10.1016/j.ijheatmasstransfer.2018.01.062

Wu, W., Yang, X., Zhang, G., Chen, K., & Wang, S. (2017). Experimental investigation on the thermal performance of heat pipe-assisted phase change material based battery thermal management system. *Energy Conversion and Management*, *138*, 486–492. https://doi.org/10.1016/j.enconman.2017.02.022

Yang, X. H., Tan, S. C., & Liu, J. (2016). Thermal management of Li-ion battery with liquid metal. *Energy Conversion and Management*, *117*, 577–585. https://doi.org/10.1016/j.enconman.2016.03.054

Ye, Y., Saw, L. H., Shi, Y., & Tay, A. A. O. (2015). Numerical analyses on optimizing a heat pipe thermal management system for lithium-ion batteries during fast charging. *Applied Thermal Engineering, 86*, 281–291. https://doi.org/10.1016/J.APPLTHERMALENG. 2015.04.066

Yu, X., Lu, Z., Zhang, L., Wei, L., Cui, X., & Jin, L. (2019). Experimental study on transient thermal characteristics of stagger-arranged lithium-ion battery pack with air cooling strategy. *International Journal of Heat and Mass Transfer, 143*, 118576. https://doi. org/10.1016/J.IJHEATMASSTRANSFER.2019.118576

Zhang, X., Liu, C., & Rao, Z. (2018). Experimental investigation on thermal management performance of electric vehicle power battery using composite phase change material. *Journal of Cleaner Production, 201*, 916–924. https://doi.org/10.1016/j.jclepro.2018.08.076

Zhao, D., & Tan, G. (2014). A review of thermoelectric cooling: Materials, modeling and applications. *Applied Thermal Engineering, 66*(1–2), 15–24. https://doi.org/10.1016/J. APPLTHERMALENG.2014.01.074

Zhao, R., Zhang, S., Liu, J., & Gu, J. (2015). A review of thermal performance improving methods of lithium ion battery: Electrode modification and thermal management system. *Journal of Power Sources, 299*, 557–577. https://doi.org/10.1016/j. jpowsour.2015.09.001

Index

For Product Safety Concerns and Information please contact our EU
representative GPSR@taylorandfrancis.com
Taylor & Francis Verlag GmbH, Kaufingerstraße 24, 80331 München, Germany

www.ingramcontent.com/pod-product-compliance
Lightning Source LLC
Chambersburg PA
CBHW060814220326
41598CB00022B/2612

9 781032 385440